INDUCTANCE

INDUCTANCE

Loop and Partial

CLAYTON R. PAUL
Professor of Electrical and Computer Engineering
Mercer University
Macon, Georgia
and
Emeritus Professor of Electrical Engineering
University of Kentucky
Lexington, Kentucky

Celebrating 125 Years
of Engineering the Future

WILEY

A JOHN WILEY & SONS, INC., PUBLICATION

Published by John Wiley & Sons, Inc., Hoboken, New Jersey.
Published simultaneously in Canada.

For general information on our other products and services or for technical support, please contact our Customer Care Department within the United States at (800) 762-2974, outside the United States at (317) 572-3993 or fax (317) 572-4002.

Wiley also publishes its books in a variety of electronic formats. Some content that appears in print may not be available in electronic formats. For more information about Wiley products, visit our web site at www.wiley.com.

Library of Congress Cataloging-in-Publication Data:

Paul, Clayton R.
   Inductance : loop and partial / Clayton R. Paul.
      p. cm.
   Includes bibliographical references and index.
   ISBN 978-0-470-46188-4
 1. Inductance. 2. Induction coils. I. Title.
   QC638.P38 2010
   621.37'42–dc22   2009031434

Printed in the United States of America

10 9 8 7 6 5 4 3 2 1

This book is dedicated to the memory of my Father and my Mother

Oscar Paul

and

Louise Paul

CONTENTS

PREFACE

This book has been written to provide a thorough and complete discussion of virtually all aspects of inductance: both "loop" and "partial." There is considerable misunderstanding and misapplication of the important concepts of inductance. Undergraduate electrical engineering curricula generally discuss "loop" inductance only very briefly and only in one undergraduate course at the beginning of the junior year in a four-year curriculum. However, that curriculum is replete with the analysis of electric circuits containing the inductance symbol. In all those electric circuit analysis courses, the *values* of the inductors are given and are not derived from physical principles. Yet in the world of industry, the analyst must somehow obtain these values as well as construct inductors having the chosen values of inductance used in the circuit analysis. This book addresses that missing link: calculation of the values of the various physical constructions of inductors, both intentional and unintentional, from basic electromagnetic principles and laws.

In addition, today's high-speed digital systems as well as high-frequency analog systems are using increasingly higher spectral content signals. Numerous "unintended" inductances such as those of the interconnection leads are becoming increasingly important in determining whether these high-speed, high-frequency systems will function properly. This is generally classified as the "signal integrity" of those systems and is an increasingly important aspect of digital system design as clock and data speeds increase at a dramatic rate. Some ten years ago the effects of interconnects such as printed circuit board lands on the function of the modules that lands interconnect were not important and could be ignored. Today, it is critical that circuit models of these interconnects be included in any analysis of the overall system. The concept of "partial inductance" is the critical link in being able to model these interconnects. Partial inductance is not covered in any undergraduate electrical engineering course but is becoming increasingly important in digital system design. A substantial portion of this book is devoted to that topic.

One of the important contributions of this book is the detailed derivation of the loop and partial inductances of numerous configurations of current-carrying conductors. Although the derivations are sometimes tedious, there is nothing we can do about it because the results are dictated by the laws of electromagnetics, and these can be complicated. Unlike other textbooks, all the details regarding derivations for the inductance of inductors are given. Although these are simplified where possible, only so much simplification can be accepted if the reader is to have a clear and unambiguous view of how the result is obtained.

In Chapter 1 we discuss inductance and show important parallels between inductance and capacitance along with some historical details. All of the derivations of the inductance of various inductors first require that we obtain their magnetic fields. Chapter 2 is devoted to this task. The fundamental laws of Biot–Savart, Gauss, and Ampère are discussed, and numerous calculations of the magnetic fields are obtained from them. In addition, the vector magnetic potential method of computing the magnetic fields is also discussed, along with the method of images and energy stored in the magnetic field. In Chapter 3 we provide a complete explanation of how the inductance, which is computed for dc currents, can be used to characterize the effect of time-varying currents. Maxwell's equations for time-varying currents are discussed in detail. An iterative solution of them is given which shows why and when the inductor, derived for dc currents, can be used to characterize the effects of time-varying currents.

All aspects of the derivation of the "loop" inductance of various current-carrying loops are covered in Chapter 4. The flux linkage method, the vector magnetic potential method, and the Neumann integral for determining the "loop" inductance are used, and the "loop" inductances are calculated from all three methods. The proximity effect for closely spaced conductors is discussed along with the loop inductance of various transmission lines.

In Chapter 5 we provide details for computation of the "partial" inductances of wires. Both the self-partial inductance of wires and the mutual partial inductances between wires are derived. These generic results can then be used to "build" a model for other current-carrying structures. Chapter 6 contains all corresponding details about the derivation of the partial inductances of conductors of rectangular cross section, referred to as "lands." The concept of geometric mean distance as an aid to the calculation of partial inductances is discussed and derived for various structures.

The final chapter of the book, Chapter 7, provides a focus on when one should use "loop" inductance and when one should use "partial" inductance for determining the effect of current-carrying conductors. This chapter is meant to provide a simple discussion of this in order to focus the results of previous chapters. The chapter concludes with the solution of a problem involving coupling between two circuit loops using the "loop" inductance method and then using the "partial" inductance method. Both methods yield the same answer, as expected. This example clearly shows the advantages of using "partial" inductance to characterize "unintentional inductors" such as wires and lands.

With the present and increasing emphasis on high-speed digital systems and high-frequency analog systems, it is imperative that system designers develop an intimate understanding of the concepts and methods in this book. No longer can we rely on low-speed, low-frequency systems to keep us from needing to learn these new concepts and analysis skills.

The author would like to acknowledge Dr. Albert E. Ruehli of the IBM T.J. Watson Research Center for many helpful discussions of partial inductance over the years.

CLAYTON R. PAUL

Macon, Georgia

1

INTRODUCTION

The concept of *inductance* is simple and straightforward. However, actual computation of the inductance of various physical structures and its implementation in an electric circuit model of that structure is often fraught with misconceptions and mistakes that prevent its correct calculation and use. This book is intended to ensure the correct understanding, calculation, and implementation of inductance.

1.1 HISTORICAL BACKGROUND

Knowledge of magnetism has a long history [3]. A type of iron ore called *lodestone* had been discovered in Magnesia in Asia. This material had some interesting properties of magnetic attraction at a distance of other ferromagnetic substances and was known to Plato and Socrates. In the sixteenth century, William Gilbert first postulated that Earth was a giant spherical magnet, and A. Kirchner, in the seventeenth century, demonstrated that the two poles of a magnet have equal strength. Pierre de Marricourt constructed a compass in 1629 that allowed the determination of the direction of the North Pole of the Earth. In 1750, John Mitchell determined the universal principle that force at a distance depends on the inverse square of the distance. At the beginning of the nineteenth century, Alessandro Volta developed a battery (called a *pile*).

This allowed the production of a current in a conducting material such as a wire. In 1820, Hans Christian Oersted showed that a current in a wire caused the needle of a compass to deflect. Around the same time, André Ampère conducted a set of experiments, resulting in his famous law. At about the same time, Jean-Baptiste Biot and Felix Savart formulated their important law governing the magnetic fields produced by currents: the Biot–Savart law. So up to this time it was known that in addition to permanent magnets, a current would produce a magnetic field. In 1831, Michael Faraday discovered that a time-changing magnetic field would also produce a current in a closed loop of wire. This discovery formed the essential idea of the inductance of a current loop. James Clerk Maxwell unified all this knowledge of the magnetic field as well as the knowledge of the electric field in 1873 in his renowned set of equations.

Extensive work on the calculation of the magnetic field of various current distributions and the associated concept of inductance dates back to the late nineteenth and early twentieth centuries. In fact, Maxwell in his famous treatise discussed inductance in 1873 [23]. An enormous amount of work was published on the determination of inductance from 1900 to 1920. (See the extensive list of references on magnetic fields in the book by Weber [11] and on inductance in the book by Grover [14].) This early work on inductance at the turn of the century was spurred by the introduction of 60-Hz ac power and its generation, distribution, and use. Some books, particularly those of the early twentieth century, tended to give only formulas for the magnetic fields of various distributions of currents and their inductance with little or no detail about the derivation of formulas. In that era, computers did not exist, so that many of the books and papers simply gave tables of values for the magnetic field and inductance as a function of certain parameters. Another important purpose of this book is to show, in considerable detail, how the results for the magnetic fields and the inductance are derived. All details of each derivation are shown. At the end of the book is a list of significant references and further readings on the subject of the computation of magnetic fields and inductance of various current-carrying structures. References to these are denoted in brackets.

1.2 FUNDAMENTAL CONCEPTS OF LUMPED CIRCUITS

We construct lumped-circuit models of electrical structures using the concepts and models of resistance, capacitance, and inductance [1,2]. We then solve for the resulting voltages and currents of that particular interconnection of circuit elements using Kirchhoff's voltage law (KVL) (which relates the various voltages of the particular interconnection of circuit elements), Kirchhoff's current law (KCL) (which relates the various currents of the particular interconnection of circuit elements), and the laws of the circuit elements (which relate the voltages of each circuit element to its currents) [1,2]. It is important to keep in mind that these lumped-circuit models are *valid only if the largest physical dimension of the circuit is "electrically short"* (e.g., L < $\frac{1}{10}$), where a wavelength λ is defined as the ratio of the velocity of wave propagation (along the component attachment leads), v, and the frequency of the wave, f [3–6]:

$$\lambda = \frac{v}{f} \qquad (1.1)$$

If the medium in which the circuit is immersed and through which the waves propagate along the connection leads is free space (essentially, air), the velocity of propagation of those waves is the speed of light, which is approximately $v_0 \cong 3 \times 10^8$ m/s. For a printed circuit board (PCB), the velocity of propagation of the waves traveling along the lands on that board is about 60% of that of free space, due to the interaction of the fields with the board substrate, and the wavelengths are consequently shorter than in free space. Hence, circuit dimensions on a PCB are electrically longer than in free space. For a sinusoidal wave in free space at a frequency of 300 MHz, a wavelength is 1 m. At frequencies below this, the wavelength is proportionately larger than 1 m, and for frequencies above this, the wavelength is proportionately smaller. For example, at a frequency of 3 MHz a wavelength in free space is 100 m, and at a frequency of 3 GHz a wavelength in free space is 10 cm. Hence, for lumped-circuit concepts to be valid for a circuit having a sinusoidal source of frequency 3 MHz, the maximum physical dimension of the circuit must be less than about 10 m or about 30 ft. Similarly, for a circuit having a sinusoidal source of frequency 3 GHz, the maximum physical dimension must be less than about 1 cm or about 0.4 inch for it to be modeled as a lumped circuit. Today's digital electronics have clock and data rates on the order of 300 MHz to 3 GHz. But these digital waveforms have a spectral content consisting of harmonics (integer multiples) of the basic repetition rate, which are generally significant up to at least the fifth harmonic. Hence, a 300-MHz clock rate has spectral content up to at least 1.5 GHz, and a 3-GHz clock rate has spectral content up to at least 15 GHz! So the lumped-circuit models (and their constituent components of capacitance and inductance) that were so reliable some 10 years ago are becoming less valid today. This trend will no doubt continue in the future as the requirement for higher clock and data speeds continues to increase, and the reader should keep in mind this fundamental limitation of inductance, capacitance, and the lumped-circuit models that use these elements.

The laws governing the calculation of resistance, capacitance, and inductance are written in terms of the *vectors* of the five basic electromagnetic field vectors, which are summarized in Table 1.1. Therefore, if we are to correctly

TABLE 1.1. Electromagnetic Field Vectors

Symbol	Vector	Units
J	Current density	A/m^2
	Electric field vectors	
E	Electric field intensity	V/meter
D	Electric flux density	C/m^2
	Magnetic field vectors	
H	Magnetic field intensity	A/meter
B	Magnetic flux density	$Wb/m^2 = T$

calculate and understand the ideas of capacitance and inductance of a physical structure as well as use them correctly to construct a lumped-circuit model of that structure, we must understand some elementary properties of vectors and some basic vector calculus concepts. Trying to circumvent the use of vector calculus ideas by relying on one's life experiences to compute and interpret the meaning of the capacitance and inductance of a structure properly has caused many of the incorrect results and misunderstanding, as well as the numerous erroneous applications that are seen throughout the literature and in conversations with engineering professionals. References [3–6] give extensive details on vector algebra and vector calculus. The Appendix of this book contains a review of the vector algebra and vector calculus concepts that are required to understand and compute the inductance of all physical structures.

The lumped-circuit elements of resistance, capacitance, and inductance are derived fundamentally for *static conditions*. Capacitance is derived for conductors that are supporting charges whose positions on those conductors are fixed. Resistance as well as inductance are derived for currents that are *not varying with time*: that is, *direct* (dc) *currents*. For charge distributions and currents that do not vary with time, the electromagnetic field equations (Maxwell's equations) that govern the field vectors simplify considerably. However, the resulting electrical elements of resistance, capacitance, and inductance can be used to construct lumped-circuit models of a structure whose currents and charge distributions vary with time. This is valid as long as the sources driving the circuit have frequency content such that the largest physical dimension of the circuit is electrically small (see Section 3.4).

To understand the computation of inductance (the main subject of this book), it is useful to understand the dual concept of capacitance and its calculation. The basic idea of the capacitance of a two-conductor structure is summarized in Fig. 1.1(a). If we apply a dc voltage V between two

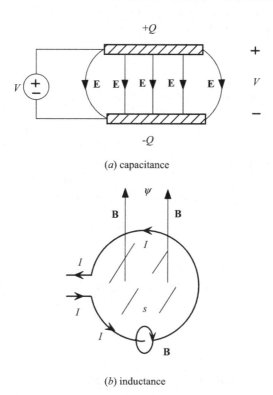

(*a*) capacitance

(*b*) inductance

FIGURE 1.1. Capacitance and inductance.

conductors, a charge Q is transferred to and stored on those conductors (equal magnitude on both conductors, but opposite polarity). This charge induces an electric field intensity $\mathbf{E}$ between the two conductors that is directed from the conductor containing the positive charge to the conductor containing the negative charge. Alternatively, we could look at this process in a different way. Place a charge on the two conductors (equal magnitude on both conductors but opposite polarity). This charge will result in an electric field $\mathbf{E}$ between the two conductors which when integrated with a line integral (see the Appendix) gives the resulting voltage between the two conductors:

$$V = -\int_{-}^{+} \mathbf{E} \cdot d\mathbf{l} \tag{1.2}$$

where the path for integration is from a point on the negatively charged conductor to a point on the positively charged conductor [3–6]. In either case, the *capacitance* of the structure is the ratio of the charge stored on the two conductors and the voltage between them [1–6]:

$$C = \frac{Q}{V} \tag{1.3}$$

Hence, the capacitance of a structure represents the ability of that structure to *store charge*. However, the capacitance of the structure is independent of the values of the voltage V and the charge Q and depends *only* on their ratio. Hence, the capacitance C of a structure depends only on its dimensions, its shape, and the properties of the medium surrounding the conductors (e.g., free space, Teflon). There is energy stored in the electric field in the space around the two conductors. That stored energy is [3–6]

$$W_E = \frac{1}{2} \int_v \mathbf{D} \cdot \mathbf{E} \, dv = \frac{1}{2} \varepsilon \int_v E^2 \, dv \tag{1.4}$$

where v is the volume of the entire space surrounding the conductors, ε is the *permittivity* of the surrounding medium, and we have used the relation $D = \varepsilon E$. In terms of capacitance this stored energy is [1,2]

$$W_E = \frac{1}{2} C V^2 \tag{1.5}$$

The dual concept is that of inductance, illustrated in Fig. 1.1(b). If we pass a steady (dc) current I around a conducting loop of wire, the current will produce a magnetic flux density $\mathbf{B}$ that circulates about the wire with its direction about the wire determined by the *right-hand rule*: If we place the thumb of the right hand in the direction of the current, the fingers will show the direction of the resulting magnetic field that is circumferential about the current. This causes a magnetic field $\mathbf{B}$ to penetrate the surface that is enclosed by the loop of current. The total magnetic flux penetrating the surface enclosed by the current loop is obtained with a surface integral (see the Appendix) as [3–6],

$$\psi = \int_s \mathbf{B} \cdot d\mathbf{s} \tag{1.6}$$

where s is the surface of the loop that is surrounded by the current. The inductance of the loop is the ratio of the total magnetic flux penetrating the loop and the current that produced it [1–6]:

$$L = \frac{\psi}{I} \tag{1.7}$$

If the surrounding medium is not ferromagnetic (iron is an example of a ferromagnetic material), that is, is not magnetizeable, the inductance is independent of the values of the flux and the current and depends only on the dimension of the loop, its shape, and the properties of the medium surrounding the conductor (e.g., free space). There is energy stored in the magnetic field in the space around the conductor loop. That stored energy is [3–6]

$$W_M = \frac{1}{2} \int_v \mathbf{B} \cdot \mathbf{H} \, dv = \frac{1}{2} \mu \int_v H^2 \, dv \qquad (1.8)$$

where v is the volume of the entire space surrounding the conductors, μ is the *permeability* of the surrounding medium, and we have used the relation $\mathbf{B} = \mu \mathbf{H}$. In terms of inductance, the stored energy is [1,2]

$$W_M = \frac{1}{2} L I^2 \qquad (1.9)$$

The duality between the concept of capacitance and the corresponding concept of inductance is striking. However, the methods and techniques for computing them are generally different in both concept and method. Visualizing how to go about calculating the capacitance of a particular structure is usually much easier to understand than is the visualization of how to go about calculating inductance.

1.3   OUTLINE OF THE BOOK

In Chapter 2 we summarize the fundamental electromagnetic field laws governing the magnetic field, those of Gauss, Ampère, and Biot–Savart, on which the inductance calculation is based. The magnetic fields $\mathbf{B}$ of various configurations carrying a dc current are derived from these laws. This is a necessary first step in computing the inductance of a structure since the magnetic flux ψ penetrating the surface that comprises the inductance must be computed from $\mathbf{B}$ via (1.6). The inductance of the structure is then obtained as the ratio of the flux and the current producing it via (1.7). The derivation of the $\mathbf{B}$ field for a particular structure that carries a dc current generally involves the setting up and evaluation of somewhat complicated integrals. An extensive table of integrals is given by Dwight [7]. Furthermore, the next step in calculation of the inductance of a structure requires a further integration of $\mathbf{B}$ as in (1.6). An alternative way of computing the $\mathbf{B}$ field of a current-carrying structure is obtained using the vector magnetic potential $\mathbf{A}$. In some cases it is easier to compute $\mathbf{A}$ directly and from this obtain $\mathbf{B}$ by differentiation. The method of images for simplifying problems involving currents over large "ground planes" is also discussed. The commonly assumed fact that all dc currents must "return to their source" and therefore must comprise closed loops is proven.

The important ideas that arise when the currents are, instead of dc, varying with time are discussed in Chapter 3. The fundamental law that provides an understanding of how an inductance produces a voltage between its two terminals is Faraday's law of induction, which is examined in detail. The important notion of displacement current in Ampère's law, which affords an understanding of how a capacitance can conduct a time-varying current through it, is also discussed. The important concepts of waves, wavelength, time delay, and electrical dimensions that allow these static ideas of capacitance and inductance to be incorporated into lumped circuits which have time-varying sources driving them are examined. The important ability of being able to use a quantity that is derived for static (dc) currents (e.g., inductance and capacitance) in a circuit where the currents vary with time is shown in terms of an iterative expansion of the electromagnetic fields. Finally, conservation of energy in the electromagnetic field and Poynting's theorem are reviewed.

With this requisite background, we are able to understand how to calculate and interpret the meaning of the "loop inductance" of a closed loop of current, which is given in Chapter 4. The "loop inductances" for various structures are also derived in Chapter 4 using several methods.

The remaining chapters are devoted to the concept of "partial inductance," which is rapidly becoming important in today's high-speed digital electronics. The general concept of "partial inductance" is examined in Chapter 5, and the self and mutual partial inductances of straight wire segments are determined. The self and mutual partial inductances of conductors of rectangular cross section, which the "lands" on printed circuit boards (PCBs) represent, are determined in Chapter 6. Chapter 7 is devoted to a critical examination of the relative merits of using loop inductances to characterize current loops versus the use of partial inductances. A fairly complex structure is analyzed by first characterizing it with loop inductances and then characterizing it with partial inductances. This example is quite useful in bringing together all the concepts of the previous chapters and in comparing their relative merits and deficiencies.

1.4 "LOOP" INDUCTANCE VS. "PARTIAL" INDUCTANCE

It is critically important that the reader understand the following two distinctions with regard to inductance. In undergraduate electrical engineering courses, only the concept of the inductance of a complete loop of current is studied. (It is shown in Section 2.9 that dc currents must form closed loops.) This "loop inductance" is given in (1.7) and requires that we be able to compute the magnetic flux ψ that passes through the enclosed surface of a closed current loop, as illustrated in Fig. 1.1(b). Therefore, computation of the loop inductance of a structure requires that we be able to identify the *complete current loop*. For "intentional" inductors this current loop is rather obvious. For example, if we wind several turns of wire around a ferromagnetic toroid core, the loop area of the current that the magnetic flux passes through is evident. Hence, the concept of loop inductance of intentional inductors is useful in that it allows us to characterize those as lumped-circuit elements.

On the other hand, if we want to assign an "inductance" to segments of a conductor on a printed circuit board (referred to as lands), there are several problems in trying to use the concept of loop inductance to do so. The first problem is that we must be able to determine the *complete current loop path* in order to calculate loop inductance of that current loop. In other words, we must be able to identify not only the "going down" path from the source to the load (which is relatively easy to do) but also the "return current" path of the current back to the source in order to determine the *complete current loop*. In today's densely packed integrated circuits and printed circuit boards carrying currents having ever-increasing spectral content, this has become virtually impossible to do! Furthermore, the complete path for the current depends on the frequency of the current. For one frequency, the return current will take a particular path, but for a higher frequency the path of the return current may be entirely different!

So the first problem with using loop inductance to model the conductors of a loop is that at different frequencies, the return path of the loop current may be different. This is best illustrated by the situation of a coaxial cable above a ground plane shown in Fig. 1.2. (See [5] for an analysis of this problem.) At dc and low frequencies, the current I takes its return path, I_G, through the massive ground plane. However, at higher frequencies, the current I takes its return path up through the shield, I_S. Therefore, the return paths and hence the complete current loops are different for different frequencies. So if we were to compute the loop inductance it would appear that we would have two different values, depending on the frequency of the current.

The final and most important problem in trying to use loop inductance to allocate inductances to the individual lands on a PCB is that *the total loop inductance of a current loop cannot be placed in any unique position in that loop*. For example, the current loop in Fig. 1.1(b) is said to present an inductance at its input terminals. But that is a loop inductance that cannot,

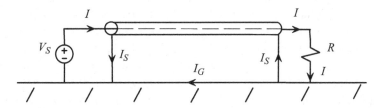

FIGURE 1.2. "Return currents" of different frequencies may take different paths.

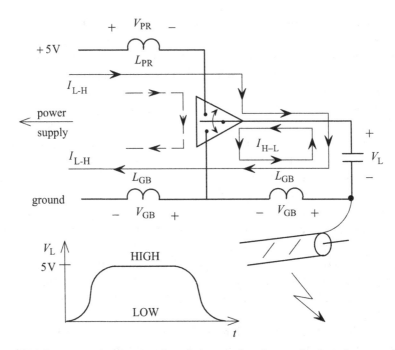

FIGURE 1.3. The problem in using "loop inductance" to characterize the inductance of PCB lands.

a priori, be divided into portions that are associated with segments of that loop!

Figure 1.3 illustrates this problem of trying to use loop inductance to model the inductance of portions of the PCB lands. We have shown a CMOS inverter that is attached to a capacitive load (perhaps representing the input to another CMOS inverter). The +5-V output of the power supply is attached to the +5-V power pin of the CMOS module via a land on the PCB. Similarly, the ground terminal of the power supply is attached to the ground pin of the CMOS module with another land on the PCB. As the inverter switches from the LOW to HIGH state, the current, I_{L-H}, is drawn from the power supply through the +5-V land and through the inverter to charge the capacitor to put the load voltage, V_L, in the HIGH state and returns to the power supply through the ground land. When the inverter switches, the load capacitor then discharges via current I_{H-L} through the inverter via a different loop: from the capacitor, through the inverter, and back to the capacitor. We have shown the conductors as each having associated individual inductances. The land connecting the +5-V output of the power supply to the +5-V pin of the inverter is shown as having an inductance L_{PR}. The land connecting the ground of the power supply to the ground pin of the inverter and the land connecting the bottom of the capacitor to the ground pin of the inverter are also shown as having inductances L_{GB}. (Although these two inductances have the same symbol, they obviously have different values, due to the different lengths of these return paths.) The current I_{L-H} clearly forms a loop: from the power supply through the inverter, through the capacitor, and back to the power supply. When the inverter is switching from LOW to HIGH, the current I_{L-H} through the +5-V land increases in value in order to charge the capacitor. Hence, a voltage is developed across L_{PR} of

$$V_{PR} = L_{PR} \frac{dI_{L-H}}{dt}$$

This is referred to as *power rail collapse*, since the voltage of the power pin of the inverter is $5 - V_{PR}$, and hence the voltage of the +5-V pin of the inverter module drops in value from +5-V. Similarly, the voltage at the ground pin of the inverter goes from zero to V_{GB}:

$$V_{GB} = L_{GB} \frac{dI_{L-H}}{dt}$$

This is referred to as *ground bounce*. On the other hand, when the load voltage is transitioning from HIGH to LOW state, a voltage is developed across the V_{GB} of the other ground land as the current I_{H-L} from the capacitor discharges through the inverter and returns to the capacitor.

Although at first glance this seems to be a straightforward characterization of the individual lands with an inductance, it is not. What do we mean by the inductances of the two lands, L_{PR} and L_{GB}? These certainly are not loop inductances because the total inductance of a loop cannot be placed uniquely in any segment of the loop. In fact, these are "partial inductances." But the type of diagram shown in Fig. 1.3 is seen throughout the literature. The problem here is that few people know how to compute L_{PR} and L_{GB}. Even worse, they often mistakenly compute L_{PR} and L_{GB} using a formula for a loop inductance they find in a handbook that does not apply to these inductances, thereby giving erroneous results for the magnitudes of V_{PR} and V_{GB}!

So loop inductance is not useful in modeling an "unintended inductance" to obtain the voltage developed between its two ends, due to a rate of change of current through it. However, using the concept of loop inductance to model "intended" physical inductors such as a toroid or a solenoid is a useful application of that concept. On the other hand, the concept of partial inductance allows us to represent the lands on a PCB as well as other types of conductors with inductances and to compute the values of those inductances uniquely to determine the correct voltage drop between two ends of the conductor. Unlike loop inductances, we can compute partial inductances *without the necessity of having to be able to identify the return paths for the currents*! We simply model all conductors with their partial inductances (self and mutual between this and other conductors in the circuit), build a lumped-circuit model using these partial inductances, and "turn the crank" (analyze the resulting lumped-circuit model) to find the return paths for the currents rather than trying to guess their paths a priori. There is a dual concept to partial inductance that is referred to by the author as *generalized capacitance* (see references [5,8] for a discussion).

Prior to a decade ago, when the digital clock and data speeds and their spectral content were below about 100 MHz and the density of electronic circuits was not what it is today, the concept of partial inductance was not as important. Today, it is a virtual necessity if we are to cope with the rapidly escalating densities of electronic circuits whose conductors carry currents having increasingly higher spectral content.

The units of the quantities are named for the great scientists who made major contributions to the discovery of these phenomena. Throughout this book, we use the abbreviations A, Wb, H, and T, respectively, for the units amperes, webers, henrys, and tesla. The standard is to use lowercase for the first letter of each of the names of these units and capital letters in their abbreviations.

2

MAGNETIC FIELDS OF DC CURRENTS (STEADY FLOW OF CHARGE)

As discussed in Chapter 1, inductance is intimately related to a closed loop of dc current which produces magnetic flux through the surface surrounded by the current loop. So our first priority is to understand the computation of the magnetic fields of steady (dc) currents that do not vary with time for various configurations of those currents.

2.1 MAGNETIC FIELD VECTORS AND PROPERTIES OF MATERIALS

The fundamental magnetic field vectors are the *magnetic field intensity* **H**, whose units are A/m, and *magnetic flux density* **B**, whose units are Wb/m^2 = T. In a simple (but very common) linear, homogeneous, and isotropic medium, **B** and **H** are related as [3–6]

$$\boxed{\mathbf{B} = \mu\mathbf{H} = \mu_0\mu_r\mathbf{H}} \tag{2.1}$$

where μ is the *permeability* of the medium. The permeability can be written as the product of the *relative permeability* μ_r and the permeability of free space (essentially, air), $\mu_0 = 4\pi \times 10^{-7}$ H/m:

$$\mu = \mu_r \mu_0$$ (2.2)

The units of permeability are named for Joseph Henry of Albany, New York, who essentially discovered Faraday's law at about the time Faraday did but did not publish his results until much later. Hence, for linear, homogeneous, and isotropic media, **B** and **H** can be freely interchanged according to (2.1). Dielectrics and metals that are not magnetizeable, such as copper, aluminum, and brass, are linear and isotropic with regard to magnetic fields and have $\mu_r = 1$. However, materials that are magnetizeable have $\mu_r > 1$ and are generally nonlinear.

There are common materials such as iron and steel that are magnetizable. These are said to be nonlinear with respect to magnetic fields. Ferromagnetic materials such as iron and steel have **B** and **H** related by the common "hysteresis curve" shown in Fig. 2.1. Suppose that we wind N turns of wire around a toroid of nonlinear magnetic material such as iron and pass a current I through the turns of wire as illustrated in Fig. 2.2. The turns of wire produce a magnetic field intensity of approximately $H = NI$. Starting from an unmagnetized toroid, $B = H = 0$, we start at the origin in Fig. 2.1. Increasing the current I we move up and to the right, reaching a point where further increases in I (and H) cause little or no change in B. At this point the material is said to be in *saturation*. Upon reducing I we proceed to a point where $I = 0$ (and $H = 0$) but the magnetic flux density B has not decreased to zero. Further reductions of I for negative values reduces B to zero where H is negative. This process continues as we cycle around the hysteresis curve.

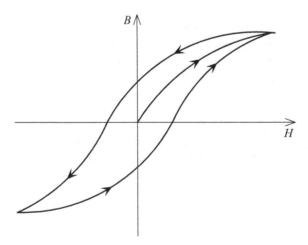

FIGURE 2.1. "Hysteresis curve" for nonlinear magnetic media.

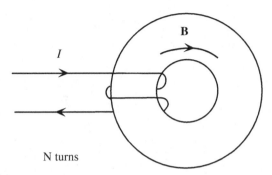

FIGURE 2.2. Toroid.

The instanteous slope at a point on the hysteresis curve is the *incremental permeability* of the material:

$$\Delta\mu = \frac{B}{H} \tag{2.3}$$

The slope of the hysteresis curve at $B = H = 0$ when the material is unmagnetized is called the *initial permeability* and is typically stated in the brochures of manufacturers of the material. Clearly, this material is nonlinear and there can be no numerical value for a "permeability" stated for it. Because of this difficulty we deal only with materials such as air and copper, for which $\mu_r = 1$, or nonlinear magnetic materials where the applied currents and consequently the levels of H are sufficiently small that we can consider the material to be linear, having its initial permeability. We could also deal with situations where, for example, the sinusoidal variations of the current and consequentially of H are sufficiently small that we can use an incremental permeability to characterize this nonlinear magnetic material. Common ferromagnetic materials are steel ($\mu_r = 2000$), iron ($\mu_r = 1000$), and nickel ($\mu_r = 600$), as well as certain powdered ferrites such as nickel–zinc ($\mu_r \cong 600$) and manganese–zinc ($\mu_r \cong 1200$). Certain exotic materials such as Mu-metal ($\mu_r = 30,000$) have very large relative permeabilities (at low frequencies, e.g., 1 kHz, and low values of H).

2.2   GAUSS'S LAW FOR THE MAGNETIC FIELD AND THE SURFACE INTEGRAL

In the case of fixed distributions of charge, electric field lines that begin on a positive charge must end on a negative charge, as illustrated in Fig. 2.3. So

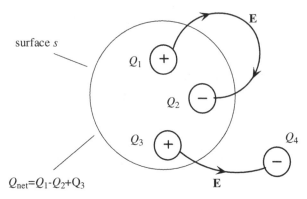

FIGURE 2.3. Static electric field of fixed distributions of charge.

we can view charges as a source of the electric field intensity vector **E**, whose units are V/m. Gauss's law for the electric field is stated as [3–6]

$$\oint_s \mathbf{D} \cdot d\mathbf{s} = Q_{\text{enclosed}} \tag{2.4}$$

where **D** is the electric flux density whose units are C/m^2 and $\mathbf{D} = \varepsilon \mathbf{E}$, where ε is the *permittivity* of the surrounding medium. The *surface integral* in (2.4) gives the *net flux* of the **D** field out of the closed surface *s* (see the Appendix for a discussion of the surface integral). Hence, Gauss's law in (2.4) simply provides that if we take the products of the differential surface elements *ds* and the components of **D** that are *perpendicular to the surface s* and add them over the closed surface *s*, we will obtain the *net* positive charge enclosed by the closed surface *s*. This is a sensible result because there are two components of **D** and **E** at a point on the surface *s*: One component is parallel to the surface and the other is perpendicular to the surface. Only the component *perpendicular* to the surface contributes to the net flux of the electric field entering or leaving the surface *s*.

However, in the case of magnetic fields, there are no known sources or sinks for the magnetic field, so that *the magnetic field lines must form closed loops*. If we cut a permanent magnet into two pieces, we do not create isolated sources of the magnetic field, as illustrated in Fig. 2.4. Gauss's law for the magnetic field states this important fact in terms of a surface integral [3–6]:

$$\psi = \oint_s \mathbf{B} \cdot d\mathbf{s} = 0 \tag{2.5}$$

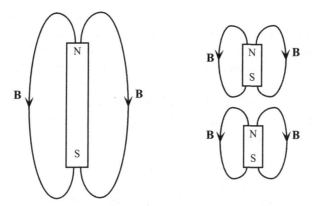

FIGURE 2.4. Permanent magnets.

This law provides that if we take the surface integral of the magnetic flux density **B** over a *closed surface s* as illustrated in Fig. 2.5 giving the net magnetic *flux*, ψ, out of the closed surface, we will obtain a result of zero for any closed surface: There is no *net* magnetic flux entering or leaving a closed surface. The units of **B** are $Wb/m^2 = T$. Hence, the units of the magnetic flux ψ *leaving the closed surface s* are webers. The surface integral in (2.5) simply provides that if we take the products of the differential surface elements *ds* and the components of **B** that are *perpendicular to the surface s* and add them over the closed surface, we will obtain a result of zero. This is a sensible

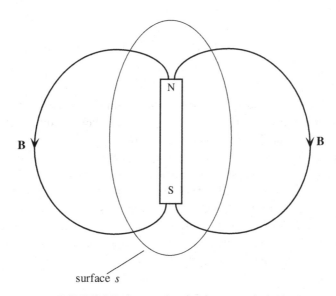

surface *s*

FIGURE 2.5. Gauss's law for the magnetic field.

result if the magnetic field lines must close on themselves since there are two components of **B** at a point on the surface: One component is parallel to the surface and the other is perpendicular to the surface. Only the component *perpendicular* to the surface contributes to the net flux of the magnetic field out of the surface *s*.

The laws of Gauss in (2.4) and (2.5) are said to be in *integral form*; that is, they apply to broad regions of space. The *point forms* of these laws apply to specific points in space and are [3–6]

$$\boxed{\nabla \cdot \mathbf{D} = \rho}$$ (2.6)

for the electric field, where ρ is the volume charge density at the point whose units are C/m^3 and

$$\boxed{\nabla \cdot \mathbf{B} = 0}$$ (2.7)

for the magnetic field. The notation $\nabla \cdot \mathbf{F}$ denotes the divergence of the vector field **F** (see the Appendix). These point forms can be derived from the integral forms using the divergence theorem (see the Appendix):

$$\oint_s \mathbf{D} \cdot d\mathbf{s} = \int_v (\nabla \cdot \mathbf{D})\, dv$$

$$= Q_{\text{enclosed}}$$

$$= \int_v \rho\, dv$$

and

$$\oint_s \mathbf{B} \cdot d\mathbf{s} = \int_v (\nabla \cdot \mathbf{B})\, dv$$

$$= 0$$

where the closed surface *s* encloses the volume *v*. Comparing both sides gives the point forms in (2.6) and (2.7). Gauss's law for the electric field in (2.6) provides that the *divergence* or net outflow of the electric field lines from a point equals the net positive volume charge density at the point (see the Appendix for a discussion of divergence). Gauss's law for the magnetic field in (2.7) simply provides that there is no *divergence* of the magnetic field lines: There are no isolated sources or sinks for the magnetic field, and the magnetic field lines must therefore form closed loops. In a rectangular coordinate system consisting of mutually orthogonal axes *x*, *y*, and *z*, we may write the "del operator" as (see the Appendix)

$$\nabla = \mathbf{a}_x \frac{\partial}{\partial x} + \mathbf{a}_y \frac{\partial}{\partial y} + \mathbf{a}_z \frac{\partial}{\partial z}$$ (2.8)

and Gauss's laws become

$$\nabla \cdot \mathbf{D} = \frac{\partial D_x}{\partial x} + \frac{\partial D_y}{\partial y} + \frac{\partial D_z}{\partial z} = \rho \qquad (2.9)$$

$$\nabla \cdot \mathbf{B} = \frac{\partial B_x}{\partial x} + \frac{\partial B_y}{\partial y} + \frac{\partial B_z}{\partial z} = 0 \qquad (2.10)$$

Note that if the vector components of **B** are *independent of* x, y, and z, respectively [i.e., $B_x(y, z)$, $B_y(x, z)$, and $B_z(x, y)$], the divergence of **B** will automatically be zero. But there are obviously many cases where the vector components of **B** are functions of some or all of the axis variables x, y, and z, yet the divergence of **B** is still zero.

2.3 THE BIOT–SAVART LAW

Perhaps the most fundamental law that allows computation of the magnetic field due to a dc current is the Biot–Savart law [3–6,9–11]:

$$\mathbf{B} = \frac{\mu_0}{4\pi} \int_v \frac{\mathbf{J} \times \mathbf{a}_R}{R^2} dv \qquad (2.11)$$

The dc current density vector is denoted as **J**, whose units are A/m^2, and v is the volume containing this current. A differential segment or "chunk" of this current density vector contains $\mathbf{J}\,dv$ ampere-meters, and the distance from this chunk of current (the source of the **B** field) to the point at which we are computing the magnetic field **B** is denoted as R. The unit vector $\mathbf{a}_R$ is directed *from* this differential chunk of current *to* the point at which we are computing **B**. The resulting **B** field is perpendicular to the plane containing **J** and $\mathbf{a}_R$ according to the right-hand rule (see the Appendix). Note that the Biot–Savart law is an inverse-square law like Coulomb's law and the law of gravity since it depends on the inverse of the square of the distance between **B** and the differential segment of the current density vector (the source of the field).

Throughout this book we generally concentrate on line currents, denoted as I, whose units are amperes. Considering a differential length of these currents as a small cylinder of length dl and cross-sectional area ds with current density distributed uniformly over the cross section (as will be the case for dc currents [3]), $J\,ds = I$, so that $J\,dv = J\,ds\,dl = I\,dl$. In this case the

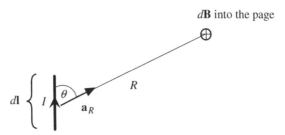

FIGURE 2.6. Biot–Savart law.

differential contribution of this current to the magnetic flux density vector at a point is

$$d\mathbf{B} = \frac{\mu_0 I}{4\pi R^2} d\mathbf{l} \times \mathbf{a}_R \qquad (2.12a)$$

as illustrated in Fig. 2.6. The direction of the vector differential length $d\mathbf{l}$ of this filamentary current segment is in the direction of the current.

Note that the magnetic field depends on the cross product $d\mathbf{l} \times \mathbf{a}_R$, where the unit vector from the current element to the point $\mathbf{a}_R$ is directed *from* the current *to* the point (see the Appendix for a review of the cross product). Hence, the magnetic field is directed into the page (perpendicular to the plane containing $d\mathbf{l}$ and the unit vector $\mathbf{a}_R$ according to the *right-hand rule*). Hence, in terms of the angle θ between these two vectors, we can write the Biot–Savart law as

$$d\mathbf{B} = \frac{\mu_0 I dl}{4\pi R^2} \sin\theta \, \mathbf{a}_n \qquad (2.12b)$$

where $\mathbf{a}_n$ is a unit vector perpendicular to the plane containing $d\mathbf{l}$ and $\mathbf{a}_R$ according to the right-hand rule in the order $d\mathbf{l} \, \mathbf{a}_n = d\mathbf{l} \times \mathbf{a}_R$ (i.e., pointing into the page). So the magnetic field is a maximum along a line perpendicular to the current element and is zero off the ends of the current element. If we place the current element along the z axis of a cylindrical coordinate system (see the Appendix), the magnetic field will be directed circumferentially around the current in the ϕ direction at *all* points around the current.

EXAMPLE

As an example, we use the Biot–Savart law to determine the magnetic field about a current of finite length L. Since dc currents must form closed loops

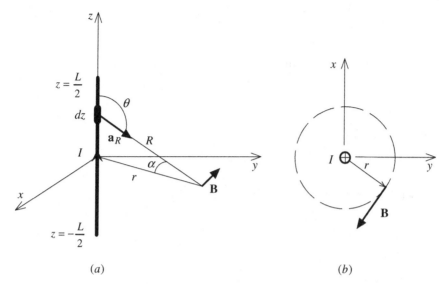

FIGURE 2.7. Current of length L and the magnetic field about it.

(see Section 2.9), we use the magnetic fields of finite lengths of current to construct the fields of closed current loops by the superposition of the fields of the current segments of the closed current loop. Hence, determining the magnetic fields of finite lengths of current is useful from that standpoint.

First, set up a rectangular coordinate system and orient the current along the z axis and centered on the origin with the current directed in the positive z direction, as shown in Fig. 2.7(a). We will determine the magnetic flux density at a point that is a distance $r = \sqrt{x^2 + y^2}$ *from the midpoint of the current* and along a line that is *perpendicular to the current*. The contribution to the magnetic field at a distance r from the origin of the coordinate system that is due to a differential length of the current dz which is at a distance R from the point is

$$dB = \frac{\mu_0 I dz}{4\pi R^2} \sin\theta$$

The direction of this B field is, according to the Biot–Savart law, perpendicular to the plane containing the positive z axis and the unit vector from the current element $I\,dz$ to the point according to the right-hand rule: $\mathbf{a}_z \times \mathbf{a}_R$. Hence, it is directed circumferentially about the current. This is in the ϕ direction in a cylindrical coordinate system, $\mathbf{a}_\phi$ (see the Appendix for a discussion of the cylindrical coordinate system). The sine of the angle involved in the cross product is

$$\sin\theta = \sin(\alpha + 90°) = \cos\alpha$$
$$= \frac{r}{R}$$

and the distance R is

$$R = \sqrt{z^2 + r^2}$$

Hence, the total magnetic field at the point is

$$
\begin{aligned}
B &= \frac{\mu_0 I}{4\pi} \int_{z=-L/2}^{L/2} \frac{r}{R^3}\,dz \\
&= \frac{\mu_0 I}{4\pi} \int_{z=-L/2}^{L/2} \frac{r}{\left(r^2 + z^2\right)^{3/2}}\,dz \\
&= \frac{\mu_0 I r}{4\pi} \left[\frac{z}{r^2\sqrt{r^2 + z^2}} \right]_{z=-L/2}^{L/2} \\
&= \frac{\mu_0 I r}{4\pi} \left[\frac{L/2}{r^2\sqrt{r^2 + (L/2)^2}} - \frac{-L/2}{r^2\sqrt{r^2 + (-L/2)^2}} \right] \\
&= \frac{\mu_0 I r}{4\pi} \frac{L}{r^2\sqrt{r^2 + (L/2)^2}} \\
&= \frac{\mu_0 I}{4\pi r} \frac{L}{\sqrt{r^2 + L^2/4}} \\
&= \frac{\mu_0 I}{2\pi r} \frac{L}{\sqrt{4r^2 + L^2}}
\end{aligned}
$$

We have used integral 200.03 from the table of integrals by Dwight [7]:

$$\int \frac{1}{\left(a^2 + x^2\right)^{3/2}}\,dx = \frac{x}{a^2\sqrt{a^2 + x^2}} \qquad \text{(D200.03)}$$

(*Note:* Throughout this book we evaluate the somewhat complicated integrals we encounter using the extensive table of integrals by Dwight [7]. These integrals will be denoted as (Dxxx.xx) according to the integral number in Dwight.) The magnetic flux density vector is directed in the circumferential direction about the wire which corresponds to the ϕ coordinate of the cylindrical coordinate system. Hence, the result can be written as a vector:

$$\mathbf{B} = \frac{\mu_0 I}{4\pi r} \frac{L}{\sqrt{r^2 + (L/2)^2}} \mathbf{a}_\phi$$

$$= \frac{\mu_0 I}{2\pi r} \frac{L}{\sqrt{4r^2 + L^2}} \mathbf{a}_\phi \qquad (2.13)$$

where $\mathbf{a}_\phi = \mathbf{a}_z \times \mathbf{a}_r$.

For an infinite length of current, $L \to \infty$, (2.13) reduces to a very fundamental result that we will use on numerous occasions:

$$\boxed{\mathbf{B} = \frac{\mu_0 I}{2\pi r} \mathbf{a}_\phi \qquad L \to \infty} \qquad (2.14)$$

Determination of the direction of the magnetic field of a current is obtained with the famous *right-hand rule*. [The official symbol of the Institute of Electrical and Electronics Engineers (IEEE) memorializes this very fundamental rule.] According to the Biot–Savart law, if we place the thumb of our right hand in the direction of the current I, the fingers of that hand will give the resulting direction of the $\mathbf{B}$ field, which is perpendicular to the plane containing (1) the current I and (2) the vector pointing from the current to the point at which we desire to determine the $\mathbf{B}$ field; $\mathbf{a}_\phi = \mathbf{a}_z \times \mathbf{a}_r$. Hence, *the magnetic field about an infinitely long current is directed circumferentially about the wire at all points along it according to the right-hand rule, decays inversely with distance from the wire, and is constant in magnitude at distances r from the wire* as illustrated in Fig. 2.7(b). An important difference between the magnetic fields of a wire of infinite length and one of finite length is that the latter has fringing fields at its endpoints.

EXAMPLE

In the preceding example we centered the current on the origin of the coordinate system and determined the $\mathbf{B}$ field at a radial distance r from that center *and on a line perpendicular to the midpoint of the current*. We next generalize this result to obtain the magnetic field of a current that is of finite length but at *any point* about the current which is *not necessarily on a line perpendicular to its midpoint*, as shown in Fig. 2.8.

We again orient the current along the z axis and center the current on the origin of that coordinate system, but the $\mathbf{B}$ field is determined at a general point that is at a horizontal distance r (the cylindrical coordinate system variable)

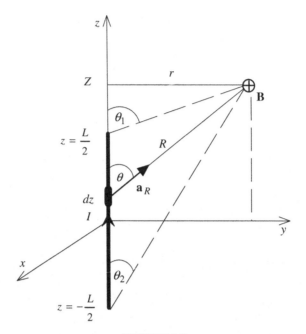

FIGURE 2.8

from the z axis and is located at an arbitrary value of $z = Z$. Using the Biot–Savart law we see that the **B** field is again circumferential about the current and, for example, is perpendicular to the yz plane. The **B** field is again in the $\mathbf{a}_\phi = \mathbf{a}_z \times \mathbf{a}_R$ direction. The Biot–Savart law again gives

$$dB = \frac{\mu_0 I}{4\pi R^2} \sin \theta \, dz$$

The distance R from the current element to the point is

$$R = \sqrt{(Z - z)^2 + r^2}$$

and

$$\sin \theta = \frac{r}{R}$$

Hence, the integral to be evaluated is

$$B = \frac{\mu_0 I r}{4\pi} \int_{z=-L/2}^{L/2} \frac{1}{\left[(Z - z)^2 + r^2\right]^{3/2}} \, dz$$

Using a change of variables, $Z - z = \lambda$, $d\lambda = -dz$ gives

$$B = \frac{\mu_0 I r}{4\pi} \int_{\lambda=Z-L/2}^{Z+L/2} \frac{1}{(\lambda^2 + r^2)^{3/2}} \, dz$$

Using Dwight [7] (D200.03) again gives

$$\mathbf{B} = \frac{\mu_0 I}{4\pi r} \left[\frac{Z + L/2}{\sqrt{(Z + L/2)^2 + r^2}} - \frac{Z - L/2}{\sqrt{(Z - L/2)^2 + r^2}} \right] \mathbf{a}_\phi \qquad (2.15)$$

which, of course, reduces to (2.13) for $Z = 0$. For a current of infinite length, $L \to \infty$, (2.15) reduces to (2.14). In terms of the angles θ_1 and θ_2 between the z axis and lines drawn from the ends of the current to the point, this becomes

$$\mathbf{B} = \frac{\mu_0 I}{4\pi r} (\cos \theta_2 - \cos \theta_1) \, \mathbf{a}_\phi \qquad (2.16)$$

In the case of a current of infinite length, $L \to \infty$, $\theta_1 \to \pi$, and $\theta_2 \to 0$ and (2.16) reduces to (2.14).

EXAMPLE

The principle of *superimposing* the contributions of several currents to give the total field at a point is a powerful technique for linear media. We show in Section 2.9 that *steady (dc) currents must form closed loops*. Hence, we use this principle of superimposing the contributions of the segments of the current of a closed current loop to obtain the total magnetic field of closed loops of current. In this example we determine the total magnetic field at a distance d from the center of a rectangular loop of current having sides of length w and l and *along a line that is perpendicular to the loop at a distance d from its center*, as shown in Fig. 2.9. We restrict this solution to a point along a line from the center of the loop because the equation for the $\mathbf{B}$ field at any other point about the loop is very difficult to derive and the result is extraordinarily complicated (see [9], p. 286). Treat this as four currents whose $\mathbf{B}$ fields are given by (2.13) and superimpose the fields. The $\mathbf{B}$ field due to each side is perpendicular to a line drawn from the center of each current to the point. Considering two pairs of opposite sides, we see from Fig. 2.9 that the horizontal contributions (in the xy plane) cancel and we are left with the total in the z direction. (Use the right-hand rule to determine the direction of the magnetic field that is due to each current.) Hence, the

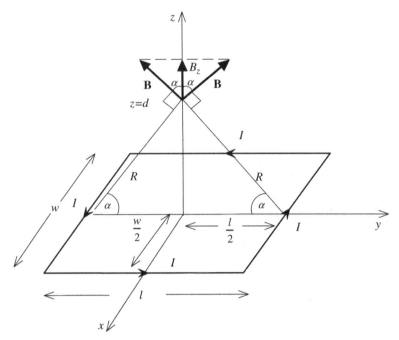

FIGURE 2.9. Magnetic field at a distance d along a line perpendicular to the center of a rectangular loop.

magnetic field at the point due to two of the opposite sides each of length w is, using (2.13),

$$\mathbf{B} = 2\,\frac{\mu_0 I}{2\pi}\,\frac{w}{R\sqrt{4R^2 + w^2}}\,\cos\alpha\,\mathbf{a}_z$$

where

$$R = \sqrt{\left(\frac{l}{2}\right)^2 + d^2}$$

$$= \frac{1}{2}\sqrt{l^2 + 4d^2}$$

and

$$\cos\alpha = \frac{l/2}{R}$$

Hence, the total from two of the opposite sides is

$$\mathbf{B} = 2\frac{\mu_0 I}{\pi}\,\frac{wl}{\left(l^2 + 4d^2\right)\sqrt{l^2 + w^2 + 4d^2}}\,\mathbf{a}_z$$

Adding the contributions from the other two opposite sides gives the total as

$$\mathbf{B} = 2\frac{\mu_0 I}{\pi} \left[\frac{wl}{\left(l^2 + 4d^2\right)\sqrt{l^2 + w^2 + 4d^2}} \right.$$

$$\left. + \frac{lw}{\left(w^2 + 4d^2\right)\sqrt{w^2 + l^2 + 4d^2}} \right] \mathbf{a}_z \qquad (2.17)$$

At the center of the loop, $d = 0$, this reduces to

$$\mathbf{B} = 2\frac{\mu_0 I}{\pi} \left(\frac{w}{l\sqrt{l^2 + w^2}} + \frac{l}{w\sqrt{w^2 + l^2}} \right) \mathbf{a}_z \qquad d = 0$$

$$= 2\frac{\mu_0 I}{\pi} \frac{\sqrt{l^2 + w^2}}{wl} \mathbf{a}_z \qquad d = 0 \qquad (2.18)$$

For a square loop, $w = l$, (2.17) reduces to

$$\mathbf{B} = 2\sqrt{2}\frac{\mu_0 I}{\pi} \frac{l^2}{\left(l^2 + 4d^2\right)\sqrt{l^2 + 2d^2}} \mathbf{a}_z \qquad l = w \qquad (2.19)$$

At the center of a square loop, $w = l$ and $d = 0$, (2.18) becomes

$$\mathbf{B} = 2\sqrt{2}\frac{\mu_0 I}{\pi l} \mathbf{a}_z \qquad w = l,\ d = 0 \qquad (2.20)$$

EXAMPLE

Consider a sheet of current lying in the yz plane as shown in Fig. 2.10(a). The sheet carries a surface current K whose units are A/m that is parallel to the yz plane and directed in the z direction. The sheet extends to infinity in all directions. Viewing this as currents of infinite length directed in the z direction whose values are $I = K\,dy$, we can superimpose their $\mathbf{B}$ fields using the results for an infinite current obtained in (2.14). The $\mathbf{B}$ field due to one of the currents at a point along the $+x$ axis at $x = d$ (perpendicular to the plane containing the surface current) as shown in Fig. 2.10(b) is

$$dB = \frac{\mu_0 K}{2\pi R} dy$$

where

$$R = \sqrt{d^2 + y^2}$$

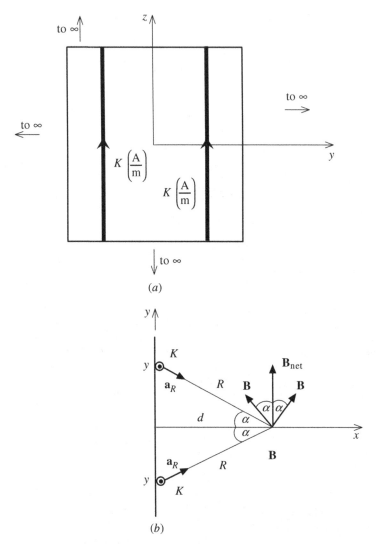

FIGURE 2.10. Infinite current sheet.

The direction of the **B** field from each current is, according to the Biot–Savart law, perpendicular to the plane containing the current and a unit vector directed from the current to the point: $\mathbf{a}_z \times \mathbf{a}_R$. The x components of the fields of two symmetrically disposed currents cancel as shown in Fig. 2.10(b), giving the net field in the y direction as

$$\mathbf{B}_{\text{net}} = 2\,\frac{\mu_0 K}{2\pi}\int_{y=0}^{\infty}\frac{1}{R}\cos\alpha\,dy\,\mathbf{a}_y$$

where

$$\cos \alpha = \frac{d}{R}$$

Substituting gives

$$\mathbf{B}_{\text{net}} = 2 \frac{\mu_0 K}{2\pi} \int_{y=0}^{\infty} \frac{1}{R} \cos \alpha \, dy \, \mathbf{a}_y$$

$$= \frac{\mu_0 K d}{\pi} \int_{y=0}^{\infty} \frac{1}{d^2 + y^2} \, dy \, \mathbf{a}_y$$

$$= \begin{cases} \dfrac{\mu_0 K}{2} \, \mathbf{a}_y & \text{for } x > 0 \\ -\dfrac{\mu_0 K}{2} \, \mathbf{a}_y & \text{for } x < 0 \end{cases} \tag{2.21}$$

and we have used integral 120.1 from Dwight [7]:

$$\int \frac{1}{a^2 + x^2} \, dx = \frac{1}{a} \tan^{-1} \frac{x}{a} \tag{D120.1}$$

Hence, the magnetic flux density at any distance from a current sheet is directed parallel to the sheet and is independent of distance from the sheet. This result applies also to the field on the other side of the sheet, but the direction of the field is in the $-y$ direction on that side.

EXAMPLE

We can generalize the result for an infinite current sheet obtained in the preceding example to one that has a finite width W and finite length L. We will determine the magnetic flux density vector $\mathbf{B}$ at a point that is a distance $x = d$ from *the center of the sheet*, as illustrated in Fig. 2.11. Again viewing this result as a superposition of the fields due to two symmetrically disposed but *finite-length* currents $I = K \, dy$ of length L that are parallel to the z axis, we can use the result obtained in (2.13) for the field at a point a distance r from the midpoint of a finite-length current and write the net $\mathbf{B}$ field as

$$\mathbf{B}_{\text{net}} = 2 \frac{\mu_0 K}{4\pi} \int_{y=0}^{W/2} \frac{L}{R\sqrt{R^2 + (L/2)^2}} \cos \alpha \, dy \, \mathbf{a}_y$$

Substituting $R = \sqrt{d^2 + y^2}$ and $\cos \alpha = d/R$ gives

$$\mathbf{B}_{\text{net}} = \frac{\mu_0 K L d}{2\pi} \int_{y=0}^{W/2} \frac{1}{(y^2 + d^2)\sqrt{y^2 + d^2 + (L/2)^2}} \, dy \, \mathbf{a}_y$$

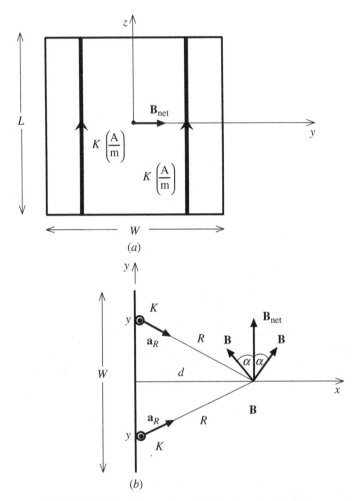

FIGURE 2.11. Current sheet of finite length and width.

Using integral 387 from Dwight [7],

$$\int \frac{dx}{(ax^2 + b)\sqrt{fx^2 + g}} = \frac{1}{\sqrt{b}\sqrt{ag - bf}} \tan^{-1} \frac{x\sqrt{ag - bf}}{\sqrt{b}\sqrt{fx^2 + g}}$$

$$\text{(D387)}$$

gives

$$\mathbf{B}_{net} = \frac{\mu_0 K}{\pi} \tan^{-1} \frac{(W/2)(L/2)}{d\sqrt{d^2 + (W/2)^2 + (L/2)^2}} \, \mathbf{a}_y \qquad x > 0 \quad (2.22)$$

On the back side of the plate, $x < 0$, the result in (2.22) must be negated according to the right-hand rule.

Taking the limit as $L \to \infty$ gives the result for an infinitely long current strip of width W at a distance d from its center and perpendicular to the strip surface as

$$\mathbf{B}_{\text{net}} = \frac{\mu_0 K}{\pi} \tan^{-1} \frac{W}{2d} \, \mathbf{a}_y \qquad L \to \infty \qquad (2.23)$$

This result in (2.23) can be derived directly by using the result for an infinite current in (2.14), $\mathbf{B} = (\mu_0 I / 2\pi r) \, \mathbf{a}_\phi$:

$$\mathbf{B} = 2 \frac{\mu_0 K}{2\pi} \int_{y=0}^{W/2} \frac{1}{R} \cos \alpha \, dy \, \mathbf{a}_y \qquad L \to \infty$$

$$= \frac{\mu_0 K}{\pi} \int_{y=0}^{W/2} \frac{d}{d^2 + y^2} \, dy \, \mathbf{a}_y$$

$$= \frac{\mu_0 K}{\pi} \tan^{-1} \frac{W}{2d} \, \mathbf{a}_y$$

where we again used integral 120.1 from Dwight [7]. Taking the limit of this as $W \to \infty$ gives an infinite current sheet and the result derived directly in (2.21).

EXAMPLE

A loop of current of radius a is centered on the origin of a rectangular coordinate system and lies in the xy plane as shown in Fig. 2.12. Determine the $\mathbf{B}$ field at a point $z = d$ on the z axis. A segment of the current loop is of length $a \, d\phi$, where ϕ is the cylindrical coordinate system variable. The distance R from the differential segment to the point is $R = \sqrt{a^2 + d^2}$, and the cosine of the angle between the differential contribution $d\mathbf{B}$ and the z axis is

$$\cos \alpha = \frac{a}{R}$$

As ϕ varies from $\phi = 0$ to $\phi = 2\pi$ the horizontal components (in the xy plane) of $d\mathbf{B}$ cancel, leaving the $\mathbf{B}$ field along the z axis in the positive z direction as

$$\mathbf{B} = \frac{\mu_0 I}{4\pi} \int_{\phi=0}^{2\pi} \frac{1}{R^2} \cos \alpha \, a \, d\phi \, \mathbf{a}_z$$

$$= \frac{\mu_0 I}{4\pi} \int_{\phi=0}^{2\pi} \frac{a^2}{(a^2 + d^2)^{3/2}} \, d\phi \, \mathbf{a}_z$$

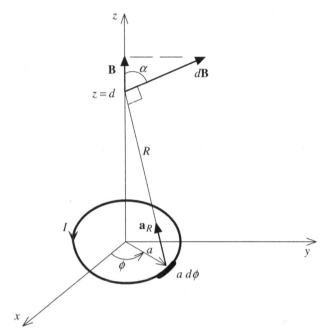

FIGURE 2.12. Current loop.

But this is a simple integral since the integrand does not depend on ϕ:

$$\mathbf{B} = \frac{\mu_0 I}{2} \frac{a^2}{(a^2 + d^2)^{3/2}} \mathbf{a}_z \qquad z \geq 0 \qquad (2.24)$$

At the center of the loop, the field is

$$\mathbf{B} = \frac{\mu_0 I}{2a} \mathbf{a}_z \qquad d = 0 \qquad (2.25)$$

At very large distances from the loop compared to the loop radius, $d \gg a$, (2.24) simplifies to

$$\mathbf{B} = \frac{\mu_0 I a^2}{2d^3} \mathbf{a}_z$$

$$= \frac{\mu_0 m}{2\pi d^3} \mathbf{a}_z \qquad d \gg a \qquad (2.26)$$

The *magnetic dipole moment m* is defined as the product of the current and the area of the current loop:

$$m = \pi a^2 I \qquad (2.27)$$

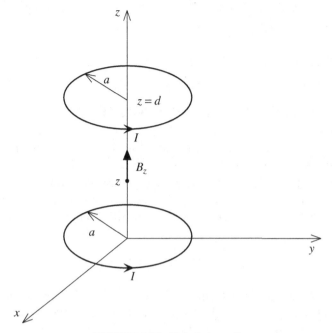

FIGURE 2.13. Helmholtz coil.

Observe that at relatively large distances from the loop, $d \gg a$, the magnetic field decays with distance as inverse-distance cubed.

A Helmholtz coil, shown in Fig. 2.13, is a pair of current loops that are used to provide a fairly uniform magnetic field. Superimposing the results for the magnetic field on the axis of each coil obtained in (2.24) gives the magnetic field along the z axis as

$$B_z = \frac{\mu_0 I a^2}{2} \left[\frac{1}{\left(a^2 + z^2\right)^{3/2}} + \frac{1}{\left(a^2 + (z-d)^2\right)^{3/2}} \right] \qquad (2.28)$$

To examine the change in the field along the z axis between these two coils, we differentiate (2.28) with respect to z to give

$$\frac{\partial B_z}{\partial z} = \frac{3\mu_0 I a^2}{2} \left[\frac{-z}{\left(a^2 + z^2\right)^{5/2}} - \frac{z-d}{\left(a^2 + (z-d)^2\right)^{5/2}} \right] \qquad (2.29)$$

This derivative is precisely zero midway between the two coils at $z = d/2$, meaning that the rate of change of the field along the z axis midway between the two coils is zero. If we take the second derivative, it can also be made zero at $z = d/2$ if we choose the separation between the two coils equal to their radii, $d = a$, thereby giving a further uniform nature of the **B** field between the two coils.

2.4   AMPÈRE'S LAW AND THE LINE INTEGRAL

In Section 2.3 we showed how to calculate the magnetic fields of currents using the Biot–Savart law. The calculations required the evaluation of certain integrals. Ampère's law allows the direct solution of many of those problems *without the evaluation of any integrals*, but the problem must exhibit a certain symmetry to be able to do so. Ampère's law for dc currents is stated as [3,6]

$$\oint_c \mathbf{H} \cdot d\mathbf{l} = I_{\text{enclosed}} \tag{2.30}$$

where **H** is the *magnetic field intensity* vector. Recall that for a linear, homogeneous, and isotropic surrounding medium, **B** and **H** can be freely interchanged using $\mathbf{B} = \mu\mathbf{H}$, and $\mu = \mu_r\mu_0$ is the permeability of the surrounding medium. Ampère's law essentially provides that if we sum the product of the differential segments of the path, dl, and the components of **H** that are *tangent* to a closed path c, we obtain the *net* current that penetrates the surface s that is enclosed by the closed path illustrated in Fig. 2.14. The direction of the closed contour c and the direction of the enclosed current are related by the right-hand rule: Place the fingers of the right hand in the direction of c and the thumb will point in the direction of I_{enclosed}. The integral on the left-hand side

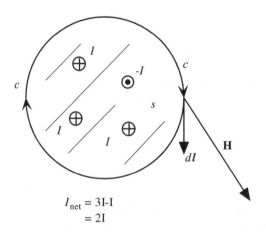

$$I_{\text{net}} = 3\text{I}\text{-}\text{I}$$
$$= 2\text{I}$$

FIGURE 2.14. Ampère's law.

of Ampère's law is said to be a *line integral* (see the Appendix for a review of the line integral). The line integral adds the products of the differential path lengths dl and the components of **H** that are *tangent* to the contour path c. There are two components of **H**: One is parallel to the path and the other is perpendicular to the path. It is sensible that only the components of **H** that are parallel to the path should contribute to the line integral.

Ampère's law is similar to Gauss's law for the electric field given in (2.4), which provides that the sum of the products of the components of **D** that are *perpendicular* to a *closed* surface s and the differential surface areas ds will give the net positive charge enclosed by the closed surface.

Ampère's law can be used to compute the **H** field for current distributions by using symmetry. To use Ampère's law to determine **H**, we must be able to *choose* a closed contour c encircling the current so that the **H** field along that contour has two properties. The first property is that **H** must be *tangent to the closed contour c at every point on it*. This allows us to remove the dot product and write Ampère's law solely in terms of the magnitudes of H and dl as

$$\oint_c H \, dl = I_{\text{enclosed}} \qquad (2.31a)$$

The second property is that H must be *constant at all points along the contour c*. This will allow us to remove H from the integral in (2.31a) and write Ampère's law as

$$H \underbrace{\oint_c dl}_{\substack{\text{total length} \\ \text{of contour } c}} = I_{\text{enclosed}} \qquad (2.31b)$$

Hence, if contour c can be chosen such that it has these two properties, H is simply the total current enclosed divided by the total length of contour c.

EXAMPLE

Determine the magnetic field intensity about a current that is infinite in length. This was solved in Section 2.3 using the Biot–Savart law, which required setting up and evaluating an integral. To solve this problem using Ampère's law, we again orient the current along the z axis with the current directed in the $+z$ direction as shown in Fig. 2.15. We observe that because of the Biot–Savart law, the **H** field will be circumferentially directed about the current *at*

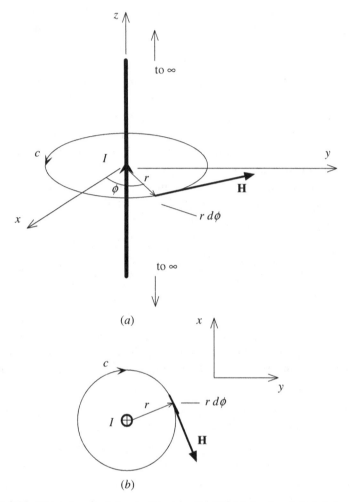

FIGURE 2.15. Using Ampère's law to determine the *H* field about an infinitely long current.

all points along it. Because of the assumption that the current is infinite in length, we may choose a closed contour *c* that is a circle of radius *r* centered on the current and place it at *any point* along the wire. The **H** field will be tangent to all points on this contour. This allows us to remove the dot product from Ampère's law:

$$\oint_c H \, dl = \int_{\phi=0}^{2\pi} Hr d\phi$$
$$= I$$

In addition, the Biot–Savart law shows that the H field magnitude will be constant in value at all points on c that are a distance r from the current, so that we may remove H from the integral and obtain

$$H \oint_c dl = H \int_{\phi=0}^{2\pi} r \, d\phi$$

$$= 2\pi r H$$

$$= I$$

giving the H field as

$$\mathbf{H} = \frac{I}{2\pi r} \, \mathbf{a}_\phi \qquad (2.32)$$

Substituting

$$\mathbf{B} = \mu_0 \mathbf{H}$$

gives the result in (2.14) that was derived with the Biot–Savart law but required the evaluation of an integral.

EXAMPLE

Currents flow through wires of circular, cylindrical cross section whose radii r_w, although small, are nonzero. If the wire is isolated from (or far from) other currents, the current inside it will be distributed symmetrically about the wire axis. (We investigate the influence of nearby currents on the current redistribution, the proximity effect, in Section 4.6). In the case of dc currents, the total current I carried by the wire will also be *uniformly distributed* over the wire cross section with a current density over the cross section of

$$J = \frac{I}{\pi r_w^2} \qquad \text{A/m}^2$$

For the purpose of determining the magnetic field external to this isolated wire, we can replace the wire and its current with a filament of current located on the axis of the wire that contains the total current I. If the wire is further assumed to be infinite in length (or very long), we can then use the basic result for a filamentary current of infinite length in (2.14) to compute the magnetic field external to the wire, and the actual radius of the wire does not enter into this. In this example we demonstrate the validity of this important principle.

Consider an isolated wire of radius r_w carrying a total current I through its cross section. Certainly because of symmetry, the current is symmetric about the wire axis. But as the frequency f of the current increases from zero (dc), it will become concentrated near the wire surface in an annulus at the wire surface having a thickness of a few skin depths [3,6], where the skin depth parameter is

$$\delta = \frac{1}{\sqrt{\pi f \mu_0 \sigma}} \quad \text{m}$$

The conductivity of the wire material is denoted as σ (copper has $\sigma = 5.8 \times 10^7 \, \text{S/m}$), and the wire material is assumed to be nonmagnetic, $\mu_r = 1$. At dc, $f = 0$, the skin depth is infinite, showing that *a dc current is distributed uniformly over the wire cross section.*

We first determine the magnetic field of an isolated wire of radius r_w and infinite length that is carrying a total dc current I in its cross section by using Ampère's law. To determine the field external to the wire using Ampère's law, we surround the wire with a circular contour of radius r as shown in Fig. 2.16(a). Since the current is distributed uniformly over the wire cross

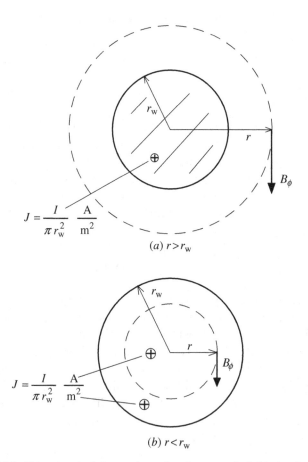

FIGURE 2.16. Using Ampère's law to determine the magnetic field of an isolated wire.

section and therefore symmetrically about the wire axis, the magnetic field intensity vector is in the circumferential or ϕ direction (in a cylindrical coordinate system) about the axis of the wire and is constant around that contour. Hence, from Ampère's law we obtain

$$\oint_c \mathbf{H} \cdot d\mathbf{l} = H_\phi 2\pi r = I$$

and we again obtain the basic result in (2.14):

$$B_\phi = \mu_0 H_\phi$$
$$= \frac{\mu_0 I}{2\pi r} \qquad r > r_w \qquad (2.33a)$$

The magnetic field external to the wire is the same as if we concentrate the entire current in a filament on the wire axis and is independent of the wire radius.

Next, we determine the magnetic field internal to the wire. Again surround an interior portion of the wire with a circular contour of radius r ($r < r_w$) centered on the wire axis as shown in Fig. 2.16(b). The current density for a total current of I that is uniformly distributed over the wire cross section is

$$J = \frac{I}{\pi r_w^2} \qquad \text{A/m}^2$$

Hence, this contour encloses a total current of

$$I_{\text{enclosed}} = \frac{I}{\pi r_w^2} \pi r^2$$
$$= I \frac{r^2}{r_w^2} \qquad \text{A}$$

Again, by symmetry about the wire axis, the magnetic field is directed circumferentially around this contour and is constant at points on it. Hence, by Ampère's law we obtain

$$B_\phi = \mu_0 H_\phi$$
$$= \frac{\mu_0}{2\pi r} I \frac{r^2}{r_w^2}$$
$$= \frac{\mu_0 I r}{2\pi r_w^2} \qquad r < r_w \qquad (2.33b)$$

The magnetic field is plotted versus the radius of the contour in Fig. 2.17.

Finally, we determine these results directly by integration as shown in Fig. 2.18. The current density over the wire cross section is again

$$J = \frac{I}{\pi r_w^2} \qquad \text{A/m}^2$$

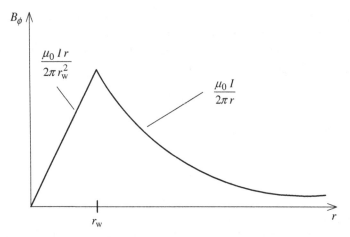

FIGURE 2.17. Plot of the magnetic field for an isolated wire.

At a radius r' and angle ϕ, a differential area is $r'\,d\phi\,dr'$, which contains a differential current of

$$\frac{I}{\pi r_w^2}\,r'\,d\phi\,dr' \qquad A$$

Treat this as an infinite-length filament of current and use the result in (2.14) to determine the differential contribution to the magnetic field at a radius r

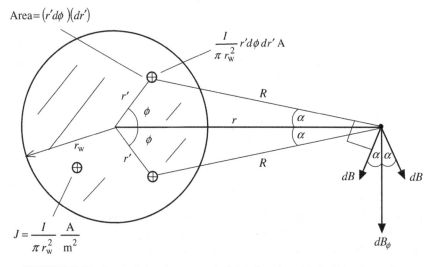

FIGURE 2.18. Determining the magnetic field of a wire with the Biot–Savart law.

from the wire axis as shown in Fig. 2.18:

$$dB = \frac{\mu_0}{2\pi R} \frac{I}{\pi r_w^2} r' \, d\phi \, dr'$$

where the distance from this differential current to the point where we desire to compute the field is (using the law of cosines)

$$R = \sqrt{r^2 + r'^2 - 2rr' \cos\phi}$$

The horizontal components of dB from symmetrically disposed elements cancel, leaving the net field in the ϕ direction as

$$dB_\phi = 2 \frac{\mu_0}{2\pi R} \frac{I}{\pi r_w^2} r' \, d\phi \, dr' \, \cos\alpha$$

$$= \frac{\mu_0 I}{\pi \left(\pi r_w^2\right)} \frac{r' \left(r - r' \cos\phi\right) d\phi \, dr'}{r^2 + r'^2 - 2rr' \cos\phi}$$

and

$$\cos\alpha = \frac{r - r' \cos\phi}{R}$$

Integrating this from $r' = 0$ to $r' = r_w$ and from $\phi = 0$ to $\phi = \pi$ gives the total magnetic field of the wire:

$$B_\phi = \frac{\mu_0 I}{\pi \left(\pi r_w^2\right)} \int_{r'=0}^{r_w} r' \left[\int_{\phi=0}^{\pi} \frac{r - r' \cos\phi}{r^2 + r'^2 - 2rr' \cos\phi} d\phi\right] dr'$$

The interior integral can be evaluated using the remarkable integral 859.124 of Dwight [7]:

$$\int_0^\pi \frac{(a - b\cos x)\, dx}{a^2 + b^2 - 2ab\cos x} = \begin{cases} \dfrac{\pi}{a} & a > b > 0 \\ 0 & b > a > 0 \end{cases} \qquad \text{(D859.124)}$$

For $r > r_w$ the result is

$$B_\phi = \frac{\mu_0 I}{\pi \left(\pi r_w^2\right)} \int_{r'=0}^{r_w} r' \left[\int_{\phi=0}^{\pi} \frac{r - r' \cos\phi}{r^2 + r'^2 - 2rr' \cos\phi} d\phi\right] dr'$$

$$= \frac{\mu_0 I}{\pi \left(\pi r_w^2\right)} \int_{r'=0}^{r_w} r' \frac{\pi}{r} dr'$$

$$= \frac{\mu_0 I}{\pi \left(\pi r_w^2\right)} \frac{\pi}{r} \frac{r_w^2}{2}$$

$$= \frac{\mu_0 I}{2\pi r} \qquad r > r_w \qquad\qquad (2.34a)$$

again giving the fundamental result in (2.14). For $r < r_w$ we break the integral into two pieces with respect to r' in order to use (D859.124):

$$
\begin{aligned}
B_\phi &= \frac{\mu_0 I}{\pi \left(\pi r_w^2\right)} \int_{r'=0}^{r_w} r' \left[\int_{\phi=0}^{\pi} \frac{r - r' \cos \phi}{r^2 + r'^2 - 2rr' \cos \phi} d\phi \right] dr' \\
&= \frac{\mu_0 I}{\pi \left(\pi\, r_w^2\right)} \left[\int_{r'=0}^{r} \frac{\pi}{r} r'\ dr' + \int_{r'=r}^{r_w} (0)\, r'\ dr' \right] \\
&= \frac{\mu_0 I}{\pi \left(\pi r_w^2\right)} \frac{\pi}{r} \frac{r^2}{2} \\
&= \frac{\mu_0 I r}{2\pi r_w^2} \qquad r < r_w
\end{aligned}
\tag{2.34b}
$$

as was derived using Ampère's law.

EXAMPLE

Next we derive the magnetic field of a coaxial cable. The coaxial cable has an infinite (or very long) length and consists of an inner wire of radius r_w contained within an overall shield of inner radius r_s and thickness t, as shown in Fig. 2.19(a). A current I is passed down the inner wire and returns in the shield.

To determine the magnetic field, we surround the inner wire with a circular contour of radius r as shown in Fig. 2.19(b). The dc current I is uniformly distributed over the cross section of the wire and over the cross section of the shield. Hence, the magnetic field is in the circumferential or ϕ direction tangent to the contour and is constant around that contour. In the region between the wire and the shield, $r_w < r < r_s$, the total current enclosed by the contour is I. Hence, Ampère's law gives for $r_w < r < r_s$,

$$
\oint_c \mathbf{H} \cdot d\mathbf{l} = H_\phi 2\pi r
$$

$$
= I \qquad r_w < r < r_s
$$

and the magnetic flux density in the region between the wire and the shield is

$$
B_\phi = \frac{\mu_0 I}{2\pi r} \qquad r_w < r < r_s
\tag{2.35a}
$$

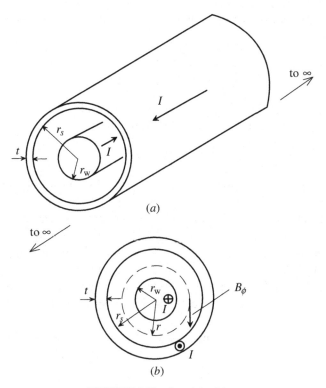

FIGURE 2.19. Coaxial cable.

The magnetic flux density inside the wire is, as in the preceding example,

$$B_\phi = \frac{\mu_0 I r}{2\pi r_w^2} \qquad r < r_w \qquad (2.35b)$$

The dc current $-I$ in the shield is also uniformly distributed over the shield cross section and has a current density of

$$J_{\text{shield}} = -\frac{I}{\pi \left[(r_s + t)^2 - r_s^2\right]} \qquad \text{A/m}^2 \qquad r_s < r < r_s + t$$

Hence, expanding the contour to within the shield, $r_s < r < r_s + t$, encloses a total current of

$$\begin{aligned} I_{\text{enclosed}} &= I - I\frac{\pi r^2 - \pi r_s^2}{\pi \left[(r_s + t)^2 - r_s^2\right]} \\ &= I\frac{(r_s + t)^2 - r^2}{(r_s + t)^2 - r_s^2} \qquad \text{A} \qquad r_s < r < r_s + t \end{aligned}$$

By Ampère's law, the magnetic field in the shield is circumferentially directed and becomes

$$B_\phi = \frac{\mu_0 I}{2\pi r} \frac{(r_s + t)^2 - r^2}{(r_s + t)^2 - r_s^2} \qquad r_s < r < r_s + t \qquad (2.35c)$$

Expanding the contour to enclose the entire coaxial cable, $r > r_s + t$, shows, by Ampère's law, that the magnetic field is

$$B_\phi = 0 \qquad r > r_s + t \qquad (2.35d)$$

since the total current enclosed is zero because of the equal but oppositely directed currents.

These results, easily obtained using Ampère's law, can also be obtained using direct integration in the same fashion as in the preceding example. The results for the fields in a coaxial cable obtained by using Ampère's law in this example and given in (2.35b) for $r < r_w$, and in (2.35a) for $r_w < r < r_s$, were obtained by direct integration in the preceding example. The final result in (2.35c) for $r_s < r < r_s + t$ can also be obtained by direct integration by reference to Fig. 2.20.

Again we set up the integration as in the preceding example. The differential currents in the shield are

$$\underbrace{\frac{-I}{\pi \left[(r_s + t)^2 - r_s^2\right]}}_{J_{\text{shield}}\,\text{A/m}^2} r'\, d\phi\, dr' \qquad \text{A}$$

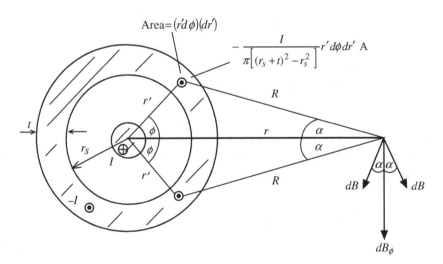

FIGURE 2.20. Determining the magnetic field within the shield by direct integration.

Hence, the magnetic field in the shield for $r_s < r < r_s + t$ *due to the currents in the shield* is

$$B_\phi = 2\frac{\mu_0}{2\pi} \underbrace{\frac{-I}{\pi \left[(r_s + t)^2 - r_s^2\right]}}_{J_{\text{shield}} \text{ A/m}^2} \int_{r'=r_s}^{r_s+t} r' \left[\int_{\phi=0}^{\pi} \frac{r - r' \cos \phi}{r^2 + r'^2 - 2rr' \cos \phi} d\phi\right] dr'$$

$$= -\frac{\mu_0 I}{\pi} \frac{1}{\pi \left[(r_s + t)^2 - r_s^2\right]} \left[\int_{r'=r_s}^{r} \frac{\pi}{r} r' \, dr' + \int_{r'=r}^{r_s+t} (0) \, r' \, dr'\right]$$

$$= -\frac{\mu_0 I}{\pi r \left[(r_s + t)^2 - r_s^2\right]} \left[\frac{r'^2}{2}\right]_{r'=r_s}^{r}$$

$$= -\frac{\mu_0 I}{2\pi r \left[(r_s + t)^2 - r_s^2\right]} \left(r^2 - r_s^2\right) \qquad r_s < r < r_s + t$$

where we have again separated the integration from $r' = r_s$ to $r' = r_s + t$ into two parts in order to use integral 859.124:

$$\int_0^\pi \frac{(a - b\cos x) \, dx}{a^2 + b^2 - 2ab \cos x} = \begin{cases} \dfrac{\pi}{a} & a > b > 0 \\ 0 & b > a > 0 \end{cases} \qquad \text{(D859.124)}$$

To this we add the contribution to the field within the shield due to the current of the interior wire:

$$B_\phi = \frac{\mu_0 I}{2\pi r} \qquad r_s < r < r_s + t$$

Combining these two contributions yields (2.35c)

$$B_\phi = \frac{\mu_0 I}{2\pi r} - \frac{\mu_0 I}{2\pi r \left[(r_s + t)^2 - r_s^2\right]} \left(r^2 - r_s^2\right)$$

$$= \frac{\mu_0 I}{2\pi r} \frac{(r_s + t)^2 - r^2}{(r_s + t)^2 - r_s^2} \qquad r_s < r < r_s + t \qquad (2.35c)$$

For the fields external to the cable, $r > r_s + t$, the integral above is, according to (D859.124),

$$B_\phi = 2\frac{\mu_0}{2\pi} \underbrace{\frac{-I}{\pi \left[(r_s + t)^2 - r_s^2\right]}}_{J_{\text{shield}} \text{ A/m}^2} \int_{r'=r_s}^{r_s+t} r' \left[\int_{\phi=0}^{\pi} \frac{r - r' \cos \phi}{r^2 + r'^2 - 2rr' \cos \phi} d\phi\right] dr'$$

$$= -\frac{\mu_0 I}{\pi} \frac{1}{\pi \left[(r_s + t)^2 - r_s^2\right]} \left[\int_{r'=r_s}^{r_s+t} \frac{\pi}{r} r' \, dr'\right]$$

$$= -\frac{\mu_0 I}{\pi r \left[(r_s + t)^2 - r_s^2\right]} \left[\frac{r'^2}{2}\right]_{r'=r_s}^{(r_s+t)}$$

$$= -\frac{\mu_0 I}{2\pi r \left[(r_s + t)^2 - r_s^2\right]} \left[(r_s + t)^2 - r_s^2\right]$$

$$= -\frac{\mu_0 I}{2\pi r} \qquad r_s + t < r$$

which, combined with the field due to the interior wire, gives a result of zero, which is (2.35d).

EXAMPLE

Use Ampère's law to determine the H field of the infinite current sheet shown in Fig. 2.10. View the sheet from the top in the xy plane and construct a *rectangular* closed contour c as shown in Fig. 2.21. By symmetry we see that the **H** field must be directed parallel to the sheet. Hence, the **H** field is

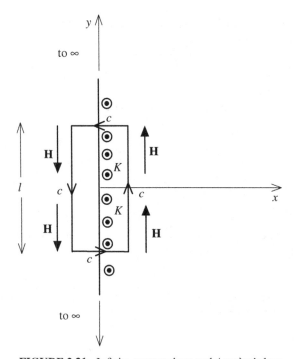

FIGURE 2.21. Infinite current sheet and Ampère's law.

parallel to the sides and perpendicular to the tops (and contributes nothing to Ampère's law along the tops of contour c). If the contour has tops of width w and sides of length l, the total current enclosed by the contour is

$$I_{\text{enclosed}} = lK \qquad \text{m} \times \text{A/m} = \text{A}$$

Hence, Ampère's law around the entire contour is

$$\oint_c \mathbf{H} \cdot d\mathbf{l} = 2 \underbrace{\int_w \mathbf{H} \cdot d\mathbf{l}}_{0} + 2 \int_l \mathbf{H} \cdot d\mathbf{l}$$

$$= 2 \int_l H \, dl$$

$$= 2lH$$

$$= I_{\text{enclosed}} = lK$$

Hence, the H field is

$$\mathbf{H} = \begin{cases} \dfrac{K}{2} \mathbf{a}_y & \text{for } x > 0 \\[2mm] -\dfrac{K}{2} \mathbf{a}_y & \text{for } x < 0 \end{cases} \qquad (2.36)$$

Substituting

$$\mathbf{B} = \mu_0 \mathbf{H}$$

gives the result in (2.21) that was derived by the Biot–Savart law but required evaluation of an integral.

2.5 VECTOR MAGNETIC POTENTIAL

Since the magnetic field has no sources or sinks, it must form closed loops everywhere. Hence, the *divergence* of the magnetic field, according to Gauss's law for the magnetic field, is *zero*:

$$\nabla \cdot \mathbf{B} = 0 \qquad (2.37)$$

In the Appendix it is shown that *the divergence of the curl of any vector field is zero*:

$$\nabla \cdot \nabla \times \mathbf{A} = 0 \qquad (2.38)$$

The divergence $\nabla \cdot \mathbf{F}$ represents the *net flux or outflow of the vector field* $\mathbf{F}$ *from a point*, whereas the curl $\nabla \times \mathbf{F}$ represents the *circulation of the vector field* **F** *about a point*. Although the identity in (2.38) is proven directly in the Appendix, it is a sensible identity. If the vector field **A** has nonzero circulation at a point, $\nabla \times \mathbf{A} \neq 0$, its curl $\nabla \times \mathbf{A}$ should have no outflow from the point and the identity is satisfied. On the other hand, suppose that the vector field has no circulation at a point, $\nabla \times \mathbf{A} = 0$. Then the divergence of this is clearly zero.

The identity in (2.38), combined with Gauss's law for the magnetic field in (2.37), $\nabla \cdot \mathbf{B} = 0$, allows us to define another, auxiliary field as

$$\boxed{\mathbf{B} = \nabla \times \mathbf{A}} \tag{2.39}$$

This new vector field **A** is called the *vector magnetic potential*. This is very similar to defining the scalar electric potential or voltage ϕ for a static (dc) electric field **E** from $\nabla \times \mathbf{E} = 0$ and using the identity from the Appendix of $\nabla \times \nabla\phi = 0$ to define the electric field in terms of the scalar potential function ϕ as $\mathbf{E} = -\nabla\phi$ [3,6]. It turns out (see [3]) that to define a vector field completely, we must define the curl of that field as well as its divergence. Equation (2.39) has defined the curl of **A**. It also turns out that we can, without any contradiction in doing so, define the divergence of **A** as zero:

$$\nabla \cdot \mathbf{A} = 0 \tag{2.40}$$

thereby completely defining this new magnetic potential vector **A**.

To determine an equation relating the vector magnetic potential to the currents that produce it, we employ Ampère's law given in (2.30):

$$\oint_c \mathbf{H} \cdot d\mathbf{l} = I_{\text{enclosed}} \tag{2.30}$$

Using Stokes's theorem (see the Appendix), we can write Ampère's law as

$$\oint_c \mathbf{H} \cdot d\mathbf{l} = \int_s (\nabla \times \mathbf{H}) \cdot d\mathbf{s}$$

$$= I_{\text{enclosed}}$$

$$= \int_s \mathbf{J} \cdot d\mathbf{s} \tag{2.41}$$

where s is the *open surface* surrounded by the closed contour c as illustrated in Fig. 2.22. The *current density vector* throught the surface s is denoted as **J**, whose units are A/m^2. The direction of the normal to the surface s as well as the direction of the contour c that encloses s are related by the right-hand rule. Place the fingers of the right hand in the direction of c and the thumb will point in the direction of $d\mathbf{s}$. Comparing both sides of (2.41) gives Ampère's law in *point form* as

$$\boxed{\nabla \times \mathbf{H} = \mathbf{J}} \tag{2.42}$$

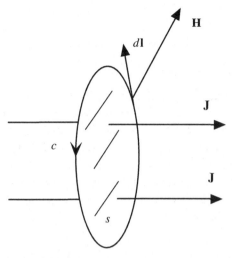

FIGURE 2.22. Ampère's law and Stokes's theorem.

Substituting the relation between **B** and **H** using the permeability of the surrounding medium (assumed not to be ferromagnetic), $\mathbf{B} = \mu_0\mathbf{H}$, gives

$$\nabla \times \mathbf{B} = \mu_0\mathbf{J} \tag{2.43}$$

Substituting the definition of the vector magnetic potential given in (2.39) gives

$$\nabla \times (\nabla \times \mathbf{A}) = \mu_0\mathbf{J} \tag{2.44}$$

The curl of the curl of a vector field can be written as [3,6]

$$\nabla \times (\nabla \times \mathbf{A}) = \nabla(\nabla \cdot \mathbf{A}) - \nabla^2\mathbf{A} \tag{2.45}$$

We defined the divergence of **A** as zero in (2.40), $\nabla \cdot \mathbf{A} = 0$, to complete the definition of the vector magnetic potential **A**. Hence, (2.44) becomes

$$\nabla^2\mathbf{A} = -\mu_0\mathbf{J} \tag{2.46}$$

The solution to (2.46) is [3,6]

$$\mathbf{A} = \frac{\mu_0}{4\pi} \int_v \frac{\mathbf{J}\,dv}{R} \tag{2.47a}$$

where v is the volume enclosing the current density **J** (which is the source of **A**). The distance between the point where we are determining **A** and a differential volume of the current that contains $\mathbf{J}\,dv$ ampere-meters is denoted as R. If the current is confined to a surface, this reduces to

$$\mathbf{A} = \frac{\mu_0}{4\pi} \int_s \frac{\mathbf{K}\,ds}{R}$$ (2.47b)

where s is the surface containing the surface current density $\mathbf{K}$ whose units are A/m. The distance between the point where we are determining $\mathbf{A}$ and a differential surface of the current that contains $\mathbf{K}\,ds$ ampere-meters is denoted as R. For line currents we consider the current I to be contained in a differential cylinder of length dl and cross-sectional area ds. If the current density is uniformly distributed over the cylinder cross section (as will be the case for dc currents [3]), the total current is $J\,ds = I$, so that $J\,dv = J\,ds\,dl = I\,dl$, and the result becomes

$$\mathbf{A} = \frac{\mu_0}{4\pi} \int_l \frac{I}{R}\,d\mathbf{l}$$ (2.47c)

where a vector differential length of the line in the direction of the current is denoted as $d\mathbf{l}$ and contains $I\,d\mathbf{l}$ ampere-meters. Again R is the distance between the point where we are determining $\mathbf{A}$ and the differential current segment. The units of the vector magnetic potential are Wb/m, magnetic flux per length. We will learn the reason for these units in Chapter 4.

It should be noted that we obtained the basic result for the computation of $\mathbf{A}$ in (2.47a) from the solution of (2.46), $\nabla^2 \mathbf{A} = -\mu_0\,\mathbf{J}$. But we obtained (2.46) by defining the divergence of $\mathbf{A}$ in (2.45) as zero (i.e., $\nabla \cdot \mathbf{A} = 0$). However, the basic result for computing $\mathbf{A}$ in (2.47a) does not require that $\nabla \cdot \mathbf{A} = 0$. The Helmholtz theorem [3,6] establishes the fact that to define a vector field such as $\mathbf{A}$ completely requires that its curl *and* its divergence be defined. But the choices for these are arbitrary and are not related. We can show this by demonstrating that taking the curl of (2.47a) gives $\mathbf{B} = \nabla \times \mathbf{A}$, where the resulting $\mathbf{B}$ is the Biot–Savart law given in (2.11). To show this we take the curl of (2.47a):

$$\mathbf{B} = \nabla \times \mathbf{A}$$
$$= \frac{\mu_0}{4\pi} \nabla \times \int_v \frac{\mathbf{J}}{R}\,dv$$
$$= \frac{\mu_0}{4\pi} \int_v \nabla \times \frac{\mathbf{J}}{R}\,dv$$ (2.48)

We can interchange the order of differentiation and integration using Leibnitz's rule [12] since the ∇ operator takes derivatives with respect to the coordinates of the location of $\mathbf{B}$ and $\mathbf{A}$, whereas the volume integral is with respect to the coordinates of the current $\mathbf{J}$. Using a vector identity [3], we can write the curl of the integrand as

$$\nabla \times \frac{\mathbf{J}}{R} = \nabla \left(\frac{1}{R}\right) \times \mathbf{J} + \frac{1}{R} \underbrace{(\nabla \times \mathbf{J})}_{0}$$

$$= \nabla \left(\frac{1}{R}\right) \times \mathbf{J}$$

$$= -\mathbf{J} \times \nabla \left(\frac{1}{R}\right) \tag{2.49}$$

The del operator takes the derivatives with respect to the coordinates of the location of **B** and **A**: the field point. Hence, the curl of **J** is zero here since **J** involves only the coordinates of the location of the source current. You can verify in spherical coordinates (see the Appendix) that

$$\nabla \left(\frac{1}{R}\right) = -\frac{1}{R^2} \mathbf{a}_R \tag{2.50}$$

where $\mathbf{a}_R$ is a unit vector pointing from the current to the field point. The vector identity in (2.49) is, of course, sensible since it is the vector counterpart to the scalar result using the chain rule. Therefore, we obtain

$$\mathbf{B} = \nabla \times \mathbf{A}$$

$$= \frac{\mu_0}{4\pi} \nabla \times \int_v \frac{\mathbf{J}}{R} \, dv$$

$$= \frac{\mu_0}{4\pi} \int_v \nabla \times \frac{\mathbf{J}}{R} \, dv$$

$$= \frac{\mu_0}{4\pi} \int_v \frac{\mathbf{J} \times \mathbf{a}_R}{R^2} \, dv \tag{2.51}$$

which is the Biot–Savart law for determining **B** given in (2.11). If the current forms a closed loop (as all dc currents must), $\nabla \cdot \mathbf{A} = 0$, but it is not necessary to define the divergence of **A** to be zero in order to obtain (2.47).

The solutions in (2.47) apply to any coordinate system. If we specialize them to a *rectangular coordinate system*, we obtain

$$A_x = \frac{\mu_0}{4\pi} \int_v \frac{J_x}{R} \, dv$$

$$A_y = \frac{\mu_0}{4\pi} \int_v \frac{J_y}{R} \, dv$$

$$A_z = \frac{\mu_0}{4\pi} \int_v \frac{J_z}{R} \, dv \tag{2.52a}$$

This is a significant result because it says that (1) each component of **J** produces the corresponding component of **A**, and (2) the direction of the resulting

A_x, A_y, A_z is the same as the direction of the corresponding J_x, J_y, J_z that produced it! In other words, a current that is directed solely in the z direction will produce a vector magnetic potential that is solely in the z direction *parallel to the J_z at all points* in the space around the current! For a current distributed over a surface s, these results become

$$A_x = \frac{\mu_0}{4\pi} \int_s \frac{K_x}{R}\, ds$$

$$A_y = \frac{\mu_0}{4\pi} \int_s \frac{K_y}{R}\, ds$$

$$A_z = \frac{\mu_0}{4\pi} \int_s \frac{K_z}{R}\, ds \qquad (2.52b)$$

If the current is a line current whose contour is l such as in a wire, these become

$$A_x = \frac{\mu_0}{4\pi} \int_l \frac{I_x}{R}\, dl = \frac{\mu_0}{4\pi} \int_l \frac{I}{R}\, dl_x$$

$$A_y = \frac{\mu_0}{4\pi} \int_l \frac{I_y}{R}\, dl = \frac{\mu_0}{4\pi} \int_l \frac{I}{R}\, dl_y$$

$$A_z = \frac{\mu_0}{4\pi} \int_l \frac{I_z}{R}\, dl = \frac{\mu_0}{4\pi} \int_l \frac{I}{R}\, dl_z \qquad (2.52c)$$

Note that unlike the Biot–Savart law, which is an inverse-square law where **B** depends on the square of the inverse distance between the current and the **B** field, the magnetic vector potential simply depends on the inverse of the distance R between the current and the component of **A** that it produces. In some problems it is simpler to determine the components of **A** from (2.47), which for rectangular coordinates are given in (2.52), and then simply determine **B** by computing its curl mechanically from $\mathbf{B} = \nabla \times \mathbf{A}$ in the appropriate coordinate system, than it is to compute **B** directly using the Biot–Savart law.

One of the main advantages of first computing the three components of the vector magnetic potential **A** via (2.47) or, in rectangular coordinates, from (2.52) and then determining **B** by differentiation via $\mathbf{B} = \nabla \times \mathbf{A}$ is that *we do not need to integrate vector quantities to obtain* **A**! The expansion of (2.47) for rectangular coordinates given in (2.52) shows this. Determining **B** directly by integration by applying the Biot–Savart law may require that we integrate vector quantities. To avoid integrating vector quantities, we utilize symmetry and resolve **B** into components, thereby restricting the solution only to points about the current where we can exploit symmetry (see, e.g., Figures 2.10 and 2.11).

Observe that the vector magnetic potential in (2.47) and (2.52) requires the integral of an inverse distance, 1/R. Hence, the resulting vector magnetic potential typically involves a natural logarithm (ln) function as the result. Thus, we expect to see these natural log functions in the following results for various configurations. The **B** field obtained from (2.39) then requires the derivatives of these natural log functions.

In some of the earlier problems we solved for the magnetic field of a very idealized case: a current of infinite length that is directed in the z direction. The magnetic flux density of a current of infinite length is finite and given by

$$\mathbf{B} = \frac{\mu_0 I}{2\pi r} \mathbf{a}_\phi$$

From the relation $\mathbf{B} = \nabla \times \mathbf{A}$ in cylindrical coordinates, **A** is related to **B** as

$$\mathbf{B} = \left(\frac{\partial A_r}{\partial z} - \frac{\partial A_z}{\partial r} \right) \mathbf{a}_\phi$$

$$= -\frac{\partial A_z}{\partial r} \mathbf{a}_\phi$$

This is because the **B** field has only a ϕ component, and the current and resulting vector magnetic potential is directed in the z direction. Integrating this, we obtain

$$A_z = -\frac{\mu_0 I}{2\pi} \ln r + C \qquad \text{current of infinite length}$$

where C is a constant of integration. However, when we are using $\mathbf{B} = \nabla \times \mathbf{A}$ to determine the **B** field for an infinite-length current by first determining **A**, we can ignore the integration constant C because we differentiate A_z in order to determine B_ϕ. Hence, for a current of infinite length we may assume a form of A_z to be

$$A_z = -\frac{\mu_0 I}{2\pi} \ln r \qquad \text{current of infinite length} \qquad (2.53)$$

EXAMPLE

Determine the vector magnetic potential at a distance r from the center of a current of length L and on a line perpendicular to its midpoint as shown in Fig. 2.23. Then determine the magnetic field **B** from that result. Although steady (dc) currents *must form closed loops*, we again use the solutions for currents of *finite length* to construct, by superposition, the magnetic fields of

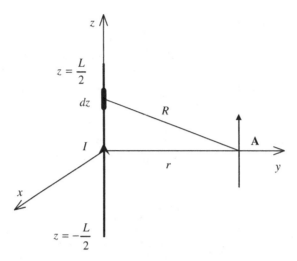

FIGURE 2.23. Vector magnetic potential of a current.

closed current loops (see the discussion in Section 2.9). Hence, determining the vector magnetic potential for currents of finite length is useful for that purpose.

The problem again fits a cylindrical coordinate system (see the Appendix for a discussion of the cylindrical coordinate system). From (2.52c) and Fig. 2.23 we see that since the current is directed solely in the z direction, the vector magnetic potential will be parallel to it at all points and directed in the z direction also. From (2.52c),

$$
\begin{aligned}
A_z &= \frac{\mu_0}{4\pi} \int_{z=-L/2}^{L/2} \frac{I}{R}\, dz \\
&= \frac{\mu_0 I}{4\pi} \int_{-L/2}^{L/2} \frac{1}{\sqrt{r^2 + z^2}}\, dz \\
&= \frac{\mu_0 I}{4\pi} \left[\ln\left(z + \sqrt{z^2 + r^2} \right) \right]_{-L/2}^{L/2} \\
&= \frac{\mu_0 I}{4\pi} \ln \frac{L/2 + \sqrt{(L/2)^2 + r^2}}{-(L/2) + \sqrt{(L/2)^2 + r^2}}
\end{aligned}
\qquad (2.54a)
$$

and we have used integral 200.01 from Dwight [7]:

$$
\int \frac{1}{\sqrt{x^2 + a^2}}\, dx = \ln\left(x + \sqrt{x^2 + a^2} \right)
\qquad (\text{D200.01})
$$

This result can be simplified to

$$A_z = \frac{\mu_0 I}{4\pi} \ln \frac{(L/2r) + \sqrt{(L/2r)^2 + 1}}{-(L/2r) + \sqrt{(L/2r)^2 + 1}}$$

$$= \frac{\mu_0 I}{2\pi} \ln \left[\frac{L}{2r} + \sqrt{\left(\frac{L}{2r}\right)^2 + 1} \right] \qquad (2.54b)$$

and we have used $\log(A/B) = \log A - \log B$ and an important natural logarithm identity:

$$\ln\left(x + \sqrt{x^2 + 1}\right) = -\ln\left(-x + \sqrt{x^2 + 1}\right)$$

which can be proven directly ($\log AB = \log A + \log B$):

$$\ln\left(x + \sqrt{x^2 + 1}\right) + \ln\left(-x + \sqrt{x^2 + 1}\right) = \ln(1) = 0$$

The result in (2.54b) can be written in an alternative form using the inverse hyperbolic sine function:

$$\sinh^{-1}x \equiv \ln\left(x + \sqrt{x^2 + 1}\right)$$

$$= -\sinh^{-1}(-x) \qquad (D700.1)$$

Hence, we could write (2.54b) as

$$A_z = \frac{\mu_0 I}{2\pi} \sinh^{-1}\frac{L}{2r} \qquad (2.54c)$$

The simplified result in (2.54b) could also have been obtained directly by utilizing symmetry and integrating from $z = 0$ to $z = L/2$ and doubling the result:

$$A_z = 2\frac{\mu_0}{4\pi} \int_{z=0}^{L/2} \frac{I}{R} \, dz$$

$$= \frac{\mu_0 I}{2\pi} \left[\ln\left(z + \sqrt{z^2 + r^2}\right) \right]_0^{L/2}$$

$$= \frac{\mu_0 I}{2\pi} \left[\ln\left(\frac{L}{2} + \sqrt{\left(\frac{L}{2}\right)^2 + r^2}\right) - \ln r \right]$$

$$= \frac{\mu_0 I}{2\pi} \ln \left[\frac{L}{2r} + \sqrt{\left(\frac{L}{2r}\right)^2 + 1} \right]$$

$$= \frac{\mu_0 I}{2\pi} \sinh^{-1}\frac{L}{2r}$$

Taking the curl of $\mathbf{A}$ to give $\mathbf{B}$ via (2.39) gives (see the Appendix for the curl in cylindrical coordinates)

$$\mathbf{B} = \nabla \times \mathbf{A}(r, \phi, z)$$

$$= \left(\frac{1}{r} \frac{\partial A_z}{\partial \phi} - \frac{\partial A_\phi}{\partial z} \right) \mathbf{a}_r + \left(\frac{\partial A_r}{\partial z} - \frac{\partial A_z}{\partial r} \right) \mathbf{a}_\phi + \left[\frac{1}{r} \frac{\partial (rA_\phi)}{\partial r} - \frac{1}{r} \frac{\partial A_r}{\partial \phi} \right] \mathbf{a}_z$$

$$= -\frac{\partial A_z}{\partial r} \mathbf{a}_\phi$$

Substituting A_z from (2.54b) and performing the differentiation gives

$$\mathbf{B} = -\frac{\partial A_z}{\partial r} \mathbf{a}_\phi$$

$$= \frac{\mu_0 I}{2\pi r} \frac{L/2}{\sqrt{(L/2)^2 + r^2}} \mathbf{a}_\phi$$

$$= \frac{\mu_0 I}{2\pi r} \frac{L}{\sqrt{4r^2 + L^2}} \mathbf{a}_\phi \qquad (2.55)$$

which is the same as (2.13), which was obtained with the Biot–Savart law. We have used A_z from (2.54b) and the derivative

$$\frac{d}{dr} \ln \left[\frac{a}{r} + \sqrt{\left(\frac{a}{r} \right)^2 + 1} \right] = -\frac{a}{r\sqrt{a^2 + r^2}}$$

Alternatively,

$$\frac{d}{dx} \sinh^{-1} \frac{a}{x} = \frac{d}{dx} \operatorname{csch}^{-1} \frac{x}{a}$$

$$= \frac{-a}{|x|\sqrt{x^2 + a^2}} \qquad (D728.8)$$

For an infinitely long current, $L \to \infty$, (2.55) evaluates to

$$\mathbf{B} = \frac{\mu_0 I}{2\pi r} \mathbf{a}_\phi \qquad L \to \infty \qquad (2.56)$$

which is (2.14) again.

EXAMPLE

Determine the vector magnetic potential of a current of length L at some general point that is at a distance r from it (the cylindrical coordinate system variable) and at $z = Z$, as shown in Fig. 2.24, and then determine the $\mathbf{B}$ field.

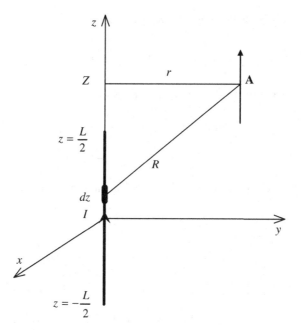

FIGURE 2.24. Vector magnetic potential of a current of finite length.

From Fig. 2.24 and (2.52c), we again see that since the current is directed solely in the z direction, the vector magnetic potential will be parallel to it at all points and directed in the z direction also. From (2.52c),

$$A_z = \frac{\mu_0 I}{4\pi} \int_{z=-L/2}^{L/2} \frac{1}{R} dz$$

where $R = \sqrt{(Z - z)^2 + r^2}$. Making a change of variables as $\lambda = Z - z$ and $d\lambda = -dz$ gives

$$
\begin{aligned}
A_z &= \frac{\mu_0 I}{4\pi} \int_{\lambda=Z-L/2}^{Z+L/2} \frac{1}{\sqrt{\lambda^2 + r^2}} d\lambda \\
&= \frac{\mu_0 I}{4\pi} \ln \frac{(Z + L/2) + \sqrt{(Z + L/2)^2 + r^2}}{(Z - L/2) + \sqrt{(Z - L/2)^2 + r^2}} \\
&= \frac{\mu_0 I}{4\pi} \left(\sinh^{-1} \frac{Z + L/2}{r} - \sinh^{-1} \frac{Z - L/2}{r} \right) \\
&= \frac{\mu_0 I}{4\pi} \left(\sinh^{-1} \frac{Z + L/2}{r} + \sinh^{-1} \frac{L/2 - Z}{r} \right)
\end{aligned}
$$

(2.57)

and we have again used integral 200.01 of Dwight [7]:

$$\int \frac{dx}{\sqrt{x^2 + a^2}} = \ln\left(x + \sqrt{x^2 + a^2}\right) \qquad \text{(D200.01)}$$

Also, we have again used the alternative relation

$$\sinh^{-1} x \equiv \ln\left(x + \sqrt{x^2 + 1}\right)$$
$$= -\sinh^{-1}(-x) \qquad \text{(D700.1)}$$

Obtain **B** by taking the curl of **A** in cylindrical coordinates, giving

$$\mathbf{B} = -\frac{\partial A_z}{\partial r}\, \mathbf{a}_\phi$$

$$= \frac{\mu_0 I}{4\pi r}\left[\frac{Z + L/2}{\sqrt{(Z + L/2)^2 + r^2}} - \frac{Z - L/2}{\sqrt{(Z - L/2)^2 + r^2}}\right] \mathbf{a}_\phi \qquad \text{(2.58)}$$

which is the same as (2.15) obtained with the Biot–Savart law. We have used the derivative

$$\frac{d}{dr}\ln\left(a + \sqrt{a^2 + r^2}\right) = -\frac{a}{r\sqrt{a^2 + r^2}} + \frac{1}{r}$$

and $\log(A/B) = \log A - \log B$. Alternatively, we could use (D728.8).

EXAMPLE

Determine the vector magnetic potential **A** due to a current loop of radius a that is centered on the origin of a rectangular coordinate system and lying in the xy plane as shown in Fig. 2.25. From that, determine **B**.

For the following computation, we determine **A** at a point located in the xz plane at $x = \rho$, $y = 0$, and z. Because of the symmetry of the current loop, the field will then be determined at any general point located at (ρ, ϕ, z) in a cylindrical coordinate system or, equivalently, at any general point (r, θ, ϕ) in a spherical coordinate system and will be independent of ϕ in either case. If we pair off current segments at $\pm\phi$ measured with respect to the x axis, from symmetry we see that **A** is in the y direction or, equivalently, in the ϕ direction at this point. At the point of interest we obtain, from (2.47c),

$$A_\phi = \frac{\mu_0 I}{4\pi} \oint \frac{dl_\phi}{R}$$

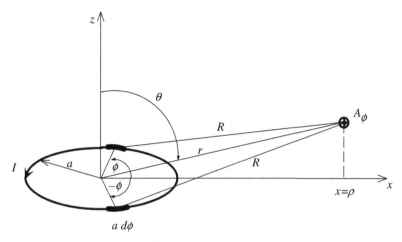

FIGURE 2.25. Current loop.

The component of a differential length of the loop, $dl = a \, d\phi$, in the direction of $\mathbf{A}$ at this point is

$$dl_\phi = a \, d\phi \, \cos \phi$$

Using the law of cosines, we obtain

$$R^2 = a^2 + r^2 - 2ar \sin \theta \cos \phi$$

Hence, at the field point

$$
\begin{aligned}
A_\phi &= 2 \frac{\mu_0 I}{4\pi} \int_{\phi=0}^{\pi} \frac{a \cos \phi}{\sqrt{a^2 + r^2 - 2ar \sin \theta \cos \phi}} \, d\phi \\
&= \frac{\mu_0 I}{2\pi} \int_{\phi=0}^{\pi} \frac{a \cos \phi}{\sqrt{a^2 + \rho^2 + z^2 - 2a\rho \cos \phi}} \, d\phi
\end{aligned}
$$

(2.59)

since $r^2 = z^2 + \rho^2$ and $\rho = r \sin \theta$.

The integral in (2.59) is difficult to evaluate. We first restrict the result to the case where the point is at a distance that is far away with respect to the current loop radius, $r \gg a$, and later will obtain the exact solution. Evaluating the denominator using the binomial theorem gives

$$
\begin{aligned}
\frac{1}{R} &\cong \left(\frac{1}{r^2 - 2ar \sin \theta \cos \phi} \right)^{1/2} = \frac{1}{r} \left(1 - \frac{2a}{r} \sin \theta \cos \phi \right)^{-1/2} \\
&\cong \frac{1}{r} \left(1 + \frac{a}{r} \sin \theta \cos \phi \right)
\end{aligned}
$$

Hence, we obtain

$$A_\phi \cong \frac{\mu_0 I}{2\pi} \int_{\phi=0}^{\pi} \frac{a \cos \phi}{r} \left(1 + \frac{a}{r} \sin \theta \cos \phi \right) d\phi$$

$$= \frac{\mu_0 I}{2\pi} \frac{a^2}{r^2} \sin \theta \frac{\pi}{2}$$

$$= \frac{\mu_0 I a^2 \sin \theta}{4 r^2}$$

Taking the curl of **A** in spherical coordinates (see the Appendix) to obtain **B** gives

$$\mathbf{B} = \nabla \times \mathbf{A}$$

$$= \frac{1}{r \sin \theta} \frac{\partial (\sin \theta \, A_\phi)}{\partial \theta} \mathbf{a}_r - \frac{1}{r} \frac{\partial (r A_\phi)}{\partial r} \mathbf{a}_\theta$$

$$= \frac{\mu_0 a^2 I}{4 r^3} (2 \cos \theta \, \mathbf{a}_r + \sin \theta \, \mathbf{a}_\theta) \qquad (2.60)$$

The *magnetic dipole moment* is defined in (2.27) as the product of the current and the area of the current loop:

$$m = \pi a^2 I \qquad (2.27)$$

Substituting (2.27) into (2.60) gives the components of **B** as

$$B_r = \frac{\mu_0 m}{2\pi r^3} \cos \theta \qquad (2.61a)$$

$$B_\theta = \frac{\mu_0 m}{4\pi r^3} \sin \theta \qquad (2.61b)$$

$$B_\phi = 0 \qquad (2.61c)$$

Note that the magnetic field of a current loop at large distances from the loop falls off inversely with the cube of distance.

The exact solution of the integral in (2.59) can be obtained in terms of the complete elliptic integrals [10,11]. To do so, we make a change of variables $\phi = \pi - 2\zeta$ and $d\phi = -2 \, d\zeta$ so that $\cos \phi = -\cos 2\zeta = 2 \sin^2 \zeta - 1$. Hence, (2.59) becomes

$$A_\phi = \frac{\mu_0 I a}{\pi} \int_{\zeta=0}^{\pi/2} \frac{2 \sin^2 \zeta - 1}{\sqrt{(a + \rho)^2 + z^2 - 4a\rho \sin^2 \zeta}} \, d\zeta \qquad (2.62)$$

Defining

$$k^2 = \frac{4a\rho}{(a + \rho)^2 + z^2} \qquad (2.63)$$

(2.62) reduces to

$$A_\phi = \frac{\mu_0 I}{\pi k} \sqrt{\frac{a}{\rho}} \left[\left(1 - \frac{k^2}{2}\right) K - E \right] \tag{2.64}$$

where the complete elliptic integrals of the first and second kind are [7]

$$K(k) = \int_{\zeta=0}^{\pi/2} \frac{d\zeta}{\sqrt{1 - k^2 \sin^2 \zeta}} \tag{D773.1}$$

and

$$E(k) = \int_{\zeta=0}^{\pi/2} \sqrt{1 - k^2 \sin^2 \zeta} \, d\zeta \tag{D774.1}$$

and are tabulated in Dwight, Tables 1040 and 1041 [7].

The magnetic flux density is obtained in cylindrical coordinates (see the Appendix) from $\mathbf{B} = \nabla \times \mathbf{A}$ as [10,11]

$$B_\rho = -\frac{\partial A_\phi}{\partial z}$$

$$= \frac{\mu_0 I}{2\pi} \frac{z}{\rho \sqrt{(a+\rho)^2 + z^2}} \left[-K + \frac{a^2 + \rho^2 + z^2}{(a - \rho)^2 + z^2} E \right] \tag{2.65a}$$

$$B_\phi = 0 \tag{2.65b}$$

$$B_z = \frac{1}{\rho} \frac{\partial (\rho A_\phi)}{\partial \rho}$$

$$= \frac{\mu_0 I}{2\pi} \frac{1}{\sqrt{(a+\rho)^2 + z^2}} \left[K + \frac{a^2 - \rho^2 - z^2}{(a - \rho)^2 + z^2} E \right] \tag{2.65c}$$

Along the z axis, $\rho = 0, k = 0$, so that $K(0) = E(0) = \pi/2$ and these general results reduce to

$$B_\rho \to 0 \qquad\qquad \rho = 0 \tag{2.66a}$$

$$B_\phi = 0 \qquad\qquad \rho = 0 \tag{2.66b}$$

$$B_z = \frac{\mu_0 I}{2} \frac{a^2}{(a^2 + z^2)^{3/2}} \qquad \rho = 0 \tag{2.66c}$$

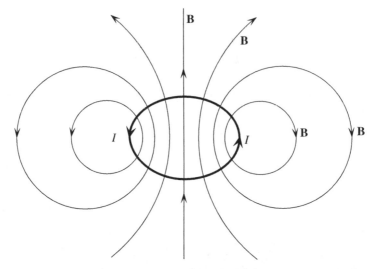

FIGURE 2.26. Magnetic field of a loop.

which is the result obtained in (2.24). In the plane of the loop, $z = 0$, (2.65) reduces to

$$B_\rho = 0 \qquad\qquad z = 0 \qquad (2.67a)$$

$$B_\phi = 0 \qquad\qquad z = 0 \qquad (2.67b)$$

$$
\begin{aligned}
B_z &= \frac{\mu_0 I}{2\pi} \left(\frac{1}{a + \rho} K + \frac{1}{a - \rho} E \right) \\
&= \frac{\mu_0 I a}{2\pi} \int_0^\pi \frac{a - \rho\cos\phi}{(a^2 - \rho^2)\sqrt{a^2 + \rho^2 - 2a\rho\cos\phi}}\, d\phi \qquad z = 0 \qquad (2.67c)
\end{aligned}
$$

where k^2 in (2.63) becomes, for $z = 0$,

$$k^2 = \frac{4a\rho}{(a + \rho)^2} \qquad z = 0 \qquad\qquad (2.67d)$$

The alternative result for B_z at $z = 0$ in (2.67c) was obtained by substituting the change of variables $\phi = \pi - 2\zeta$ and $d\phi = -2d\zeta$, so that $\cos\phi = -\cos 2\zeta = 2\sin^2\zeta - 1$ into K and E. At the center of the loop ($z = 0$, $\rho = 0$), $k = 0$, and $K = E = \pi/2$, so that (2.67c) reduces to $B_z = \mu_0 I/2a$, which is (2.25). The magnetic fields of a current loop are illustrated in Fig. 2.26.

EXAMPLE

Derive the **B** field at any point about a current sheet of finite width W and infinite length as shown in Fig. 2.27 using the vector magnetic potential.

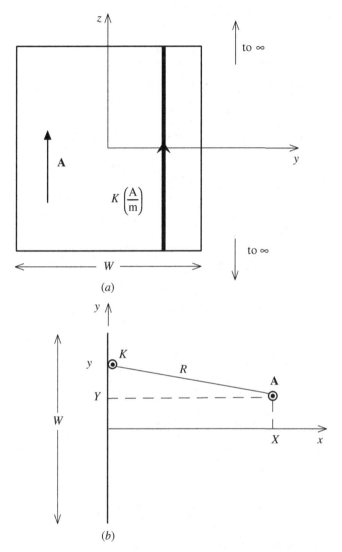

FIGURE 2.27. Current sheet of finite width and infinite length.

Again the sheet surface current can be viewed as currents $I = K \, dy$, which are infinite in length. We can assume a form for the differential contribution to A_z given in (2.53):

$$dA_z = -\frac{\mu_0 K \, dy}{2\pi} \ln R \qquad \text{current of infinite length}$$

Here

$$R = \sqrt{(y - Y)^2 + X^2}$$

and hence

$$A_z = -\frac{\mu_0 K}{2\pi} \int_{y=-W/2}^{W/2} \ln R \, dy$$

$$= -\frac{\mu_0 K}{4\pi} \int_{y=-W/2}^{W/2} \ln \left[(y-Y)^2 + X^2 \right] dy$$

Making a change of variables $\lambda = y - Y$, this becomes

$$A_z = -\frac{\mu_0 K}{4\pi} \int_{\lambda=-(W/2+Y)}^{W/2-Y} \ln \left(\lambda^2 + X^2 \right) d\lambda$$

$$= -\frac{\mu_0 K}{4\pi} \left[\lambda \ln \left(\lambda^2 + X^2 \right) - 2\lambda + 2X \tan^{-1} \frac{\lambda}{X} \right]_{-(W/2+Y)}^{W/2-Y}$$

$$= -\frac{\mu_0 K}{4\pi} \left\{ (W/2-Y) \ln \left[(W/2-Y)^2 + X^2 \right] \right.$$

$$-2(W/2-Y) + 2X \tan^{-1} \frac{W/2-Y}{X}$$

$$+ (W/2+Y) \ln \left[(W/2+Y)^2 + X^2 \right]$$

$$\left. -2(W/2+Y) + 2X \tan^{-1} \frac{W/2+Y}{X} \right\}$$

$$= -\frac{\mu_0 K}{4\pi} \left\{ (W/2-Y) \ln \left[(W/2-Y)^2 + X^2 \right] \right.$$

$$+ (W/2+Y) \ln \left[(W/2+Y)^2 + X^2 \right]$$

$$\left. -2W + 2X \tan^{-1} \frac{WX}{X^2 + Y^2 - (W/2)^2} \right\} \qquad (2.68)$$

This was evaluated using integral 623 from Dwight [7]:

$$\int \ln \left(x^2 + a^2 \right) dx = x \ln \left(x^2 + a^2 \right) - 2x + 2a \tan^{-1} \frac{x}{a} \qquad (D623)$$

We also used the trigonometric identity

$$\tan^{-1} \theta_1 \pm \tan^{-1} \theta_2 = \tan^{-1} \frac{\theta_1 \pm \theta_2}{1 \mp \theta_1 \theta_2} \qquad \theta_1, \theta_2 \geq 0 \qquad (2.69a)$$

giving

$$\tan^{-1}(x+y) + \tan^{-1}(x-y) = \tan^{-1} \frac{2x}{1 - x^2 + y^2} \qquad (2.69b)$$

and

$$\tan^{-1}(x+y) - \tan^{-1}(x-y) = \tan^{-1}\frac{2y}{1+x^2-y^2} \qquad (2.69c)$$

This gives $[x = (W/2)/X$ and $y = Y/X]$

$$\tan^{-1}\frac{W/2+Y}{X} + \tan^{-1}\frac{W/2-Y}{X} = \tan^{-1}\frac{W/X}{1-[(W/2)/X]^2 + (Y/X)^2}$$

$$= \tan^{-1}\frac{WX}{X^2+Y^2-(W/2)^2}$$

The magnetic field is determined from $\mathbf{B} = \nabla \times \mathbf{A}$ in rectangular coordinates as

$$\mathbf{B} = \frac{\partial A_z}{\partial Y}\mathbf{a}_x - \frac{\partial A_z}{\partial X}\mathbf{a}_y$$

$$= \frac{\mu_0 K}{4\pi}\left\{-\ln\frac{(W/2+Y)^2+X^2}{(W/2-Y)^2+X^2}\mathbf{a}_x + 2\tan^{-1}\frac{WX}{X^2+Y^2-(W/2)^2}\mathbf{a}_y\right\}$$

$$(2.70)$$

and we have used

$$\frac{\partial}{\partial u}\tan^{-1}u = \frac{1}{1+u^2} \qquad (D512.4)$$

Letting $Y \to 0$ in (2.70) gives the field along a line perpendicular to the strip:

$$\mathbf{B} = \frac{\mu_0 K}{2\pi}\tan^{-1}\frac{WX}{X^2-(W/2)^2}\mathbf{a}_y$$

$$= \frac{\mu_0 K}{\pi}\tan^{-1}\frac{W}{2X}\mathbf{a}_y \qquad Y=0 \qquad (2.71)$$

and we have used the identity in (2.69). But (2.71) is the same as (2.23) $(X=d)$, which was derived directly for the $\mathbf{B}$ field using the Biot–Savart law.

This result can be derived directly from the Biot–Savart law. A cross-sectional view of the problem is shown in Fig. 2.28. The differential contribution to the $\mathbf{B}$ field at a general point $x=X$ and $y=Y$ can be obtained by again considering the sheet to be composed of infinitely long filaments of currents $I = K\,dy$ and using the fundamental result in (2.14):

$$d\mathbf{B} = \frac{\mu_0 K dy}{2\pi R}(-\cos\theta\,\mathbf{a}_x + \sin\theta\,\mathbf{a}_y)$$

where

$$R = \sqrt{(Y-y)^2+X^2}$$

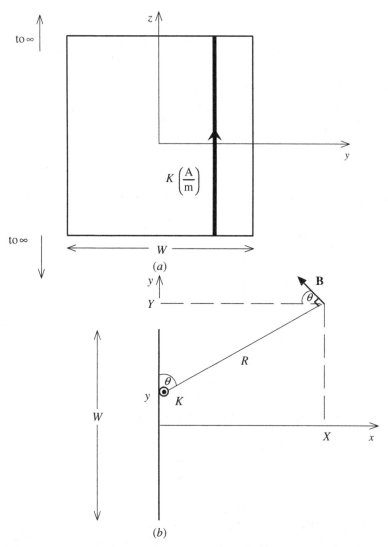

FIGURE 2.28. Magnetic field of a current sheet of finite width and infinite length using the Biot–Savart law.

and

$$\cos \theta = \frac{Y - y}{R}$$

$$\sin \theta = \frac{X}{R}$$

Integrating gives the total **B** field as

$$\mathbf{B} = \int_{y=-W/2}^{W/2} d\mathbf{B} = B_x \mathbf{a}_x + B_y \mathbf{a}_y$$

Evaluating these gives

$$B_x = -\frac{\mu_0 K}{2\pi} \int_{y=-W/2}^{W/2} \frac{Y-y}{(Y-y)^2 + X^2} \, dz$$

Making a change of variables, $\lambda = Y - y$, $d\lambda = -dy$ gives

$$B_x = -\frac{\mu_0 K}{2\pi} \int_{\lambda=Y-W/2}^{Y+W/2} \frac{\lambda}{\lambda^2 + X^2} \, d\lambda$$

Using integral 121.1 from Dwight [7],

$$\int \frac{x}{a^2 + x^2} dx = \frac{1}{2} \ln \left(a^2 + x^2 \right) \qquad\qquad \text{(D121.1)}$$

gives

$$B_x = -\frac{\mu_0 K}{4\pi} \ln \frac{(Y+W/2)^2 + X^2}{(Y-W/2)^2 + X^2}$$

which is the x component given in (2.70). The y component becomes

$$B_y = \frac{\mu_0 K}{2\pi} \int_{y=-W/2}^{W/2} \frac{X}{(Y-y)^2 + X^2} \, dy$$

Again making a change of variables, $\lambda = Y - y$, $d\lambda = -dy$ gives

$$B_y = \frac{\mu_0 K}{2\pi} \int_{\lambda=Y-W/2}^{Y+W/2} \frac{X}{\lambda^2 + X^2} \, d\lambda$$

Using integral 120.1 from Dwight [7],

$$\int \frac{1}{a^2 + x^2} dx = \frac{1}{a} \tan^{-1} \frac{x}{a} \qquad\qquad \text{(D120.1)}$$

gives

$$B_y = \frac{\mu_0 K}{2\pi} \left(\tan^{-1} \frac{Y+W/2}{X} - \tan^{-1} \frac{Y-W/2}{X} \right)$$

Using the identity in (2.69c) gives

$$B_y = \frac{\mu_0 K}{2\pi} \tan^{-1} \frac{WX}{X^2 + Y^2 - (W/2)^2}$$

which is the y component given in (2.70).

2.5.1 Leibnitz's Rule: Differentiate Before You Integrate

Solving for the **B** field by first obtaining the vector magnetic potential **A** via (2.47) or (2.52) avoids the integration of vector quantities that occurs by a direct solution for **B** using the Biot–Savart law. However, the final step of differentiating that to give $\mathbf{B} = \nabla \times \mathbf{A}$ may involve some rather complicated differentiations. A convenient way of avoiding those complicated differentiations and going directly to the **B** field is by using Leibnitz's rule [12]. Leibnitz's rule allows us to exchange the order of differentiation and integration:

$$\frac{\partial}{\partial y} \int_a^b f(x, y)\, dx = \int_a^b \frac{\partial f(x, y)}{\partial y}\, dx$$

There are some rather mild restrictions: $f(x, y)$ must be continuous and have continuous derivatives in $a \le x \le b$ and $y_1 \le y \le y_2$ over which the result is to be obtained.

For example, consider the problem of determining the **B** field for a finite-length current as shown in Fig. 2.24. The vector magnetic potential was obtained as

$$A_z = \frac{\mu_0 I}{4\pi} \int_{\lambda = Z - L/2}^{Z + L/2} \frac{1}{\sqrt{\lambda^2 + r^2}}\, d\lambda$$

$$= \frac{\mu_0 I}{4\pi} \ln \frac{(Z + L/2) + \sqrt{(Z + L/2)^2 + r^2}}{(Z - L/2) + \sqrt{(Z - L/2)^2 + r^2}} \qquad (2.57)$$

and the **B** field was obtained from $\mathbf{B} = \nabla \times \mathbf{A}$ as

$$\mathbf{B} = -\frac{\partial A_z}{\partial r}\, \mathbf{a}_\phi$$

$$= \frac{\mu_0 I}{4\pi r} \left[\frac{Z + L/2}{\sqrt{(Z + L/2)^2 + r^2}} - \frac{Z - L/2}{\sqrt{(Z - L/2)^2 + r^2}} \right] \mathbf{a}_\phi \qquad (2.58)$$

This final step required the differentiation of a somewhat complicated natural log function. We can instead formulate this as

$$B_\phi = -\frac{\partial A_z}{\partial r}$$

$$= -\frac{\partial}{\partial r} \left[\frac{\mu_0 I}{4\pi} \int_{\lambda = Z - L/2}^{Z + L/2} \frac{1}{\sqrt{\lambda^2 + r^2}}\, d\lambda \right]$$

$$= -\frac{\mu_0 I}{4\pi} \int_{\lambda = Z - L/2}^{Z + L/2} \frac{\partial}{\partial r} \left[\frac{1}{\sqrt{\lambda^2 + r^2}} \right] d\lambda$$

$$= -\frac{\mu_0 I}{4\pi} \int_{\lambda = Z - L/2}^{Z + L/2} \left[-\frac{1}{2} \frac{2r}{(\lambda^2 + r^2)^{3/2}} \right] d\lambda$$

$$= \frac{\mu_0 I}{4\pi} r \int_{\lambda = Z - L/2}^{Z + L/2} \frac{1}{(\lambda^2 + r^2)^{3/2}}\, d\lambda$$

$$= \frac{\mu_0 I}{4\pi} r \left[\frac{1}{r^2} \frac{\lambda}{\sqrt{\lambda^2 + r^2}} \right]_{Z-L/2}^{Z+L/2}$$

$$= \frac{\mu_0 I}{4\pi r} \left[\frac{Z + L/2}{\sqrt{(Z + L/2)^2 + r^2}} - \frac{Z - L/2}{\sqrt{(Z - L/2)^2 + r^2}} \right]$$

as obtained in (2.58) by differentiation of A_z after the integration to obtain it. We have used integral 200.03 from the table of integrals by Dwight [7]:

$$\int \frac{1}{(a^2 + x^2)^{3/2}} dx = \frac{x}{a^2 \sqrt{a^2 + x^2}} \qquad \text{(D200.03)}$$

Integrating A_z and then obtaining **B** from $\mathbf{B} = \nabla \times \mathbf{A}$ required the differentiation of a natural log function.

As another example, consider the problem of a current sheet of finite width and infinite length shown in Fig. 2.27. The vector magnetic potential is in the z direction and is given by

$$A_z = -\frac{\mu_0 K}{2\pi} \int_{y=-W/2}^{W/2} \ln R \, dy$$

$$= -\frac{\mu_0 K}{4\pi} \int_{y=-W/2}^{W/2} \ln \left[(y - Y)^2 + X^2 \right] dy$$

$$= -\frac{\mu_0 K}{4\pi} \int_{\lambda=-(W/2+Y)}^{W/2-Y} \ln(\lambda^2 + X^2) \, d\lambda$$

The **B** field is obtained from $\mathbf{B} = \nabla \times \mathbf{A}$ as

$$\mathbf{B} = \frac{\partial A_z}{\partial Y} \mathbf{a}_x - \frac{\partial A_z}{\partial X} \mathbf{a}_y$$

$$= \frac{\mu_0 K}{4\pi} \left[-\ln \frac{(W/2 + Y)^2 + X^2}{(W/2 - Y)^2 + X^2} \mathbf{a}_x + 2\tan^{-1} \frac{WX}{X^2 + Y^2 - (W/2)^2} \mathbf{a}_y \right]$$

$$(2.70)$$

Instead of integrating to obtain A_z and then differentiating to obtain **B**, use Leibnitz's rule to obtain

$$B_x = \frac{\partial A_z}{\partial Y}$$

$$= \frac{\partial}{\partial Y} \left[-\frac{\mu_0 K}{4\pi} \int_{y=-W/2}^{W/2} \ln \left[(y - Y)^2 + X^2 \right] dy \right]$$

$$= -\frac{\mu_0 K}{4\pi} \int_{y=-W/2}^{W/2} \left[\frac{\partial}{\partial Y} \ln \left[(y - Y)^2 + X^2 \right] \right] dy$$

$$= -\frac{\mu_0 K}{4\pi} \int_{y=-W/2}^{W/2} \frac{2(y-Y)(-1)}{(y-Y)^2 + X^2} \, dz$$

$$= \frac{\mu_0 K}{2\pi} \int_{\lambda=-(W/2+Y)}^{W/2-Y} \frac{\lambda}{\lambda^2 + X^2} \, d\lambda$$

$$= \frac{\mu_0 K}{2\pi} \frac{1}{2} \left[\ln\left(\lambda^2 + X^2\right) \right]_{-(W/2+Y)}^{W/2-Y}$$

$$= -\frac{\mu_0 K}{4\pi} \ln \frac{(W/2+Y)^2 + X^2}{(W/2-Y)^2 + X^2}$$

as was obtained in (2.70) and we used integral 121.1 of Dwight [7]:

$$\int \frac{x}{a^2 + x^2} dx = \frac{1}{2} \ln\left(a^2 + x^2\right) \tag{D121.1}$$

Similarly, the y component of **B** is obtained as

$$B_y = -\frac{\partial A_z}{\partial X}$$

$$= -\frac{\partial}{\partial X} \left[-\frac{\mu_0 K}{4\pi} \int_{y=-W/2}^{W/2} \ln\left[(y-Y)^2 + X^2\right] dy \right]$$

$$= \frac{\mu_0 K}{4\pi} \int_{y=-W/2}^{W/2} \left[\frac{\partial}{\partial X} \ln\left[(y-Y)^2 + X^2\right] \right] dy$$

$$= \frac{\mu_0 K}{4\pi} \int_{y=-W/2}^{W/2} \frac{2X}{(y-Y)^2 + X^2} \, dy$$

$$= \frac{\mu_0 K}{2\pi} X \int_{\lambda=-(W/2+Y)}^{W/2-Y} \frac{1}{\lambda^2 + X^2} \, d\lambda$$

$$= \frac{\mu_0 K}{2\pi} X \left[\frac{1}{X} \tan^{-1} \frac{\lambda}{X} \right]_{-(W/2+Y)}^{W/2-Y}$$

$$= \frac{\mu_0 K}{2\pi} \left[\tan^{-1} \frac{W/2-Y}{X} - \tan^{-1} \frac{-(W/2+Y)}{X} \right]$$

$$= \frac{\mu_0 K}{2\pi} \tan^{-1} \frac{WX}{X^2 + Y^2 - (W/2)^2}$$

as was obtained in (2.70) and we used integral 120.1 of Dwight [7]:

$$\int \frac{1}{a^2 + x^2} dx = \frac{1}{a} \tan^{-1} \frac{x}{a} \tag{D120.1}$$

2.6 DETERMINING THE INDUCTANCE OF A CURRENT LOOP: A PRELIMINARY DISCUSSION

The process of determining the inductance of a loop formed by a current-carrying conductor was discussed briefly in Chapter 1. To determine the inductance of any loop shape we first need to determine the **B** field over the surface of the loop, s, that is bounded by the conductor. Next we must integrate that **B** field over the loop surface with a surface integral to determine the total magnetic flux through the loop surface as

$$\psi = \int_s \mathbf{B} \cdot d\mathbf{s} \tag{1.6}$$

The inductance of the loop is the ratio of this flux and the current I that created it:

$$L = \frac{\psi}{I} \tag{1.7}$$

The circular current loop shown in Fig. 2.29 will be used to illustrate the first part of that process: determining the **B** field over the surface of the loop. The loop has a radius a and is formed by a wire of radius r_w. In this section we determine the magnetic flux density **B** over the flat surface of the loop, s, that is bounded by the interior surface of the wire by using three methods. In Chapter 4, that **B** field will be integrated over the loop surface via the surface

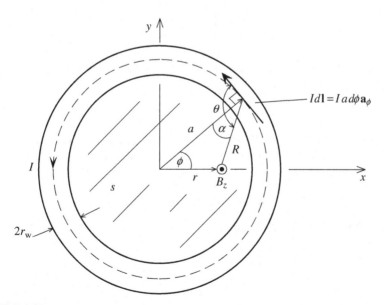

FIGURE 2.29. Determining the magnetic flux through the surface enclosed by a circular wire loop.

integral in (1.6) to give the total magnetic flux through the loop and hence the inductance of the circular loop via (1.7). The **B** field over the loop surface will, by the right-hand rule, be z-directed (out of the page), $\mathbf{B} = B_z \mathbf{a}_z$, and hence the total magnetic flux throught the surface via (1.6) will be obtained in Chapter 4 by further integration as

$$\psi = \int_{r=0}^{a-r_w} \int_{\phi'=0}^{2\pi} B_z \underbrace{r\, d\phi'\, dr}_{ds} \tag{1.6}$$

We assume that the current is *uniformly distributed over the wire cross section* so that we may compute the magnetic fields from it by replacing the wire with a filament on its axis that contains the total current.

Perhaps the most fundamental method for determining the magnetic field over the surface bounded by the wire loop is by using the Biot–Savart law. A differential length of current produces a net magnetic field over the surface bounded by the wire that is in the z direction and is therefore perpendicular to the surface of the loop. Because of symmetry, we can, without loss of generality, determine the **B** field in the plane of the loop, the xy plane, at a point that is located along the x axis at a distance r from the center of the loop as shown in Fig. 2.29. For $r < a$ this gives the field inside the loop. This result will also be valid in the plane of the loop for points outside the loop, $r > a$. From the Biot–Savart law, the differential contribution to the $\mathbf{B} = B_z \mathbf{a}_z$ field at the point from this differential current is

$$dB_z = \frac{\mu_0 I}{4\pi} \frac{1}{R^2} \underbrace{a\, d\phi}_{dl} \sin\theta \tag{2.72a}$$

where the distance R between the differential current segment and the point is, using the law of cosines,

$$R^2 = a^2 + r^2 - 2ar \cos\phi \tag{2.72b}$$

and θ is the angle between the differential current vector and the vector directed from it to the point as shown in Fig. 2.29. The sine of θ can be determined in terms of the angle α as $\theta = 90° + \alpha$ and

$$\sin\theta = \sin(90° + \alpha) = \cos\alpha$$

Using the law of cosines again gives $r^2 = a^2 + R^2 - 2aR \cos\alpha$, so that

$$\sin\theta = \cos\alpha = \frac{a^2 + R^2 - r^2}{2aR}$$

$$= \frac{a - r\cos\phi}{R} \tag{2.72c}$$

The contribution to the field at the point from the lower half of the current is the same as that from the upper half. Hence, the magnetic field at the point is determined from the Biot–Savart law as

$$B_z(r) = 2\frac{\mu_0 Ia}{4\pi} \int_{\phi=0}^{\pi} \frac{a - r\cos\phi}{(a^2 + r^2 - 2ar\cos\phi)^{3/2}} d\phi$$

$$= \frac{\mu_0 Ia}{2\pi} \int_{\phi=0}^{\pi} \frac{a - r\cos\phi}{(a^2 + r^2 - 2ar\cos\phi)^{3/2}} d\phi \qquad z = 0 \quad (2.73)$$

A second method of determining B_z is by using the vector magnetic potential and performing $B_z = \nabla \times A_\phi = (1/r)[\partial(rA_\phi)/\partial r]$ (see the Appendix). The vector magnetic potential is obtained by evaluating (2.59) in the plane of the loop (Fig. 2.25, $z = 0$, $\rho = r$, $\theta = 0$) to give

$$A_\phi = \frac{\mu_0 Ia}{2\pi} \int_{\phi=0}^{\pi} \frac{\cos\phi}{\sqrt{a^2 + r^2 - 2ar\cos\phi}} d\phi \qquad z = 0 \qquad (2.74)$$

Hence, the magnetic flux density over the loop surface is obtained from

$$\begin{aligned}
\mathbf{B} &= \nabla \times \mathbf{A} \\
&= \frac{1}{r}\frac{\partial(rA_\phi)}{\partial r}\mathbf{a}_z \\
&= \frac{\mu_0 Ia}{2\pi r}\frac{\partial}{\partial r}\left[\int_{\phi=0}^{\pi} \frac{r\cos\phi}{\sqrt{a^2 + r^2 - 2ar\cos\phi}} d\phi\right]\mathbf{a}_z \\
&= \frac{\mu_0 Ia}{2\pi r}\int_{\phi=0}^{\pi}\left[\frac{\partial}{\partial r}\frac{r\cos\phi}{\sqrt{a^2 + r^2 - 2ar\cos\phi}}\right] d\phi\,\mathbf{a}_z \\
&= \frac{\mu_0 Ia}{2\pi r}\int_{\phi=0}^{\pi} \frac{a\cos\phi(a - r\cos\phi)}{(a^2 + r^2 - 2ar\cos\phi)^{3/2}} d\phi \qquad z = 0 \qquad (2.75)
\end{aligned}$$

where we have used Leibnitz's rule to interchange the order of differentiation and integration. This result seems to be undefined at the center of the loop, $r = 0$:

$$\begin{aligned}
\lim_{r\to 0} B_z &= \frac{\mu_0 I}{2\pi}\lim_{r\to 0}\frac{\displaystyle\int_{\phi=0}^{\pi}\frac{a^2\cos\phi(a - r\cos\phi)}{(a^2 + r^2 - 2ar\cos\phi)^{3/2}}d\phi}{r} \\
&= \frac{\mu_0 I}{2\pi}\frac{\int_{\phi=0}^{\pi}[(a^3\cos\phi)/a^3]d\phi}{0} = \frac{0}{0}
\end{aligned}$$

However using l'Hôpital's rule gives a limit of

$$
\lim_{\underbrace{}_{r \to 0}} B_z = \frac{\mu_0 I}{2\pi} \lim_{\underbrace{}_{r \to 0}} \frac{\frac{\partial}{\partial r} \int_{\phi=0}^{\pi} \frac{a^2 \cos \phi \, (a - r \cos \phi)}{(a^2 + r^2 - 2ar \cos \phi)^{3/2}} d\phi}{\frac{\partial}{\partial r}(r)}
$$

$$
= \frac{\mu_0 I \int_{\phi=0}^{\pi} [(2 \cos^2 \phi)/a] d\phi = \pi/a}{2\pi \qquad\qquad 1}
$$

$$
= \frac{\mu_0 I}{2a} \qquad r = 0, z = 0
$$

which is the result derived directly by the Biot–Savart law and given in (2.25). We again used Leibnitz's rule to interchange the order of differentiation and integration in the numerator.

The third method of obtaining the B_z field in the plane of the loop is to use directly the result obtained by Smythe [10] and Weber [11] and given in (2.65). Evaluating this in the plane of the loop, $z = 0$, gives (2.67):

$$
B_z = \frac{\mu_0 I}{2\pi} \left(\frac{1}{a+r} K + \frac{1}{a-r} E \right)
$$

$$
= \frac{\mu_0 I a}{2\pi} \int_0^{\pi} \frac{a - r \cos \phi}{(a^2 - r^2) \sqrt{a^2 + r^2 - 2ar \cos \phi}} \, d\phi \qquad z = 0 \qquad (2.67c)
$$

where $K(k)$ and $E(k)$ are the complete elliptic integrals of the first and second kind, respectively [7]:

$$
K(k) = \int_{\zeta=0}^{\pi/2} \frac{d\zeta}{\sqrt{1 - k^2 \sin^2 \zeta}} \tag{D773.1}
$$

$$
E(k) = \int_{\zeta=0}^{\pi/2} \sqrt{1 - k^2 \sin^2 \zeta} \, d\zeta \tag{D774.1}
$$

and

$$
k^2 = \frac{4ar}{(a+r)^2} \tag{2.67d}
$$

As a check on these results, at the center of the loop all three evaluate to

$$
B_z = \frac{\mu_0 I}{2a} \qquad r = 0, z = 0 \tag{2.25}
$$

which is the result obtained directly and given in (2.25).

The interesting aspect of these three results for B_z is that they are all seemingly different! All three results have the form

$$
B_z(r) = \frac{\mu_0 I a}{2\pi} \int_{\phi=0}^{\pi} [\text{Integrand}] \, d\phi \qquad z = 0 \tag{2.76}
$$

but *all three integrands are different.* For example, compare the integrands of the result using the Biot–Savart law and given in (2.73):

$$\int_{\phi=0}^{\pi} \frac{a - r\cos\phi}{(a^2 + r^2 - 2ar\cos\phi)^{3/2}} \, d\phi = \frac{1}{a^2} \int_{\phi=0}^{\pi} \frac{1 - u\cos\phi}{(1 + u^2 - 2u\cos\phi)^{3/2}} \, d\phi$$

(2.77a)

the result obtained by differentiating the vector magnetic potential according to $\mathbf{B} = \nabla \times \mathbf{A}$ and given in (2.75):

$$\int_{\phi=0}^{\pi} \frac{a\cos\phi\,(a - r\cos\phi)}{r(a^2 + r^2 - 2ar\cos\phi)^{3/2}} d\phi = \frac{1}{a^2} \int_{\phi=0}^{\pi} \frac{\cos\phi\,(1 - u\cos\phi)}{u\,(1 + u^2 - 2u\cos\phi)^{3/2}} \, d\phi$$

(2.77b)

and the result obtained by Smythe and Weber in (2.67c):

$$\int_{0}^{\pi} \frac{a - r\cos\phi}{(a^2 - r^2)\sqrt{a^2 + r^2 - 2ar\cos\phi}} \, d\phi$$

$$= \frac{1}{a^2} \int_{\phi=0}^{\pi} \frac{1 - u\cos\phi}{(1 - u^2)\sqrt{1 + u^2 - 2u\cos\phi}} \, d\phi \qquad (2.77c)$$

and we have written the three integrands in terms of the ratio of the radius to the point, r, and the radius of the loop, a, as

$$u = \frac{r}{a} \qquad (2.77d)$$

Therefore, all of the results depend on the ratio of the radius to the point and the radius of the loop. For points interior to the loop, $r < a$ and $u < 1$, the magnetic field should be directed out of the page, and hence the integrals should be positive so that $B_z > 0$. For points exterior to the loop, $r > a$ and $u > 1$, the magnetic field should be directed into the page and hence the integrals should be negative, so that $B_z < 0$. This is determined using the right-hand rule.

But how can these three *different* integrands give the same result for the $B_z(r)$ field over the surface enclosed by the loop? The answer is that what is important is the result of the integration, and integrands having different curves over the limits of the integral, $0 \leq \phi \leq \pi$, are capable of enclosing the same *area*. Fortunately, it turns out that this is the case for the three results above: *All three seemingly different integrals give the same magnetic field $B_z(r)$,* as we show next using numerical integration.

Since the three integrals in (2.77) cannot be integrated in closed form, a numerical integration routine was used to perform the integration for various values of the radius to the point, r, and the radius of the loop, a. Figure 2.30(a) shows the plots of the integrands for $r = 1$ and $a = 2$ over the range of the

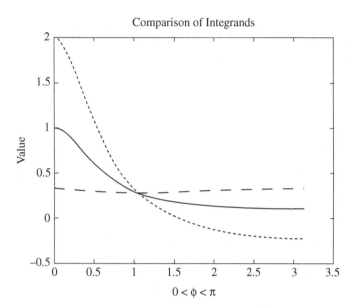

FIGURE 2.30(a). Plots of the integrands for $0 \le \phi \le \pi$: (———) by the Biot–Savart law in (2.77a); (....) by the vector potential method in (2.77b); and ($_-$ $_-$ $_-$) by the result from Smythe [10] and Weber [11] in (2.77c) for $r = 1$ and $a = 2$. All three integrals evaluate to 0.9783.

integral, $0 \le \phi \le \pi$. The curves of the three integrands are considerably different, yet the integral evaluates to 0.9783 for all three integrands.

Figure 2.30(b) shows the plots of the integrands for $r = 0.1$ and $a = 2$ over the range of the integral, $0 \le \phi \le \pi$. The curve of the integrand by the vector magnetic potential method in (2.77b) is considerably different from the other two, yet the integral evaluates to 0.7869 for all three integrands. For $r = 0$ all three integrals approach $\pi/a^2 = 0.7854$.

Figure 2.30(c) shows the plots of the integrands for $r = 1.9$ and $a = 2$ over the range of the integral, $0 \le \phi \le \pi$. The curves of the integrands by the Biot–Savart law in (2.77a) and by the vector potential method in (2.77b) are virtually identical but are different from the result by Smythe and Weber in (2.77c), yet the integral evaluates to 5.6550 for all three integrands.

The three integrals for B_z in (2.77) are also valid for points in the plane of the loop ($z = 0$) which are outside the loop, $r > a$. Figure 2.30(d) shows the plots of the integrands for $r = 2$ and $a = 1$ over the range of the integral, $0 \le \phi \le \pi$. The curves of the three integrands are considerably different over the range of the integral, $0 \le \phi \le \pi$, yet the integral evaluates to -0.2709 for all three integrands. The fact that the integral is negative and $B_z < 0$ makes sense because for points in the plane of the loop ($z = 0$) but lying outside the current loop, $r > a$, the $\mathbf{B} = B_z \mathbf{a}_z$ field is, by the right-hand rule, in the negative z direction (into the page).

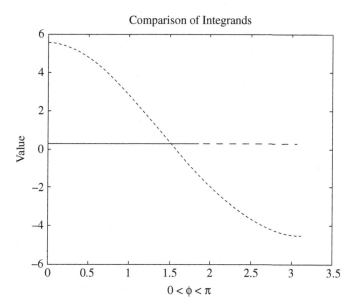

FIGURE 2.30(b). Plots of the integrands for $0 \leq \phi \leq \pi$: (——) by the Biot–Savart law in (2.77a), (....) by the vector potential method in (2.77b), and (_ _ _) by the result from Smythe [10] and Weber [11] in (2.77c) for $r = 0.1$ and $a = 2$. All three integrals evaluate to 0.7869.

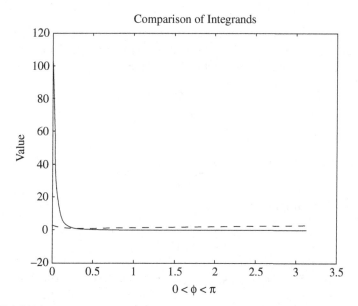

FIGURE 2.30(c). Plots of the integrands for $0 \leq \phi \leq \pi$: (——) by the Biot–Savart law in (2.77a), (....) by the vector potential method in (2.77b), and (_ _ _) by the result from Smythe [10] and Weber [11] in (2.77c) for $r = 1.9$ and $a = 2$. All three integrals evaluate to 5.6550.

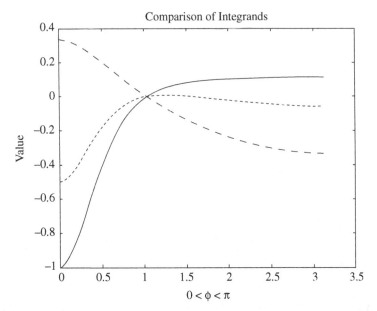

FIGURE 2.30(d). Plots of the integrands for $0 \leq \phi \leq \pi$: (——) by the Biot–Savart law in (2.77a), (....) by the vector potential method in (2.77b), and (– – –) by the result from Smythe [10] and Weber [11] in (2.77c) for $r = 2$ and $a = 1$ (points outside the loop). All three integrals evaluate to -0.2709.

The curves representing the three integrands in (2.77a), (2.77b), and (2.77c) which are plotted in Fig. 2.30 are equal for only one value of ϕ. This equality of the three integrands occurs when in (2.77b)

$$\frac{\cos \phi}{u} = 1$$

and when in (2.77c)

$$\frac{1 + u^2 - 2u \cos \phi}{1 - u^2} = 1$$

Both these conditions occur when

$$\phi = \cos^{-1} u \qquad u < 1$$

which has values only for $u \leq 1$. For the case of $r = 1$ and $a = 2$ giving $u = 1/2$ in Fig. 2.30(a), this value of ϕ is $\phi = 60° = 1.0472$ rad. For the case of $r = 0.1$ and $a = 2$ giving $u = \frac{1}{20}$ in Fig. 2.30(b), this value of ϕ is $\phi = 87.134° = 1.5208$ rad. For the case of $r = 1.9$ and $a = 2$ giving $u = 0.95$ in Fig. 2.30(c), this value of ϕ is $\phi = 18.195° = 0.3176$ rad. For points outside the current loop, $r > a$, so that $u > 1$, the term in the numerator of each integrand, $(1 - u \cos \phi)$, is zero for all three integrands at

$$\phi = \cos^{-1} \frac{1}{u} \qquad u > 1$$

Hence for the case of $r = 2$ and $a = 1$ giving $u = 2$ in Fig. 2.30(d), this value of ϕ where all three integrands are zero is $\phi = 60° = 1.0472$ rad.

2.7 ENERGY STORED IN THE MAGNETIC FIELD

The electric energy stored in a region of space of volume v due to a system of charges is [3,6]

$$\boxed{W_E = \tfrac{1}{2} \int_v \mathbf{D} \cdot \mathbf{E} \, dv \qquad \text{J}} \tag{2.78}$$

If the space surrounding the charges is linear, homogeneous, and isotropic and described by a permittivity ε, then $\mathbf{D}$ is related to $\mathbf{E}$ by $\mathbf{D} = \varepsilon \, \mathbf{E}$ and (2.78) becomes

$$W_E = \tfrac{1}{2}\varepsilon \int_v E^2 \, dv \qquad \text{J} \tag{2.79}$$

Similarly, the magnetic energy stored in a region of space of volume v due to a system of current loops is [3,6]

$$\boxed{W_M = \tfrac{1}{2} \int_v \mathbf{B} \cdot \mathbf{H} \, dv \qquad \text{J}} \tag{2.80}$$

If the space surrounding the currents is linear, homogeneous, and isotropic and described by a permeability μ, then $\mathbf{B}$ is related to $\mathbf{H}$ by $\mathbf{B} = \mu \, \mathbf{H}$ and (2.80) becomes

$$W_M = \tfrac{1}{2}\mu \int_v H^2 \, dv \qquad \text{J} \tag{2.81}$$

As noted in Chapter 1, these have a direct parallel with the energy stored in the fields of a capacitor (stored in its electric field)

$$\boxed{W_E = \tfrac{1}{2}C \, V^2 \qquad \text{J}} \tag{2.82}$$

where V is the voltage between the capacitor plates and C is its capacitance. Similarly, the energy stored in an inductor (stored in its magnetic field) is

$$W_M = \tfrac{1}{2} L I^2 \qquad J \tag{2.83}$$

where I is the current passed through the inductor and L is its inductance. Hence, we have an alternative means of determining the capacitance or inductance of a structure indirectly by, instead, determining the energy stored in the electric or magnetic fields of the element:

$$C = \frac{2W_E}{V^2} \tag{2.84a}$$

$$L = \frac{2W_M}{I^2} \tag{2.84b}$$

For some structures, (2.84b) will be a useful way of determining the inductance of the structure.

2.8   THE METHOD OF IMAGES

Problems often involve a flat sheet of metal that is very large in extent. Charges and/or currents exist above the sheet and it is desired to determine the electric and magnetic fields in the space above the sheet. This is a very difficult problem, because to solve it we have to determine the distribution of the charge/currents induced on the surface of this sheet. Fortunately, with the method of images we can replace these problems with equivalent problems that are much easier to solve. Although conductive metals have very large conductivities, it is nonetheless desirable to replace them with *perfect conductors*. A perfect electric conductor is a fictitious material that has an infinite conductivity, $\sigma \to \infty$. In the case of electric fields in the conductor, the current density in that conductor is related to the electric field by Ohm's law, $\mathbf{J} = \sigma \mathbf{E}$. For $\sigma = \infty$ we must have either $\mathbf{E} = 0$ or $\mathbf{J} = \infty$. Having an infinite current density would mean that either (1) a finite amount of charge is moved in zero time, or (2) an infinite amount of charge is moved in a finite time. Since neither of these is acceptable physically, we conclude that $\mathbf{E} = 0$ in a perfect conductor. No charge can exist in the interior of a perfect conductor and must exist only on its surface. Any charge in a very good conductor having a very large conductivity will decay to zero (move to the conductor surface) in a very short time called the *relaxation time*, $\tau = \varepsilon_0/\sigma$ [3]. Similarly, magnetic fields that vary with time cannot exist in a perfect conductor, but steady (dc) magnetic fields in a superconductor can [3].

The boundary conditions at the surface of a *perfect conductor* are that (1) the component of the total electric field intensity **E** that is tangent to the surface must be zero, and (2) the component of the total magnetic flux density **B** normal to the surface must be zero [3]. This means that on the surface of a perfect conductor (1) the total electric field must be normal to it, and (2) the total magnetic field must be tangent to it.

The electrostatic potential function or voltage V is defined such that the negative gradient of V gives the static electric field: $\mathbf{E} = -\nabla V$ (see the Appendix for the gradient function). Hence, the equipotential surfaces on which the voltage is constant are perpendicular to the lines of the **E** field and hence must be tangent to the surface of a perfect conductor. Similarly, the vector magnetic potential **A** is defined such that its curl gives the magnetic flux density: $\mathbf{B} = \nabla \times \mathbf{A}$. Since the **B** field lines must be tangent to the surface of a perfect conductor with no component perpendicular to the conductor surface, the lines of **A** must be tangent to the surface of a perfect conductor.

Ground planes consisting of a conductor of large extent whose conductance, although finite, is very large (e.g., copper) are frequently found in electronic systems either intentionally or unintentionally. The metallic frame of an airplane fuselage acts like a ground plane to the electromagnetic fields of the antennas that are mounted above it. Other metallic enclosures such as are used in constructing shielded rooms are intended to contain or exclude unwanted electromagnetic fields that may cause interference with sensitive electronic devices [5].

First consider a static charge above a perfect conductor of infinite extent shown in Fig. 2.31. Although the equivalent image problem requires a perfect conductor of infinite extent, in practice a reasonably good conductor of very large extent is usually a sufficient approximation. A positive charge Q at a height h above an infinite, perfectly conducting plane has, according to the boundary conditions on the electric field at its surface, its electric field normal to the surface of the plane [3]. If we replace the plane with a negative charge

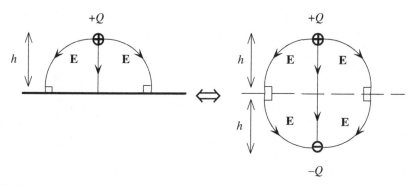

FIGURE 2.31. Static charges above a perfect conductor and the equivalent image problem.

$-Q$ at a depth h below the previous position of the plane, the electric fields *above the position of the plane* will be identical in either case [3].

In the case of currents, a similar imaging can be used. A foolproof way of getting the correct directions of the images of the currents is to recall that current is the flow of charge. So we can visualize a finite-length current as having positive and negative charge being accumulated at the ends and then image those charges as shown in Fig. 2.32. A current that is parallel to the plane is imaged at the same depth below the plane but with the current direction reversed as shown in Fig. 2.32(a). A current that is perpendicular to the plane is imaged at the same depth below the plane but with its direction the same as the current above the plane as shown in Fig. 2.32(b). In either case, the magnetic field in the space above the plane will be the same as when the plane is replaced by images. The total magnetic field will be tangent to the

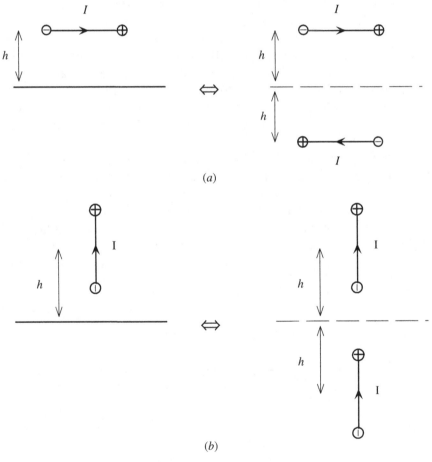

FIGURE 2.32. Imaging currents.

plane at all points on the plane in either case, thereby satisfying the boundary conditions on the magnetic field on the surface of a perfectly conducting plane [3]. Currents that are neither horizontal nor perpendicular to the plane can be imaged by resolving the current into its vertical and horizontal components and imaging those.

You should show that the total magnetic field at all points on the surface of the plane in Fig. 2.32 is tangent to the surface, and there is no component perpendicular to the surface of the plane. Do this by replacing the plane with its image and then superimposing the magnetic fields due to the original current and its image using the results for the magnetic fields of the currents (finite length or infinite length) that were derived previously.

2.9   STEADY (DC) CURRENTS MUST FORM CLOSED LOOPS

Steady currents (dc currents that do not vary with time) must form closed loops (i.e., must return to their source). This is rather simple to prove. First, we recall the *law of conservation of charge*:

$$I_{\text{leaving } s} = \oint_s \mathbf{J} \cdot d\mathbf{s} = -\frac{d}{dt} Q_{\text{enclosed}} \qquad (2.85)$$

The surface integral $\oint_s \mathbf{J} \cdot d\mathbf{s}$ gives the *net current leaving the closed surface s*, $\mathbf{J}$ is the current density in A/m^2 over that surface, and Q_{enclosed} is the *net positive charge enclosed by the surface*. This is illustrated in Fig. 2.33. This mathematical statement of the law of conservation of charge is very sensible since it says merely that *the net outflow of current out of a closed surface s equals the time rate of decrease of the charge enclosed by that surface.* Recalling that current is the rate of flow of charge, this mathematical statement

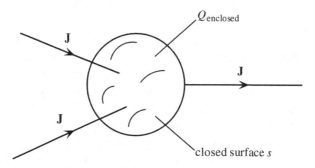

FIGURE 2.33. Conservation of charge.

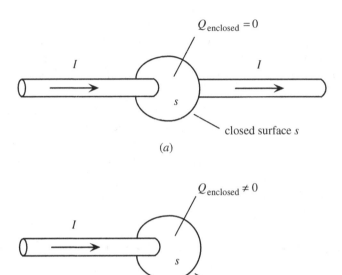

FIGURE 2.34. Finite-length currents and conservation of charge.

of the law of conservation of charge is elegantly obvious since it requires that if there is a *net* current leaving the closed surface, it must be accompanied by a decrease in the *net positive* charge contained in that surface since charge can be neither created nor destroyed inside the closed surface *s*!

Consider the case of a wire carrying a steady (dc) current I as illustrated in Fig. 2.34. Surround a point along the wire with a closed surface (a node in the vernacular of lumped-circuit theory) as shown in Fig. 2.34(a). In the case of steady or direct (dc) currents, the right-hand side of (2.85) must be zero:

$$\oint_s \mathbf{J} \cdot d\mathbf{s} = 0 \qquad \text{steady (dc) currents} \qquad (2.86)$$

According to (2.86) the dc current entering the node must equal the dc current leaving the node, and hence Kirchhoff's current law satisfies conservation of charge for steady (dc) currents [1,2]. However, this also shows that *finite lengths of dc currents cannot exist*. This is simple to show because if we surround the end of a finite length of current with a closed surface as illustrated in Fig. 2.34(b), we will have current I entering but no current leaving the closed surface, thereby violating (2.86). Hence, *steady (dc) currents must form closed loops*. In other words, *a steady (dc) current must return to its source*.

The current $\mathbf{J}$ in (2.86) is *conduction current*, which is the flow of *free charge* such as in a wire. When we add displacement current to Ampère's law for time-varying fields (and time-varying currents) in Chapter 3 we may make the statement that for time-varying currents, the sum of the conduction and the displacement current must form closed loops (i.e., must return to their source). This may be shown by taking the divergence of Ampère's law for time-varying currents in point form (see Chapter 3):

$$\nabla \cdot \nabla \times \mathbf{H} = \nabla \cdot \left(\underbrace{\mathbf{J}}_{\substack{\text{conduction} \\ \text{current}}} \right) + \nabla \cdot \left(\underbrace{\frac{\partial \mathbf{D}}{\partial t}}_{\substack{\text{displacement} \\ \text{current}}} \right)$$

$$= 0 \qquad\qquad (2.87)$$

since we have the identity $\nabla \cdot \nabla \times \mathbf{F} = 0$ for any vector field (see the Appendix). Using the divergence theorem (see the Appendix), this gives

$$\oint_s \left(\mathbf{J} + \frac{\partial \mathbf{D}}{\partial t} \right) \cdot d\mathbf{s} = 0 \qquad\qquad (2.88)$$

thereby showing that for time-varying currents, the *total current* must form closed loops. Where conduction current ends, displacement current takes over to complete the loop as in a circuit containing a capacitor.

In several of the examples illustrating the Biot–Savart law as well as in illustrating the use of the vector magnetic potential to obtain $\mathbf{B}$ via $\mathbf{B} = \nabla \times \mathbf{A}$, we employed *finite lengths of dc current*. But if these cannot exist, of what use are those results for the magnetic fields of a current of finite length? The answer is that we use solutions for the fields of finite-length current segments

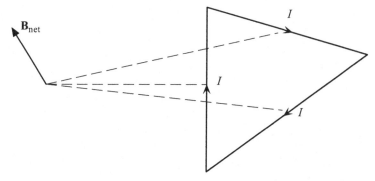

FIGURE 2.35. Combining magnetic fields of finite-length currents to determine the magnetic field of a closed loop of current.

to construct the solutions for the fields of a *closed current loop* of which these current elements are a part, as illustrated in Fig. 2.35. Clearly, we are using *superposition* and the surrounding medium must be linear, at least with regard to its magnetic field properties. So determining the magnetic fields for steady (dc) currents of finite length is useful in that regard and for that purpose.

3

FIELDS OF TIME-VARYING CURRENTS (ACCELERATED CHARGE)

In Chapter 2 we investigated the calculation of the magnetic fields produced by various configurations of static (dc) currents (the *steady* flow of charge). We discussed in Chapter 1 how the inductance of a structure will be obtained from these static magnetic fields by first obtaining the magnetic flux penetrating the surface of the current loop from (1.6):

$$\psi = \int_s \mathbf{B} \cdot d\mathbf{s} \tag{1.6}$$

and then obtaining the inductance of the loop from (1.7):

$$L = \frac{\psi}{I} \tag{1.7}$$

This inductance parameter will then be used in lumped circuits to determine its effect in circuits in which the currents vary with time (*accelerated* charge). The electromagnetics law that allows this determination is Faraday's law of induction. However, we seem to have a logical inconsistency in this process: A circuit element, inductance, that was derived for static (dc) currents will be used to evaluate its effect on time-varying currents. The ability to use a result derived for dc currents in a situation where the currents vary with time is shown in Section 3.4 to be a valid approximation using an iterative solution of the field equations. This approximation will be valid for circuits

Inductance: Loop and Partial, By Clayton R. Paul
Copyright © 2010 John Wiley & Sons, Inc.

whose maximum dimensions are "electrically small," that is, much less than a wavelength at the frequency of the driving source. The notion of electrically small dimensions is discussed in Section 3.3.

3.1 FARADAY'S FUNDAMENTAL LAW OF INDUCTION

Faraday's law is perhaps the most profound of the collective group of laws governing all macroscopic electromagnetic fields that are known as Maxwell's equations. Without Faraday's law we would not have the use of "electricity" and all its myriad implications.

To state Faraday's law in unambiguous mathematical terms, consider Fig. 3.1, which shows an open surface s that has a contour or path c surrounding it. With reference to Fig. 3.1, Faraday's law can be stated in mathematical form as [3–6]

$$\text{emf} = -\frac{d\psi}{dt} \tag{3.1}$$

where the *electromotive force* emf around the *closed loop* c is obtained with a *line integral* as

$$\text{emf} = \oint_c \mathbf{E} \cdot d\mathbf{l} \tag{3.2}$$

and the *magnetic flux* that passes through the *open surface* s is obtained with a *surface integral* as

$$\psi = \int_s \mathbf{B} \cdot d\mathbf{s} \tag{3.3}$$

Hence, Faraday's fundamental law of induction is

$$\boxed{\oint_c \mathbf{E} \cdot d\mathbf{l} = -\frac{d}{dt} \int_s \mathbf{B} \cdot d\mathbf{s}} \tag{3.4}$$

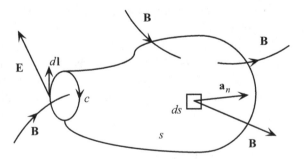

FIGURE 3.1. Faraday's law.

The contour c of the *closed loop* can be thought of as either a conducting material (as in the case of a wire) or an imaginary contour of nonconducting material (as in the case of free space) and **E** is the *electric field intensity vector* with units of V/m along that contour. The *dot product* in the integrand of the emf in (3.2), **E** $\cdot$ *d***l**, means that we take the product of the differential lengths of this contour *dl* and the electric field lines that are *tangent to the contour*. We then sum these products (with an integral) to obtain the emf around that closed path. **E** has a component parallel or tangent to this path and a component perpendicular to this path, and the components that are perpendicular to this path do not contribute to the sum. Observe that the electromotive force in (3.2) has units of volts and acts like a voltage. However, the minus sign that was present in the definition of voltage due to a charge distribution in Chapter 1 is absent here, so that instead of being a voltage *produced by charge*, the emf represents a form of *voltage source inserted in the loop*. If the electrical dimensions (in wavelengths) of the closed loop are electrically small ($\ll \lambda$), we may treat this emf as a lumped voltage source and *place it anywhere in the loop*.

The right-hand side of Faraday's law in (3.1) is the rate of *decrease* (the negative sign is referred to as *Lenz's law*) of the magnetic flux ψ given in (3.3) that passes through the surface s that the closed loop c encloses, and **B** is the *magnetic flux density* vector with units of Wb/m^2 (tesla). The result of the *surface integral* in (3.3), ψ, gives the *net magnetic flux passing through the surface that is enclosed by the contour c*. The units of that flux are webers. A vector differential surface of that surface is $d\mathbf{s} = ds\,\mathbf{a}_n$, where $\mathbf{a}_n$ is the unit normal to the surface. The dot product **B** $\cdot$ *d***s** in the integrand of (3.3) means that we take the product of the differential surface areas *ds* and the components of **B** that are *perpendicular to the surface*. Then we add (with an integral) these products to give the *net magnetic flux ψ leaving (or passing through) the open surface s*. This is again sensible since **B** has two components: one perpendicular to the surface and one that is tangent to the surface. The component of **B** that is tangent to the surface does not (and should not) contribute to the net flux *passing through the surface*.

So we may interpret Faraday's law as providing that:

> *A time-varying magnetic field passing through an open surface s will induce (produce) an electric field around the contour c that encircles the surface.*

In Section 3.4 we provide the rationale for saying that the magnetic field produces an electric field rather than the reverse, although this is still somewhat arbitrary. This is the process behind some particle accelerators that accelerate charged particles to enormous speeds and smash them into other particles in order to break those particles into their constituent pieces. A large, time-varying magnetic field creates an electric field that exerts a force on electric charge. The path here into which the electric field is induced is an imaginary contour in space. Faraday's law also makes possible electric transformers and electric motors and generators among an enormous number of other applications that are absolutely essential to our daily lives and commerce. In an electric generator, coils of wire rotate around a shaft and pass through a dc magnetic field, thereby causing a time-varying magnetic field to penetrate the surfaces enclosed by those coils of wire. Hence voltages are induced in those coils of wire by Faraday's law, thereby producing electricity.

The contour or path c in the general statement of Faraday's law can be thought of as the mouth of a balloon which can be inflated to give different surfaces s, as illustrated in Fig. 3.1. All these surfaces give the same result as long as the contour c remains the same. Magnetic field lines that enter and leave the surface and do not pass through the mouth of the balloon do not contribute to the net flux through the surface and hence do not contribute to the induced electric field. Only those magnetic field lines that pass through the mouth of the balloon contribute to the net flux exiting the balloon surface. The direction of the contour c and the direction "out of" the open surface s are again related by the *right-hand rule*. Placing the fingers of our right hand in the direction of the contour c, our thumb will point in the direction "out of the open surface."

To simplify the discussion we choose a flat surface and a circular contour enclosing that surface as shown in Fig. 3.2. Again, the components of the magnetic flux density that penetrate or pass through this surface are those that are normal (perpendicular) to the surface, $\mathbf{B} \cdot d\mathbf{s}$ and $d\mathbf{s} = ds\,\mathbf{a}_n$, where $\mathbf{a}_n$ is a unit normal to the surface. Again, this is sensible because the components of $\mathbf{B}$ that are tangent (parallel) to the surface do not "exit" the surface. Faraday's law provides that we may replace the *effect* of the magnetic flux density vector passing through the surface by inserting an equivalent voltage source whose value is

$$V = \frac{d\psi}{dt} \qquad (3.5)$$

into the contour of the loop that encloses the surface. To "lump" this induced emf in the loop, we will assume that the physical dimensions of this loop are electrically small ($\ll \lambda$). Furthermore, we consider the loop contour to be constructed of a conducting material such as a wire (a conductor having a circular, cylindrical cross section). We can lump these effects of the

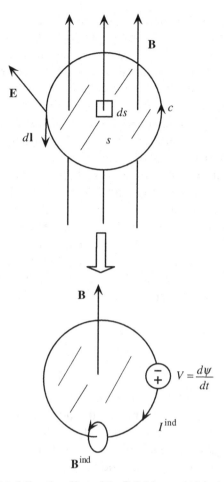

FIGURE 3.2. Modeling the effect of the **B** field as an induced voltage source.

time-changing magnetic field through the loop into a lumped voltage source whose value is given in (3.5) and *place it anywhere in the loop contour* because we assume that the loop dimensions are electrically small.

Getting the polarity of this induced source correct is critical. Faraday's law essentially provides that the voltage source representing the induced emf has a polarity such that it *opposes (Lenz's law) the rate of change of the magnetic flux through the loop.* A foolproof way of determining the correct polarity of the source is the following. The source should tend to induce or "push" a current I^{ind} around this conducting loop in a direction such that this induced current produces another induced magnetic flux $\mathbf{B}^{ind}$ that *opposes* any change in the original magnetic field **B**. This is a very sensible result because if the magnetic field induced by the source did not *oppose the original magnetic* *field*, an induced current would produce an induced magnetic flux that would increase the net magnetic flux through the loop, thereby increasing the value of the induced voltage, which produces a larger induced magnetic field, and so on without bound. As we found in Chapter 2, a current in a wire produces a magnetic field whose direction can be obtained with the *right-hand rule*. That is, if we place the thumb of our right hand in the direction of the current, the fingers will give the direction of the induced magnetic field about the wire. If the original magnetic flux through the surface enclosed by the loop is directed upward as shown in Fig. 3.2, the source should have a polarity such that it tends to push a current *out of its positive terminal* that circulates clockwise, thereby producing (by the right-hand rule) an induced magnetic field that is directed downward *through the loop surface* such that this induced magnetic field opposes the original magnetic field. Observe that the value of the induced voltage source *V* in (3.5) depends on the *time rate of change* of the magnetic flux. Hence, either a large *B* field that is slowly varying with time (such as a 60-Hz power frequency current) or a small *B* field that is rapidly varying with time (such as a 2-GHz current in a cell phone) will have a similar effect.

EXAMPLE

This example shows the utility of using an induced voltage source to model the effect of an incident magnetic field through a closed loop and also shows that the positions of the measurement leads to a voltmeter affect the reading of that voltmeter. Figure 3.3(a) shows a circuit where two resistors comprise the circuit and a uniform external magnetic field of $B = 5t^2 \, \text{Wb/m}^2$ directed out of the page threads the loop that the circuit encloses. A high-impedance voltmeter that draws negligible current is attached across a resistor. The $2 \, \text{m} \times 3 \, \text{m}$ circuit loop encloses a total magnetic flux of

$$\psi = \int_s \mathbf{B} \cdot d\mathbf{s}$$

$$= 5t^2 \left(\text{Wb/m}^2 \right) \times 6 \left(\text{m}^2 \right)$$

$$= 30t^2 \quad (\text{Wb})$$

Hence, the magnitude of the voltage source induced in the loop is

$$V = \frac{d\psi}{dt}$$

$$= 60t \quad \text{V}$$

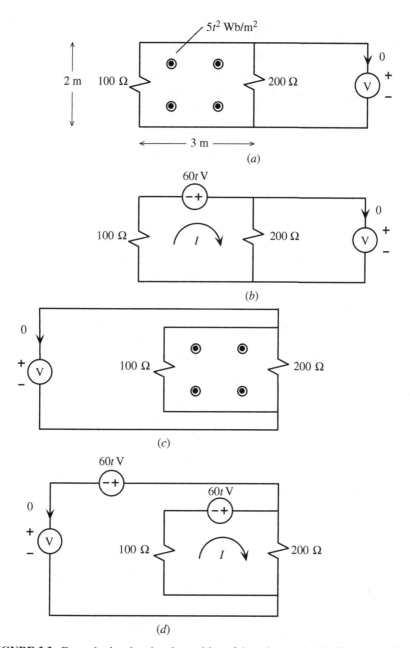

FIGURE 3.3. Example showing that the position of the voltmeter leads affects its reading.

The source representing this induced emf is inserted as shown in Fig. 3.3(b). The source has the polarity shown in order to enforce Lenz's law (it tends to produce a current that circulates around the loop in the clockwise direction so as to produce, according to the right-hand rule, an induced B field that tends to oppose the change in the original B field).

From that circuit we calculate a current flowing around the loop in the clockwise direction of

$$I = \frac{60t}{100 + 200}$$
$$= 0.2t \quad A$$

Hence, the measured voltage is

$$V = 200I$$
$$= 40t \quad V$$

In Fig. 3.3(c) the voltmeter is attached to the same two points, but the voltmeter leads are routed differently. The equivalent circuit for Fig. 3.3(c) is shown in Fig. 3.3(d). Observe that the voltmeter leads now also enclose the magnetic flux, and another voltage source must be inserted in the loop formed by those voltmeter leads as shown. Since the impedance of the voltmeter is assumed infinite, it draws neglible current and we again obtain

$$I = \frac{60t}{100 + 200}$$
$$= 0.2t \quad A$$

But its measured voltage is now

$$V = 200I - 60t$$
$$= -20t \quad V$$

This can also be obtained by summing KVL around the inner loop of that circuit to again obtain

$$V = -60t + 60t - 100I$$
$$= -20t \quad V$$

Hence, Faraday's law shows that the orientation of the voltmeter leads can influence its reading significantly.

EXAMPLE

Consider Fig. 3.4(a), where an open-circuit loop is situated near a two-wire transmission line bearing equal but oppositely directed time-varying currents $I(t)$. It is assumed that the time variation of the currents is sufficiently slow that the loop is electrically small at the significant spectral frequencies of the currents. Each current produces a component of the **B** field threading the loop as shown in Fig. 3.4(b). Assuming that the currents are very long with respect to the loop dimensions, the fundamental result in (2.14) for the magnetic field of an infinitely long current can be applied in an approximate fashion:

$$B_\phi = \frac{\mu_0 I(t)}{2\pi r}$$

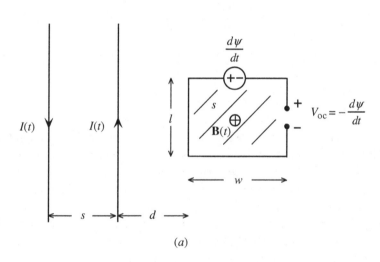

(a)

(b)

FIGURE 3.4. Example.

Hence, the net magnetic flux threading the loop [into the page in Fig. 3.4(a)] due to both currents is (use the right-hand rule)

$$
\begin{aligned}
\psi &= -\int_{z=0}^{l} \int_{r=s+d}^{s+d+w} \frac{\mu_0\, I(t)}{2\pi r}\, dr\; dz + \int_{z=0}^{l} \int_{r=d}^{d+w} \frac{\mu_0\, I(t)}{2\pi\, r}\, dr\; dz \\
&= \frac{\mu_0 l\, I(t)}{2\pi} \left(-\ln \frac{s+d+w}{s+d} + \ln \frac{d+w}{d} \right) \\
&= \frac{\mu_0 l}{2\pi} \ln \frac{(d+w)(s+d)}{d\,(s+d+w)}\, I(t)
\end{aligned}
$$

Hence, the induced voltage source in the loop has the magnitude $d\psi/dt$ and the polarity shown. Thus, the open-circuit voltage at the loop terminals with polarity shown is

$$
\begin{aligned}
V_{oc}(t) &= -\frac{d\psi}{dt} \\
&= -\frac{\mu_0 l}{2\pi} \ln \frac{(d+w)(s+d)}{d\,(s+d+w)}\, \frac{dI(t)}{dt}
\end{aligned}
$$

EXAMPLE

This final example of Faraday's law illustrates that since the magnitude of the induced voltage source is the time rate of change of the flux through the loop and the total flux through the loop is essentially the product of the **B** field and the area of the loop, the induced voltage can also be produced by a constant **B** field but a time-changing loop area. Consider a set of conducting rails across which a conducting shorting bar moves to the right with velocity v as shown in Fig. 3.5. The magnetic field threading the loop is constant (independent of time) and is uniformly distributed over the loop area. The horizontal width of the loop area is vt, so that the total area of the loop is

$$
\text{area} = lvt
$$

The induced voltage source has the polarity shown and a magnitude of

$$
\begin{aligned}
\frac{d\psi}{dt} &= B\frac{d\,\text{area}}{dt} \\
&= Blv
\end{aligned}
$$

Hence, the open-circuit voltage is

$$
V_{oc} = Blv
$$

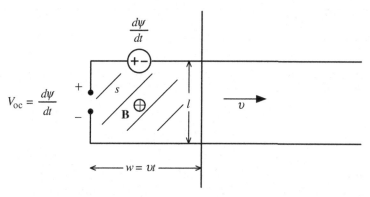

FIGURE 3.5. Example showing Faraday's law for a moving contour.

The form of Faraday's law in (3.4) is said to be its *integral form*. This form is useful for describing its meaning. The *point form* is useful for performing numerical solutions. It is obtained by applying Stokes's theorem (see the Appendix) to the left-hand side to give

$$\oint_c \mathbf{E} \cdot d\mathbf{l} = \int_s (\nabla \times \mathbf{E}) \cdot d\mathbf{s}$$

$$= -\frac{d}{dt} \int_s \mathbf{B} \cdot d\mathbf{s}$$

Comparing both sides gives the *point form* of Faraday's law:

$$\boxed{\nabla \times \mathbf{E} = -\frac{\partial \mathbf{B}}{\partial t}}$$ (3.6)

where $\nabla \times \mathbf{E}$ gives the curl or circulation of $\mathbf{E}$ at a point. Applying the general result (see the Appendix) that $\nabla \cdot \nabla \times \mathbf{F} = 0$ for any general vector field $\mathbf{F}$ to (3.6), we obtain Gauss's law for the magnetic field: $\nabla \cdot \mathbf{B} = 0$. Expanding the curl (see the Appendix) in a rectangular coordinate system and comparing both sides gives

$$\frac{\partial E_z}{\partial y} - \frac{\partial E_y}{\partial z} = -\frac{\partial B_x}{\partial t}$$ (3.7a)

$$\frac{\partial E_x}{\partial z} - \frac{\partial E_z}{\partial x} = -\frac{\partial B_y}{\partial t}$$ (3.7b)

$$\frac{\partial E_y}{\partial x} - \frac{\partial E_x}{\partial y} = -\frac{\partial B_z}{\partial t}$$ (3.7c)

EXAMPLE

A very common form of wave propagation is the *uniform plane wave* [3–6]. If the **E** field is given by

$$\mathbf{E} = E_m \cos(\omega t - \beta z)\, \mathbf{a}_x$$

determine the corresponding **B** field such that the fields satisfy Faraday's law. Since the **E** field is directed solely in the x direction, $E_y = E_z = 0$. Furthermore, the **E** field is independent of x and y, so that $\partial/\partial x = \partial/\partial y = 0$. Hence, (3.7) becomes simply

$$\frac{\partial E_x}{\partial z} = -\frac{\partial B_y}{\partial t} \qquad (3.7b)$$

Substituting the form of **E** gives

$$\beta E_m \sin(\omega t - \beta z) = -\frac{\partial B_y}{\partial t}$$

Integrating this gives the **B** field as

$$\mathbf{B} = \frac{\beta}{\omega} E_m \cos(\omega t - \beta z)\, \mathbf{a}_y$$

The **B** field is in the y direction orthogonal to the **E** field.

3.2   AMPÈRE'S LAW AND DISPLACEMENT CURRENT

We studied Ampère's law for static (dc) fields in Chapter 2:

$$\oint_c \mathbf{H} \cdot d\mathbf{l} = \underbrace{\int_s \mathbf{J} \cdot d\mathbf{s}}_{I_{\text{enclosed}}}$$

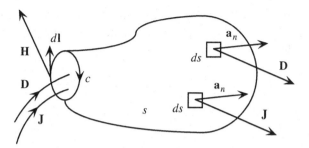

FIGURE 3.6. Ampère's law for time-varying fields.

The surface current density **J** has units of A/m^2 and represents current due to free charges such as electrons in a wire. For time-varying fields a term must be added:

$$\oint_c \mathbf{H} \cdot d\mathbf{l} = \underbrace{\int_s \mathbf{J} \cdot d\mathbf{s}}_{\substack{\text{conduction} \\ \text{current}}} + \underbrace{\frac{d}{dt} \int_s \mathbf{D} \cdot d\mathbf{s}}_{\substack{\text{displacement} \\ \text{current}}} \qquad (3.8)$$

The vector **D** is the electric flux density vector with units of C/m^2. The open surface s is enclosed by the closed contour c as for Faraday's law and the directions are related by the right-hand rule. This is illustrated in Fig. 3.6.

The displacement current is essentially a time-varying electric field. For static fields (3.8) reduces to the static field version of Ampère's law. This addition of the displacement current term to the static version of Ampère's law allows current to flow between two plates of a capacitor, thereby completing the current loop as shown in Fig. 3.7. Ampère's law for time-varying fields in

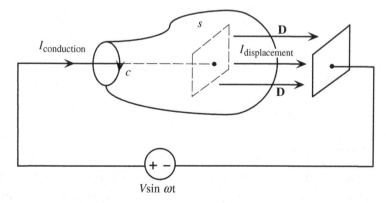

FIGURE 3.7. Displacement current flows between the two plates of a capacitor.

(3.8) shows that a time-varying electric field and its associated displacement current act exactly like conduction current, and either one can produce a magnetic field.

EXAMPLE

For the capacitor circuit of Fig. 3.7, a 1-μ F capacitor has a sinusoidal voltage source $10 \sin \omega t$ volts attached across its terminals, and the frequency of the source is 1 kHz. The conduction current is

$$I_{\text{conduction}} = \frac{10 \text{ V}}{1/\omega C}$$

$$= 62.8 \text{ mA}$$

The capacitance of a parallel-plate capacitor (neglecting fringing of the fields at the edges) is $C = \varepsilon(A/d)$, where ε is the permittivity of the dielectric between the plates, A is the plate area, and d is the separation of the plates. The electric field between the plates is (neglecting fringing of the fields at the edges) $E = 10 \text{ V}/d$. Hence, the D field is

$$D = \varepsilon E$$

$$= \frac{C}{A}(10 \text{ V})$$

$$= \frac{10^{-5}}{A}$$

and the displacement current is

$$I_{\text{displacement}} = \frac{d}{dt} \int_s \mathbf{D} \cdot d\mathbf{s}$$

$$= \omega \frac{10^{-5}}{A} A$$

$$= 62.8 \text{ mA}$$

Hence, the conduction and displacement currents are equal, as they must be.

The form of Ampère's law in (3.8) is said to be its *integral form*. Again, this form is useful for describing its meaning. The *point form* is useful for performing numerical solutions. It is obtained by applying Stokes's theorem (see the Appendix) to the left-hand side to give

$$\oint_c \mathbf{H} \cdot d\mathbf{l} = \int_s (\nabla \times \mathbf{H}) \cdot d\mathbf{s}$$

$$= \int_s \mathbf{J} \cdot d\mathbf{s} + \frac{d}{dt} \int_s \mathbf{D} \cdot d\mathbf{s}$$

Comparing both sides gives the *point form* of Ampère's law:

$$\boxed{\nabla \times \mathbf{H} = \mathbf{J} + \frac{\partial \mathbf{D}}{\partial t}} \tag{3.9}$$

Applying the general result (see the Appendix) that $\nabla \cdot \nabla \times \mathbf{F} = 0$ for any general vector field $\mathbf{F}$ to (3.9), we obtain $\nabla \cdot \mathbf{J} = -\partial(\nabla \cdot \mathbf{D})/\partial t = -\partial\rho(t)/\partial t$ by substituting Gauss's law so that (3.9) satisfies the law of conservation of charge. Expanding the curl in a rectangular coordinate system and comparing both sides gives

$$\frac{\partial H_z}{\partial y} - \frac{\partial H_y}{\partial z} = J_x + \frac{\partial D_x}{\partial t} \tag{3.10a}$$

$$\frac{\partial H_x}{\partial z} - \frac{\partial H_z}{\partial x} = J_y + \frac{\partial D_y}{\partial t} \tag{3.10b}$$

$$\frac{\partial H_y}{\partial x} - \frac{\partial H_x}{\partial y} = J_z + \frac{\partial D_z}{\partial t} \tag{3.10c}$$

EXAMPLE

Again a very common form of wave propagation is the *uniform plane wave* [3–6]. If the **H** field is given by

$$\mathbf{H} = H_m \cos(\omega t - \beta z)\, \mathbf{a}_y$$

determine the corresponding **E** field such that the fields satisfy Ampère's law. We assume that the fields are in free space so that there is no conduction current, $\mathbf{J} = 0$. Since the **H** field is directed solely in the y direction, $H_x = H_z = 0$. Furthermore, the **H** field is independent of x and y so that $\partial/\partial x = \partial/\partial y = 0$. Hence, (3.10) becomes simply

$$-\frac{\partial H_y}{\partial z} = \frac{\partial D_x}{\partial t} \tag{3.10a}$$

Substituting the form of **H** gives

$$-\beta H_m \sin(\omega t - \beta z) = \frac{\partial D_x}{\partial t}$$

Integrating this gives the **D** field as

$$\mathbf{D} = \frac{\beta}{\omega} H_m \cos{(\omega t - \beta z)}\, \mathbf{a}_x$$

Substituting $\mathbf{D} = \varepsilon_0\, \mathbf{E}$ gives the electric field:

$$\mathbf{E} = \frac{\beta}{\varepsilon_0 \omega} H_m \cos{(\omega t - \beta z)}\, \mathbf{a}_x$$

Again, the **E** field is in the x direction orthogonal to the **H** field.

3.3 WAVES, WAVELENGTH, TIME DELAY, AND ELECTRICAL DIMENSIONS

We routinely model electronic circuits with a *lumped-circuit model* which is a particular interconnection of the lumped-circuit elements of resistance, capacitance, and inductance [1,2]. We then solve these lumped-circuit models for the resulting voltages and currents of those elements using Kirchhoff's voltage and current laws. *These lumped-circuit models and the voltages and currents obtained from them are only valid as long as the largest physical dimension of the circuit is electrically small (i.e., much less than a wavelength at the frequency of excitation, f, of that circuit)* [3–6]. A wavelength is

$$\boxed{\lambda = \frac{v}{f}} \tag{3.11}$$

where v denotes the velocity of propagation of, for example, the currents along the connection leads attached to the elements. If the surrounding medium is free space (for all practical purposes air), the velocity of propagation is approximately $v = 3 \times 10^8$ m/s. If a sinusoidal source excites the circuit and has a frequency of 300 MHz, a wavelength is 1 m, and if the excitation frequency of the source is 3 GHz, a wavelength is 10 cm or approximately 4 in. In the case of a printed circuit board the velocities of propagation of the signals carried by the lands on the board are about 60% of that of free space, due to the interaction of the electromagnetic fields produced by those signals with the board substrate, and hence the wavelengths are smaller than in air.

In lumped circuits we can ignore the effects of the connection leads attached to the lumped elements because for the model to be valid, their physical lengths must be electrically small (i.e., $\ll \lambda$). If the connection leads that are attached to an element are electrically long, currents at the two endpoints of this leads *will not be the same* but will have a phase difference between them,

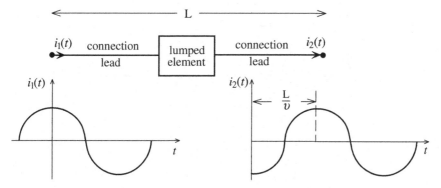

FIGURE 3.8. Effect of element interconnection leads.

as illustrated in Fig. 3.8. The current along the connection lead is, in fact, a *wave*. Suppose that the current and the associated wave are sinusoidal. Such a wave can be written as a function of time, t, and position along the lead, z, as [3–6]

$$i(z, t) = I \cos(\omega t - \beta z) \tag{3.12}$$

where β is the phase-shift constant in rad/m, and $\omega = 2\pi f$, where f is the cyclic frequency of the wave. The velocity of propagation of the wave can be found by observing that to track the movement of the wave, we must follow a point on the wave. Hence, the argument of the cosine in (3.12) must be a constant: $\omega t - \beta z = C$. Differentiating this gives the velocity of propagation of the wave:

$$v = \frac{\omega}{\beta} \tag{3.13}$$

Substituting (3.13) into (3.12) gives

$$i(z, t) = I \cos\left(\omega\left(t - \frac{z}{v}\right)\right) \tag{3.14}$$

Therefore, the phase shift in the frequency domain translates to a time delay of z/v seconds in the time domain. Hence, the currents at two ends of the leads have a *time delay* between them of

$$T_D = \frac{L}{v} \tag{3.15}$$

where L is the total length of the connection leads.

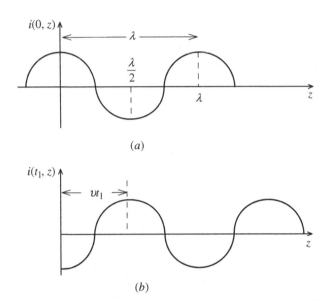

FIGURE 3.9. Wave propagation and wavelength.

A *wavelength* λ is the distance the wave must travel to shift phase by 2π radians or 360°:

$$\beta\lambda = 2\pi$$

as illustrated in Fig. 3.9. Substituting this into (3.13) again gives the wavelength in terms of the velocity of propagation and the frequency as

$$\lambda = \frac{v}{f}$$

If the total length of the connection leads is one-half wavelength, these currents at the endpoints of the lead will be 180° out of phase with each other. If the length of the connection leads is only $\lambda/100$, the phase difference between the two currents at the endpoints is an inconsequential 3.6° and can be ignored. This phase difference translates in the time domain to a time delay; the current at one end of the connection lead and the current at the other end will have a time delay between them. For a connection lead of length L this time delay (in seconds) can be written as $T_D = L/v = (L/\lambda)(1/f) = (L/\lambda)P$, where $P = 1/f$ is the period of the sinusoidal waveforms. Hence, the sinusoidal waveforms at the two ends of the connection lead will be shifted in time relative to each other by a fraction of their period, L/λ. If the connection leads are electrically short, $L \ll \lambda$, the two waveforms will be almost coincident in time and the time delay can be ignored. Otherwise, the time delay will be significant.

3.4   HOW CAN RESULTS DERIVED USING STATIC (DC) VOLTAGES AND CURRENTS BE USED IN PROBLEMS WHERE THE VOLTAGES AND CURRENTS ARE VARYING WITH TIME?

At the beginning of this chapter we alluded to the apparent contradiction that we compute the lumped elements of capacitance and inductance using static (dc) voltages and currents, yet we use these elements to investigate the effects of time-varying voltages and currents. How is this possible? The answer is, of course, as an approximation. In this section we look more closely at this approximation.

Maxwell's equations are commonly considered to be the collection of five equations: Faraday's law, Ampère's law, the two laws of Gauss, and the law of conservation of charge:

$$\oint_c \mathbf{E} \cdot d\mathbf{l} = -\frac{d}{dt} \int_s \mathbf{B} \cdot d\mathbf{s} \tag{3.16a}$$

$$\oint_c \mathbf{H} \cdot d\mathbf{l} = \int_s \mathbf{J} \cdot d\mathbf{s} + \frac{d}{dt} \int_s \mathbf{D} \cdot d\mathbf{s} \tag{3.16b}$$

$$\oint_s \mathbf{D} \cdot d\mathbf{s} = \int_v \rho_v \, dv \tag{3.16c}$$

$$\oint_s \mathbf{B} \cdot d\mathbf{s} = 0 \tag{3.16d}$$

$$\oint_s \mathbf{J} \cdot d\mathbf{s} = -\frac{d}{dt} \int_v \rho_v \, dv \tag{3.16e}$$

where ρ_v is the volume (free) charge distribution throughout volume v. The point forms of these laws were obtained from the integral forms as

$$\nabla \times \mathbf{E} = -\frac{\partial \mathbf{B}}{\partial t} \tag{3.17a}$$

$$\nabla \times \mathbf{H} = \mathbf{J} + \frac{\partial \mathbf{D}}{\partial t} \tag{3.17b}$$

$$\nabla \cdot \mathbf{D} = \rho_v \tag{3.17c}$$

$$\nabla \cdot \mathbf{B} = 0 \tag{3.17d}$$

$$\nabla \cdot \mathbf{J} = -\frac{\partial \rho_v}{\partial t} \tag{3.17e}$$

We solve these in an approximate manner by an *iterative process* [13]. First disregard all time derivatives giving the *zero-order* solutions:

$$\nabla \times \mathbf{E}_0 = 0 \tag{3.18a}$$

$$\nabla \times \mathbf{H}_0 = \mathbf{J}_0 \tag{3.18b}$$

$$\nabla \cdot \mathbf{D}_0 = \rho_{v0} \tag{3.18c}$$

$$\nabla \cdot \mathbf{B}_0 = 0 \tag{3.18d}$$

$$\nabla \cdot \mathbf{J}_0 = 0 \tag{3.18e}$$

Next, we put back the time derivatives but use the zero-order solutions in those time derivatives to obtain these *first-order* solutions for other variables not contained in time derivatives:

$$\nabla \times \mathbf{E}_1 = -\frac{\partial \mathbf{B}_0}{\partial t} \tag{3.19a}$$

$$\nabla \times \mathbf{H}_1 = \mathbf{J}_1 + \frac{\partial \mathbf{D}_0}{\partial t} \tag{3.19b}$$

$$\nabla \cdot \mathbf{D}_1 = \rho_{v1} \tag{3.19c}$$

$$\nabla \cdot \mathbf{B}_1 = 0 \tag{3.19d}$$

$$\nabla \cdot \mathbf{J}_1 = -\frac{\partial \rho_{v0}}{\partial t} \tag{3.19e}$$

Similarly, we can obtain a more refined solution known as the *second-order* solution by using the first-order solutions in the time derivatives to obtain the other variables not contained in the time derivatives:

$$\nabla \times \mathbf{E}_2 = -\frac{\partial \mathbf{B}_1}{\partial t} \tag{3.20a}$$

$$\nabla \times \mathbf{H}_2 = \mathbf{J}_2 + \frac{\partial \mathbf{D}_1}{\partial t} \tag{3.20b}$$

$$\nabla \cdot \mathbf{D}_2 = \rho_{v2} \tag{3.20c}$$

$$\nabla \cdot \mathbf{B}_2 = 0 \tag{3.20d}$$

$$\nabla \cdot \mathbf{J}_2 = -\frac{\partial \rho_{v1}}{\partial t} \tag{3.20e}$$

The zero-order solutions in (3.18) are the static (dc) solutions we obtained in Chapter 2. The zero-order solutions in (3.18) were used to obtain, for example, the solution for the first-order, induced electric field, E_1, in Faraday's law in (3.19a) by using the zero-order solution for the $\mathbf{B}$ field, $\mathbf{B}_0$. As we continue this process, we obtain a more accurate solution of Maxwell's equations. The combination of the zero-order solutions in (3.18) and the first-order solutions in (3.19) are usually referred to as the *quasistatic solution*. Generally speaking, the quasistatic solution obtained iteratively using the zero-order solutions and refining them to give the first-order solutions give adequate accuracy as long as the maximum physical dimension of the electromagnetic structure being investigated is electrically small (i.e., $L \ll \lambda$) [13]. This gives the rationale for using circuit elements such as capacitance and inductance which were derived using dc voltages and dc currents in circuits whose currents and voltages vary with time so long as the maximum dimension of the circuit is electrically small.

It is simple to show that the sums of the partial solutions in this iterative process converge to the true solution to Maxwell's equations:

$$\mathbf{E} = \mathbf{E}_0 + \mathbf{E}_1 + \mathbf{E}_2 + \cdots \qquad (3.21a)$$

$$\mathbf{H} = \mathbf{H}_0 + \mathbf{H}_1 + \mathbf{H}_2 + \cdots \qquad (3.21b)$$

$$\mathbf{D} = \mathbf{D}_0 + \mathbf{D}_1 + \mathbf{D}_2 + \cdots \qquad (3.21c)$$

$$\mathbf{B} = \mathbf{B}_0 + \mathbf{B}_1 + \mathbf{B}_2 + \cdots \qquad (3.21d)$$

$$\mathbf{J} = \mathbf{J}_0 + \mathbf{J}_1 + \mathbf{J}_2 + \cdots \qquad (3.21e)$$

$$\rho_v = \rho_{v0} + \rho_{v1} + \rho_{v2} + \cdots \qquad (3.21f)$$

Adding the zero-order equations in (3.18a), the first-order equations in (3.19a), the second-order equations in (3.20a), and so on, gives

$$\nabla \times (\mathbf{E}_0 + \mathbf{E}_1 + \mathbf{E}_2 + \cdots) = 0 - \frac{\partial}{\partial t}(\mathbf{B}_0 + \mathbf{B}_1 + \mathbf{B}_2 + \cdots) \qquad (3.22)$$

Substituting (3.21a) and (3.21d) gives the first Maxwell equation:

$$\nabla \times \mathbf{E} = -\frac{\partial \mathbf{B}}{\partial t}$$

The other equations of Maxwell are obtained in a similar fashion.

3.5   VECTOR MAGNETIC POTENTIAL FOR TIME-VARYING CURRENTS

In Chapter 2 we introduced the *vector magnetic potential* **A** for determining the magnetic field **B** for static (dc) current configurations. In this section we rederive the vector magnetic potential for time-varying currents. The main purpose in doing so is that the result will clearly demonstrate that the quasistatic solutions of the field equations, (3.18) and (3.19), are valid approximations as long as the maximum physical dimensions of the problem are much less than a wavelength ($L \ll \lambda$).

Because of Gauss's law for the magnetic field,

$$\nabla \cdot \mathbf{B} = 0 \tag{3.23}$$

and the vector identity (see the Appendix) that the divergence of the curl of *any* vector field is zero;

$$\nabla \cdot \nabla \times \mathbf{A} = 0 \tag{3.24}$$

we can define the *vector magnetic potential* $\mathbf{A}$ as

$$\mathbf{B} = \nabla \times \mathbf{A} \tag{3.25}$$

For static current distributions, $\mathbf{A}$ was obtained in Chapter 2 as

$$\mathbf{A} = \frac{\mu}{4\pi} \int_v \frac{\mathbf{J}}{R} \, dv \tag{3.26}$$

where $\mathbf{J}$ is the current distribution (A/m^2), v is the volume enclosing that current distribution, and R is the distance between a differential volume of that current distribution containing $\mathbf{J}\,dv$ and the point at which we wish to determine $\mathbf{A}$.

For time-varying currents this result obviously must be modified. To demonstrate that result, first note that substituting (3.25) into Faraday's law gives

$$\nabla \times \left(\mathbf{E} + \frac{\partial \mathbf{A}}{\partial t} \right) = 0$$

This seems to imply that the sum in parentheses is zero. But we have the identity (see the Appendix) that

$$\nabla \times \nabla \phi = 0 \tag{3.27}$$

for any scalar field ϕ. Hence, we can write, in general,

$$\boxed{\mathbf{E} = - \underbrace{\nabla \phi}_{\substack{\text{due to} \\ \text{charges}}} - \underbrace{\frac{\partial \mathbf{A}}{\partial t}}_{\substack{\text{due to} \\ \text{time--varying} \\ \text{currents}}}} \tag{3.28}$$

Hence, in general, the electric field is the result of two "sources": the charges in the system and the time-varying currents in the system. For dc, this reduces to $\mathbf{E} = -\nabla \phi$ and ϕ is said to be the *potential function* that is more commonly known as "voltage."

With these results we can now derive the result for the vector magnetic potential for time-varying currents. Proceeding in a fashion similar to that of Section 2.5 of for static fields, we substitute $\mathbf{B} = \mu\,\mathbf{H}$ and $\mathbf{D} = \varepsilon\,\mathbf{E}$ into Ampère's law to yield

$$\nabla \times \mathbf{B} = \mu\,\mathbf{J} + \mu\varepsilon\frac{\partial\mathbf{E}}{\partial t}$$

Substituting the relation for $\mathbf{B}$ in terms of $\mathbf{A}$ given in (3.25) yields

$$\nabla \times \nabla \times A = \mu\,\mathbf{J} + \mu\varepsilon\frac{\partial\mathbf{E}}{\partial t} \tag{3.29}$$

Substituting the relation for $\mathbf{E}$ given in (3.28) gives

$$\nabla \times \nabla \times \mathbf{A} = \mu\,\mathbf{J} + \mu\varepsilon\left(-\nabla\left(\frac{\partial\phi}{\partial t}\right) - \frac{\partial^2\mathbf{A}}{\partial^2 t}\right) \tag{3.30}$$

But we have the vector identity [3,6]

$$\nabla \times \nabla \times \mathbf{A} = \nabla\,(\nabla \cdot \mathbf{A}) - \nabla^2\mathbf{A} \tag{3.31}$$

Substituting (3.31) into (3.30) and collecting terms gives

$$\nabla^2\mathbf{A} - \mu\varepsilon\,\frac{\partial^2\mathbf{A}}{\partial^2 t} = -\mu\,\mathbf{J} + \nabla\left(\nabla \cdot \mathbf{A} + \mu\varepsilon\,\frac{\partial\phi}{\partial t}\right) \tag{3.32}$$

Again, the complete definition of a vector quantity requires that we define both the curl and the divergence of it. We defined the curl of $\mathbf{A}$ in (3.25). We are free to define the divergence of $\mathbf{A}$. From (3.32) a convenient way to define the divergence of $\mathbf{A}$ is so that the term in (3.32) in parentheses is rendered zero:

$$\nabla \cdot \mathbf{A} = -\mu\varepsilon\,\frac{\partial\phi}{\partial t} \tag{3.33}$$

This is commonly referred to as the *Lorentz choice of gauge*. Note that for static currents, this reduces to $\nabla \cdot \mathbf{A} = 0$, which was chosen in Chapter 2 for static (dc) currents. Hence the equation for the vector magnetic potential for time-varying currents becomes

$$\nabla^2\mathbf{A} - \mu\varepsilon\,\frac{\partial^2\mathbf{A}}{\partial^2 t} = -\mu\,\mathbf{J} \tag{3.34}$$

It can be shown [3,6] that the solution to (3.34) is

$$\boxed{\mathbf{A} = \frac{\mu}{4\pi}\int_v \frac{\mathbf{J}\left(t - \frac{R}{v}\right)}{R}\,dv} \tag{3.35a}$$

where v is a velocity of propagation:

$$v = \frac{1}{\sqrt{\mu\varepsilon}} \tag{3.35b}$$

and again, $\mathbf{J}$ is the current distribution, v is the volume enclosing that current distribution, and R is the distance between a differential volume of that current distribution containing $\mathbf{J}\,dv$ and the point at which we wish to determine $\mathbf{A}$. This shows that the vector magnetic potential at a point that is a distance R away from a current element $\mathbf{J}\,dv$ has a time delay of effect of R/v. This is refered to as *retardation* and is characteristic of all time-varying fields. If we write this result for fields and currents that are varying sinusoidally with time, the result in (3.35a) becomes the phasor vector magnetic potential $\hat{\mathbf{A}}$ [3]:

$$\hat{\mathbf{A}} = \frac{\mu}{4\pi} \int_v \frac{\hat{\mathbf{J}}\, e^{-j\beta R}}{R}\, dv \tag{3.36a}$$

in terms of the phasor current density $\hat{\mathbf{J}}$ where the *phase constant* β is again

$$\beta = \frac{\omega}{v}$$

$$= \frac{2\pi}{\lambda} \tag{3.36b}$$

and the wavelength is again

$$\lambda = \frac{v}{f}$$

$$= \frac{1}{\sqrt{\mu\varepsilon}\, f} \tag{3.36c}$$

Again, the *phase shift* term $e^{-j\beta R}$ in the frequency domain amounts to a *time delay* in the time domain. We can expand the exponential term in (3.36a) as

$$e^{-j\beta R} = 1 - j\beta R + \frac{\beta^2 R^2}{2} + \cdots \tag{3.37a}$$

It is this result that shows why quasistatic results can be used to approximate time-varying fields. Substituting (3.36b) for β gives

$$e^{-j\beta R} = 1 - j2\pi \frac{R}{\lambda} + \frac{(2\pi)^2}{2} \left(\frac{R}{\lambda}\right)^2 + \cdots \tag{3.37b}$$

Hence, the retardation term depends on powers of R/λ, which gives the physical distance to the field point in terms of its electrical distance in wavelengths. For electrically small dimensions of the problem, $R \ll \lambda$, the exponential term approximates to unity, $e^{-j\beta R} \cong 1$, and the vector magnetic potential for time-varying currents in (3.36a) reduces to the static field result for $\mathbf{A}$ that was used in Chapter 2 and is given in (2.52). Also observe that in Chapter 2 we chose the Coulomb gauge to define the divergence of $\mathbf{A}$ for static fields:

$\nabla \cdot \mathbf{A} = 0$. The more general Lorentz gauge for time-varying field problems in (3.33) reduces to the Coulomb gauge for static problems.

In the case of sinusoidal variation of the fields, the phasor form of Faraday's law and Ampère's law are obtained by replacing all time derivatives with $j\omega$ and become [3–6]

$$\nabla \times \hat{\mathbf{E}} = -j\omega\hat{\mathbf{B}}$$

(3.38a)

$$\nabla \times \hat{\mathbf{B}} = \mu\hat{\mathbf{J}} + j\omega\mu\varepsilon\hat{\mathbf{E}}$$

(3.38b)

and we have substituted $\hat{\mathbf{B}} = \mu\hat{\mathbf{H}}$ and $\hat{\mathbf{D}} = \varepsilon\hat{\mathbf{E}}$ into Ampère's law in (3.38b). Once the phasor vector magnetic potential $\hat{\mathbf{A}}$ is obtained from (3.36a), the phasor magnetic field is determined from

$$\hat{\mathbf{B}} = \nabla \times \hat{\mathbf{A}}$$

(3.39a)

The phasor electric field is determined from Ampère's law in (3.38b) in the region outside the current distribution where $\hat{\mathbf{J}} = 0$ as

$$\hat{\mathbf{E}} = \frac{1}{j\omega\mu\varepsilon}\nabla \times \hat{\mathbf{B}}$$

$$= \frac{1}{j\omega\mu\varepsilon}\nabla \times \nabla \times \hat{\mathbf{A}}$$

(3.39b)

and the solution for all the fields is determined in terms of the vector magnetic potential $\hat{\mathbf{A}}$. Using the equation for the phasor vector magnetic potential in (3.36a) and (3.37b) shows that for electrically small structures, the quasistatic fields obtained from (3.39) provide reasonable approximations.

3.6 CONSERVATION OF ENERGY AND POYNTING'S THEOREM

In this section we discuss the dissipation and storage of energy in the electromagnetic field. The product of the units of $\mathbf{E}$ and $\mathbf{H}$ is $V/m \times A/m = W/m^2$, representing a *power density in the combined field*. Hence, it is natural to define the power density vector as

$$\mathbf{S} = \mathbf{E} \times \mathbf{H} \qquad W/m^2$$

(3.40)

This is referred to as the *Poynting vector* after the English physicist John H. Poynting. The net outflow of power from a point is represented by the divergence of **S**. Using a vector identity [3] of $\nabla \cdot (\mathbf{E} \times \mathbf{H}) = \mathbf{H} \cdot (\nabla \times \mathbf{E}) - \mathbf{E} \cdot (\nabla \times \mathbf{H})$ and substituting Faraday's and Ampère's laws gives

$$
\nabla \cdot \mathbf{S} = \mathbf{H} \cdot \left(-\frac{\partial \mathbf{B}}{\partial t} \right) - \mathbf{E} \cdot \left(\mathbf{J} + \frac{\partial \mathbf{D}}{\partial t} \right)
$$

$$
= -\mathbf{E} \cdot \mathbf{J} - \mathbf{E} \cdot \frac{\partial \mathbf{D}}{\partial t} - \mathbf{H} \cdot \frac{\partial \mathbf{B}}{\partial t} \qquad (3.41)
$$

Integrating this result throughout a volume v and using the divergence theorem (see the Appendix) gives

$$
\underbrace{- \oint_s \mathbf{S} \cdot d\mathbf{s}}_{\substack{\text{power} \\ \text{entering} \\ \text{surface } S}} = \underbrace{\int_v (\mathbf{E} \cdot \mathbf{J})\, dv}_{\substack{\text{power} \\ \text{dissipated} \\ \text{in volume } v}} + \underbrace{\int_v \left(\mathbf{E} \cdot \frac{\partial \mathbf{D}}{\partial t} \right) dv}_{\substack{\text{rate of change} \\ \text{of stored} \\ \text{energy in the} \\ \text{electric field}}} + \underbrace{\int_v \left(\mathbf{H} \cdot \frac{\partial \mathbf{B}}{\partial t} \right) dv}_{\substack{\text{rate of change} \\ \text{of stored} \\ \text{energy in the} \\ \text{magnetic field}}} \qquad (3.42)
$$

where the closed surface s encloses the volume v. This indicates the expected energy balance since the left side, which represents the total power *entering* the closed surface s, equals the sum of three terms. The first term represents the ohmic power dissipation throughout the volume v, while the second and third terms represent the time rate of change of the energy stored in the electric and magnetic fields, respectively, in the volume v. The right-hand side can be rewritten, assuming that the medium is linear, homogeneous, and isotropic using the basic relations $\mathbf{J} = \sigma \mathbf{E}$, $\mathbf{B} = \mu \mathbf{H}$, and $\mathbf{D} = \varepsilon \mathbf{E}$, as

$$
\underbrace{- \oint_s \mathbf{S} \cdot d\mathbf{s}}_{\substack{\text{power} \\ \text{entering} \\ \text{surface } S}} = \underbrace{\int_v \left(\sigma\, |E|^2 \right) dv}_{\substack{\text{power} \\ \text{dissipated} \\ \text{in volume } v}} + \underbrace{\frac{1}{2} \frac{d}{dt} \int_v \left(\varepsilon\, |E|^2 \right) dv}_{\substack{\text{rate of change} \\ \text{of stored} \\ \text{energy in the} \\ \text{electric field}}} + \underbrace{\frac{1}{2} \frac{d}{dt} \int_v \left(\mu\, |H|^2 \right) dv}_{\substack{\text{rate of change} \\ \text{of stored} \\ \text{energy in the} \\ \text{magnetic field}}}
$$

$$
= \int_v (\mathbf{E} \cdot \mathbf{J})\, dv + \frac{1}{2} \frac{d}{dt} \int_v (\mathbf{D} \cdot \mathbf{E})\, dv + \frac{1}{2} \frac{d}{dt} \int_v (\mathbf{B} \cdot \mathbf{H})\, dv \qquad (3.43)
$$

We used the $\frac{1}{2}$ factor and moved the time partial derivatives outside the last two volume integrals since

$$
\mathbf{E} \cdot \frac{\partial \mathbf{D}}{\partial t} = \varepsilon \mathbf{E} \cdot \frac{\partial \mathbf{E}}{\partial t}
$$

$$
= \varepsilon \frac{1}{2} \frac{\partial}{\partial t}\, |\mathbf{E}|^2
$$

$$= \varepsilon \frac{1}{2} \frac{\partial}{\partial t} (\mathbf{E} \cdot \mathbf{E})$$

$$= \frac{1}{2} \frac{\partial}{\partial t} (\mathbf{D} \cdot \mathbf{E})$$

$$\mathbf{H} \cdot \frac{\partial \mathbf{B}}{\partial t} = \mu \, \mathbf{H} \cdot \frac{\partial \mathbf{H}}{\partial t}$$

$$= \mu \frac{1}{2} \frac{\partial}{\partial t} |\mathbf{H}|^2$$

$$= \mu \frac{1}{2} \frac{\partial}{\partial t} (\mathbf{H} \cdot \mathbf{H})$$

$$= \frac{1}{2} \frac{\partial}{\partial t} (\mathbf{B} \cdot \mathbf{H})$$

because we can write for any vector field $\mathbf{F}$, using the *chain rule*,

$$\mathbf{F} \cdot \frac{\partial \mathbf{F}}{\partial t} = F_x \frac{\partial F_x}{\partial t} + F_y \frac{\partial F_y}{\partial t} + F_z \frac{\partial F_z}{\partial t}$$

$$= \frac{1}{2} \frac{\partial F_x^2}{\partial t} + \frac{1}{2} \frac{\partial F_y^2}{\partial t} + \frac{1}{2} \frac{\partial F_z^2}{\partial t}$$

$$= \frac{1}{2} \frac{\partial |\mathbf{F}|^2}{\partial t}$$

$$= \frac{1}{2} \frac{\partial (\mathbf{F} \cdot \mathbf{F})}{\partial t}$$

The result in (3.43) suggests that for linear, homogeneous, and isotropic media, the energy stored in the electric and magnetic fields inside the volume is

$$W_E = \frac{1}{2} \int_v (\mathbf{D} \cdot \mathbf{E}) \, dv \qquad \mathrm{J} \tag{3.44a}$$

and

$$W_M = \frac{1}{2} \int_v (\mathbf{B} \cdot \mathbf{H}) \, dv \qquad \mathrm{J} \tag{3.44b}$$

respectively.

3.7   INDUCTANCE OF A CONDUCTING LOOP

In the remaining chapters we discuss the computation of the inductance of a closed loop that is constructed of a conductor such as a wire or a land on a printed circuit board. This final section will serve as a preliminary to those discussions.

Faraday's fundamental law of induction indicates that an electromotive force is induced in the perimeter of any closed loop through the enclosed surface of which a time-varying magnetic field passes. We have represented that emf in Fig. 3.2 as a lumped voltage source placed at an *indeterminate position* in the loop:

$$V = \frac{d\psi}{dt} \tag{3.45}$$

This voltage source represents the time rate of change of the total magnetic flux penetrating that loop:

$$\psi = \int_s \mathbf{B} \cdot d\mathbf{s} \tag{3.46}$$

We can also write the flux through the loop in an equivalent form in terms of the line integral of the vector magnetic potential $\mathbf{A}$ around the loop using the identity $\mathbf{B} = \nabla \times \mathbf{A}$. Hence, the magnetic flux through the loop can be written as a line integral of $\mathbf{A}$ around that loop as

$$\psi = \int_s \mathbf{B} \cdot d\mathbf{s}$$
$$= \int_s (\nabla \times \mathbf{A}) \cdot d\mathbf{s}$$
$$= \oint_c \mathbf{A} \cdot d\mathbf{l} \tag{3.47}$$

where we have used Stokes's theorem (see the Appendix) to convert a surface integral over the open surface s enclosed by the conducting loop into a line integral around the contour c enclosing the surface. Hence, Faraday's fundamental law of induction can be written in two alternative forms as

$$\oint_c \mathbf{E} \cdot d\mathbf{l} = -\frac{d}{dt} \int_s \mathbf{B} \cdot d\mathbf{s}$$
$$= -\frac{d}{dt} \oint_c \mathbf{A} \cdot d\mathbf{l} \tag{3.48}$$

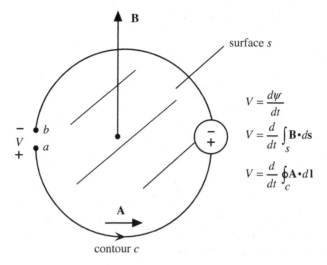

FIGURE 3.10. Conducting loop with a small gap.

Consider a conducting loop composed of a perfect conductor that has a very small gap cut in it, as shown in Fig. 3.10. Along the conductor of the loop $\mathbf{E} = 0$, so that Faraday's law gives

$$\oint_c \mathbf{E} \cdot d\mathbf{l} = \int_{\text{gap}} \mathbf{E}_{\text{gap}} \cdot d\mathbf{l} + \int_{\text{conductor}} \underbrace{\mathbf{E}_{\text{conductor}}}_{0} \cdot d\mathbf{l}$$

$$= -\frac{d}{dt} \int_s \mathbf{B} \cdot d\mathbf{l}$$

$$= -\frac{d}{dt} \oint_c \mathbf{A} \cdot d\mathbf{l} \qquad (3.49)$$

But the electric field along the perfect conductor is zero, giving

$$\int_{\text{gap}} \mathbf{E}_{\text{gap}} \cdot d\mathbf{l} = -\frac{d}{dt} \int_s \mathbf{B} \cdot d\mathbf{s}$$

$$= -\frac{d}{dt} \oint_c \mathbf{A} \cdot d\mathbf{l} \qquad (3.50)$$

Moving the minus sign to the left-hand side gives

$$
\begin{aligned}
V &= -\int_{gap} \mathbf{E}_{gap} \cdot d\mathbf{l} \\
&= \frac{d\psi}{dt} \\
&= \frac{d}{dt} \int_s \mathbf{B} \cdot d\mathbf{s} \\
&= \frac{d}{dt} \oint_c \mathbf{A} \cdot d\mathbf{l}
\end{aligned}
\tag{3.51}
$$

Hence, a voltage that is related to the time rate of change of the magnetic flux appears at the terminals of the open-circuited loop. Essentially, the time-changing magnetic flux through the loop induces an electric field in the conductor that forces the charges (electrons) in the conductor to move to the terminals of the gap, thereby creating another electric field due to this electric field across the gap induced by the charge that is accumulated at the terminals. The sum of this induced electric field caused by the charges at the gap and the original electric field combine to give a net electric field that is zero on the surface of the conductor, therby satisfying the boundary conditions that the tangential electric field on the surface of a perfect conductor must equal zero.

4

THE CONCEPT OF "LOOP" INDUCTANCE

In this chapter we examine the calculation of the "loop" inductance of various configurations of closed current loops.

4.1   SELF INDUCTANCE OF A CURRENT LOOP FROM FARADAY'S LAW OF INDUCTION

Faraday's law of induction, discussed in Chapter 3, is fundamental to the notion of inductance. For example, consider the circular loop of conducting wire shown in Fig. 4.1. Suppose that we cut a small gap in the loop and inject a current I into that gap so that the current flows around the loop in the counterclockwise direction as shown in Fig. 4.1. This current will, by the right-hand rule, produce a magnetic flux density **B** threading the surface s that the current surrounds. We have shown this surface as being flat to simplify the discussion, although any surface shape will give the same result as long as it is surrounded by the loop. For a current directed in the counterclockwise direction around the loop, the magnetic field is directed upward through the surface surrounded by the loop. The total magnetic flux penetrating the loop is obtained as

$$\psi = \int_s \mathbf{B} \cdot d\mathbf{s} \qquad (4.1)$$

Inductance: Loop and Partial, By Clayton R. Paul
Copyright © 2010 John Wiley & Sons, Inc.

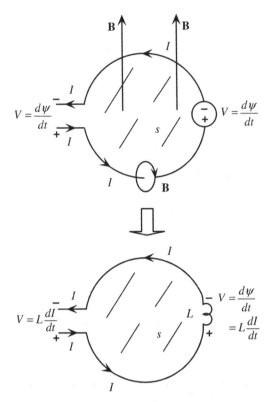

FIGURE 4.1. Loop inductance by Faraday's law.

where s is the surface the current loop surrounds. If the current and associated magnetic field varies with time, Faraday's law of induction essentially provides that the time rate of change of the magnetic flux through the loop will essentially induce an electromotive force (emf) around the loop contour:

$$\text{emf} = \oint_c \mathbf{E} \cdot d\mathbf{l} = -\frac{d\psi}{dt} \tag{4.2}$$

where c is the contour of the loop that surrounds the surface s. If the dimensions of the loop are electrically small, we may represent this emf as a lumped voltage source whose value is the time rate of change of the magnetic flux through the loop:

$$V = \frac{d\psi}{dt} \tag{4.3}$$

and place it *anywhere in the loop perimeter* as shown in Fig. 4.1. *The exact location of the voltage source in the loop perimeter cannot be determined uniquely,* nor does it need to be.

It is important to determine correctly the polarity of the induced voltage source. The minus sign in Faraday's law in (4.2) is referred to as Lenz's law. The induced voltage source should induce a current, $I^{induced}$, leaving its positive terminal such that this induced current will produce an induced magnetic field, $\mathbf{B}^{induced}$, through the loop surface that tends to oppose the rate of change of the original magnetic field, $\mathbf{B}$, produced by the original current I. Hence, the voltage source is inserted with the polarity shown in Fig. 4.1.

The inductance of the current loop is defined fundamentally, as the ratio of the magnetic flux threading the loop and the current producing it:

$$\boxed{L = \frac{\psi}{I}}$$

(4.4a)

or

$$\psi = LI$$

(4.4b)

If the surrounding medium is linear, homogeneous, and isotropic, the total magnetic flux threading the loop is directly proportional to the current I that produced it, and hence the inductance of the loop is only a function of the loop shape and its dimensions as well as the material properties of the surrounding medium. Hence, the induced voltage source is

$$V = \frac{d\psi}{dt}$$

$$= L\frac{dI}{dt}$$

(4.5)

Figure 4.1 shows that we can replace this induced source with the usual inductor symbol, and the voltage induced across this inductance is given by (4.5). This voltage appears across the terminals of the loop like a Thèvenin open-circuit voltage. If the contour of the loop (represented here as a wire) has resistance, that is represented as well by the usual resistor symbol inserted in series with the loop, thereby giving an additional voltage drop of IR around the loop.

The process of calculating the inductance of a loop is referred to as the *method of flux linkages,* since we compute the flux that "links" the current. It is a four-step process:

1. Inject a current I around the closed loop.
2. Determine the magnetic flux density $\mathbf{B}$ over the surface of the enclosed loop by the methods of Chapter 2.
3. Compute the total magnetic flux threading the loop according to (4.1).
4. Divide that flux by the current I according to (4.4a).

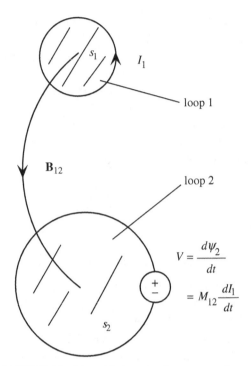

FIGURE 4.2. Mutual inductance between two loops.

The inductance so obtained is referred to as the *self inductance* of the loop. The *mutual inductance* between two loops, one of which carries a current I_1, is defined with reference to Fig. 4.2 as

$$M_{12} = \frac{\psi_2}{I_1} \qquad (4.6)$$

where ψ_2 is the flux penetrating the surface of the second loop, s_2, that is caused by the current of the first loop:

$$\psi_2 = \int_{s_2} \mathbf{B}_{12} \cdot d\mathbf{s} \qquad (4.7)$$

In Chapter 2 we found that the computation of the magnetic flux density $\mathbf{B}$ could be accomplished by various methods. But they all required that we evaluate some rather complex integrals. To complete the process of determining the inductance of the structure by the method of flux linkages, we will further have to evaluate some rather complicated integrals involving those $\mathbf{B}$ fields in order to determine the flux through the loop via (4.1) and (4.7) or by other means. However, there are other methods that we will investigate to compute the self and mutual inductances of and between current loops that avoid the direct calculation of **B** and the flux through the loop as in (4.1) or (4.7). Nevertheless, the fundamental definition of inductance is via Faraday's law.

4.1.1   Rectangular Loop

In this section we determine the inductance of the rectangular loop shown in Fig. 4.3(a), whose length is l and width is w. The conductors of the loop are

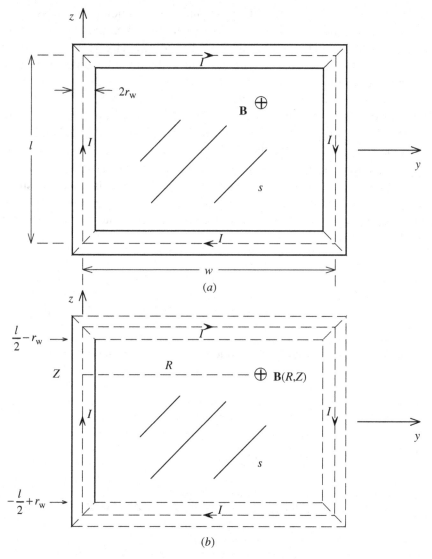

FIGURE 4.3. Rectangular loop.

wires having radii r_w. We assume that the current I is *uniformly distributed across the cross section of the wires*, so that with regard to computing the magnetic field from it, the current can be considered to be concentrated in a filament on the axes of those wires. For isolated direct currents (dc) not in proximity to other currents, the current is, in fact, uniformly distributed over the wire cross section. However, for a current that is in close proximity to other currents, the current in the wire will not be distributed uniformly over the wire cross section. Nearby currents will cause the current to be concentrated on the side of the wire nearest the neighboring current, a phenomenon known as the *proximity effect*. Proximity effect is usually not pronounced unless the two currents are within about four radii of each other (i.e., one wire will just fit between the two). This is investigated in Section 4.6. High-frequency currents will be symmetric about the wire axis but will tend to be concentrated in an annulus at the surface of thickness that is a few skin depths. High-frequency redistribution of the current is investigated in Section 6.5.

The loop through which we determine the magnetic flux is the area formed by the interior edges of the wires of the loop. To determine the total flux through that loop, we determine the flux through the loop caused by the current of each wire separately and then add the four fluxes. The flux through the loop caused by the left wire segment as shown in Fig. 4.3(b) can be obtained by using the result for the **B** field due to a length of wire given in equation (2.15). The **B** field is perpendicular to the loop surface and directed into the page according to the right-hand rule:

$$B\,(R,\,Z) = \frac{\mu_0 I}{4\pi R}\left[\frac{Z+l/2}{\sqrt{(Z+l/2)^2 + R^2}} - \frac{Z-l/2}{\sqrt{(Z-l/2)^2 + R^2}}\right] \quad (4.8)$$

Hence, the flux through the loop due to the current of the left side is

$$\psi_{\text{left side}} = \int_{Z=r_w-l/2}^{l/2-r_w} \int_{R=r_w}^{w-r_w} B(R,\,Z)\,dR\,dZ \quad (4.9)$$

Then the total flux through the loop is

$$\psi_{\text{loop}} = 2\int_{Z=r_w-l/2}^{l/2-r_w} \int_{R=r_w}^{w-r_w} B\,(R,\,Z)\,dR\,dZ$$

$$+2\int_{Z=r_w-w/2}^{w/2-r_w} \int_{R=r_w}^{l-r_w} B\,(R,\,Z)\,dR\,dZ \quad (4.10)$$

The flux through the loop surface due to the left side is evaluated as follows:

$$\psi_{\text{left side}} = \frac{\mu_0 I}{4\pi} \int_{Z=r_w-l/2}^{l/2-r_w} \int_{R=r_w}^{w-r_w} \frac{1}{R} \left[\frac{Z+l/2}{\sqrt{(Z+l/2)^2 + R^2}} \right.$$

$$\left. + \frac{l/2 - Z}{\sqrt{(l/2-Z)^2 + R^2}} \right] dR\, dZ$$

$$(4.11)$$

Using integral 221.01 from Dwight [7],

$$\int \frac{dx}{x\sqrt{x^2 + a^2}} = -\frac{1}{a} \ln \left| \frac{a + \sqrt{x^2 + a^2}}{x} \right|$$

$$(D221.01)$$

this becomes

$$\psi_{\text{left side}} = \frac{\mu_0 I}{4\pi} \int_{Z=r_w-l/2}^{l/2-r_w} \left[-\ln \frac{(Z+l/2) + \sqrt{(Z+l/2)^2 + R^2}}{R} \right.$$

$$\left. -\ln \frac{(l/2-Z) + \sqrt{(l/2-Z)^2 + R^2}}{R} \right]_{R=r_w}^{w-r_w} dZ$$

$$= \frac{\mu_0 I}{4\pi} \int_{Z=r_w-l/2}^{l/2-r_w} \left[-\sinh^{-1}\frac{Z+l/2}{R} - \sinh^{-1}\frac{l/2-Z}{R} \right]_{R=r_w}^{w-r_w} dZ$$

$$(4.12)$$

where we have written this result in terms of the inverse hyperbolic sine:

$$\sinh^{-1} x = \ln\left(x + \sqrt{x^2 + 1}\right)$$

$$(D700.1)$$

Evaluating this at the limits gives

$$\psi_{\text{left side}} = \frac{\mu_0 I}{4\pi} \int_{Z=r_w-l/2}^{l/2-r_w} \left(-\sinh^{-1}\frac{Z+l/2}{w-r_w} - \sinh^{-1}\frac{l/2-Z}{w-r_w} \right.$$

$$\left. + \sinh^{-1}\frac{Z+l/2}{r_w} + \sinh^{-1}\frac{l/2-Z}{r_w} \right) dZ$$

$$(4.13)$$

To evaluate this final integral we use a change of variables,

$$\lambda = Z + \frac{l}{2}$$

$$d\lambda = dZ$$

and

$$\zeta = \frac{l}{2} - Z$$

$$d\zeta = -dZ$$

giving

$$\psi_{\text{left side}} = \frac{\mu_0 I}{4\pi} \int_{\lambda = r_w}^{l - r_w} \left(-\sinh^{-1} \frac{\lambda}{w - r_w} + \sinh^{-1} \frac{\lambda}{r_w} d\lambda \right)$$

$$+ \frac{\mu_0 I}{4\pi} \int_{\zeta = r_w}^{l - r_w} \left(-\sinh^{-1} \frac{\zeta}{w - r_w} + \sinh^{-1} \frac{\zeta}{r_w} \right) d\zeta$$

$$= 2\frac{\mu_0 I}{4\pi} \int_{\lambda = r_w}^{l - r_w} \left(-\sinh^{-1} \frac{\lambda}{w - r_w} + \sinh^{-1} \frac{\lambda}{r_w} \right) d\lambda \qquad (4.14)$$

Evaluating this using Dwight's integral 730 [7],

$$\int \sinh^{-1} \frac{x}{a} dx = x \sinh^{-1} \frac{x}{a} - \sqrt{x^2 + a^2} \qquad a > 0 \qquad (D730)$$

gives

$$\psi_{\text{left side}} = \frac{\mu_0 I}{2\pi} \left[-\lambda \sinh^{-1} \frac{\lambda}{w - r_w} + \sqrt{\lambda^2 + (w - r_w)^2} \right.$$

$$\left. + \lambda \sinh^{-1} \frac{\lambda}{r_w} - \sqrt{\lambda^2 + (r_w)^2} \right]_{\lambda = r_w}^{l - r_w}$$

$$= \frac{\mu_0 I}{2\pi} \left[-(l - r_w)\sinh^{-1} \frac{l - r_w}{w - r_w} + \sqrt{(l - r_w)^2 + (w - r_w)^2} \right.$$

$$+ (l - r_w)\sinh^{-1} \frac{l - r_w}{r_w} - \sqrt{(l - r_w)^2 + (r_w)^2}$$

$$+ r_w \sinh^{-1} \frac{r_w}{w - r_w} - \sqrt{(r_w)^2 + (w - r_w)^2}$$

$$\left. - \underbrace{r_w \sinh^{-1} \frac{r_w}{r_w}}_{\ln(1 + \sqrt{2})} + \underbrace{\sqrt{(r_w)^2 + (r_w)^2}}_{\sqrt{2} r_w} \right] \qquad (4.15)$$

Then the total flux through the loop given by (4.10) is

$$\psi_{\text{loop}} = 2\psi_{\text{left side}}(l, w, r_w) + 2\psi_{\text{top side}}(w, l, r_w) \qquad (4.16)$$

where we simply interchange l and w in (4.15) to obtain $\psi_{\text{top side}}(w, l, r_w)$. The inductance of the loop is

$$
\begin{aligned}
L_{\text{loop}} &= 2 \frac{\psi_{\text{left side}}(l, w, r_w) + \psi_{\text{top side}}(w, l, r_w)}{I} \\
&= \frac{\mu_0}{\pi} \left[-(l - r_w) \sinh^{-1} \frac{l - r_w}{w - r_w} - (w - r_w) \sinh^{-1} \frac{w - r_w}{l - r_w} \right. \\
&\quad + (l - r_w) \sinh^{-1} \frac{l - r_w}{r_w} + (w - r_w) \sinh^{-1} \frac{w - r_w}{r_w} \\
&\quad + r_w \sinh^{-1} \frac{r_w}{w - r_w} + r_w \sinh^{-1} \frac{r_w}{l - r_w} \\
&\quad + 2\sqrt{(l - r_w)^2 + (w - r_w)^2} - 2\sqrt{(w - r_w)^2 + (r_w)^2} \\
&\quad \left. - 2\sqrt{(l - r_w)^2 + (r_w)^2} - 2 r_w \ln(1 + \sqrt{2}) + 2\sqrt{2}\, r_w \right]
\end{aligned}
$$

(4.17)

If the loop dimensions are much larger than the wire radius, $l, w \gg r_w$, the result in (4.17) simplifies to

$$
\begin{aligned}
L_{\text{loop}} &\cong \frac{\mu_0}{\pi} \left(-l \sinh^{-1} \frac{l}{w} - w \sinh^{-1} \frac{w}{l} \right. \\
&\quad \left. + l \sinh^{-1} \frac{l}{r_w} + w \sinh^{-1} \frac{w}{r_w} + 2\sqrt{l^2 + w^2} - 2w - 2l \right) \\
&= \frac{\mu_0}{\pi} \left[-l \ln\left(\frac{l}{w} + \sqrt{\left(\frac{l}{w}\right)^2 + 1} \right) - w \ln\left(\frac{w}{l} + \sqrt{\left(\frac{w}{l}\right)^2 + 1} \right) \right. \\
&\quad \left. + l \ln\left(\frac{2l}{r_w}\right) + w \ln\left(\frac{2w}{r_w}\right) + 2\sqrt{l^2 + w^2} - 2w - 2l \right] \\
&= \frac{\mu_0}{\pi} \left[-l \ln\left(1 + \sqrt{1 + \left(\frac{w}{l}\right)^2} \right) - w \ln\left(1 + \sqrt{1 + \left(\frac{l}{w}\right)^2} \right) \right. \\
&\quad \left. + l \ln \frac{2w}{r_w} + w \ln \frac{2l}{r_w} + 2\sqrt{l^2 + w^2} - 2w - 2l \right] \quad l, w \gg r_w
\end{aligned}
$$

(4.18)

This result for the inductance of a rectangular loop in 4.17 simplifies considerably if the loop is square (i.e., $l = w$). The loop inductance of a square loop becomes

$$L_{\text{square loop}} = 2\frac{\mu_0}{\pi}\left[(l - r_{\text{w}})\,\sinh^{-1}\frac{l - r_{\text{w}}}{r_{\text{w}}} - l\ln\left(1 + \sqrt{2}\right) + l\sqrt{2}\right.$$
$$\left. + r_{\text{w}}\,\sinh^{-1}\frac{r_{\text{w}}}{l - r_{\text{w}}} - 2\sqrt{(l - r_{\text{w}})^2 + (r_{\text{w}})^2}\right]\qquad l = w$$

(4.19)

In practical situations, the wire radius is much smaller than the side length of the loop (i.e., $l \gg r_{\text{w}}$), and this simplifies to

$$L_{\text{square loop}} \cong 2\frac{\mu_0}{\pi}\left[l\,\sinh^{-1}\frac{l}{r_{\text{w}}} - l\ln\left(1 + \sqrt{2}\right) + l\sqrt{2} - 2l\right]$$
$$\cong 2\frac{\mu_0}{\pi}l\left[\ln\left(2\frac{l}{r_{\text{w}}}\right) - \ln\left(1 + \sqrt{2}\right) + \sqrt{2} - 2\right]$$
$$= 2\frac{\mu_0}{\pi}l\left[\ln\frac{l}{r_{\text{w}}} - 0.774\right]\qquad l = w \gg r_{\text{w}}$$

(4.20)

This tedious derivation will be obtained in a simple and straightforward manner using the concept of partial inductance in Chapter 5.

4.1.2   Circular Loop

Next, we determine the loop inductance of a circular loop of radius a lying in the xy plane which is composed of a wire of radius r_{w}, as shown in Fig. 4.4. Again we assume that the (dc) current is uniformly distributed over the cross section of the wire so that for the purposes of computing the flux through the loop surface, we can consider the current I to be contained in a filament at the center of the wire. The magnetic flux density is directed solely in the z direction over the loop, $\mathbf{B} = B_z \mathbf{a}_z$, and is therefore perpendicular to the surface s that is surrounded by the wire. Once the $\mathbf{B}$ field over the surface s is computed, we next determine the total magnetic flux through the surface of the loop with a surface integral as

$$\psi = \int_s \mathbf{B} \cdot d\mathbf{s}$$
$$= \int_{r=0}^{a-r_{\text{w}}} \int_{\phi'=0}^{2\pi} B_z \underbrace{r\,d\phi'\,dr}_{ds}$$

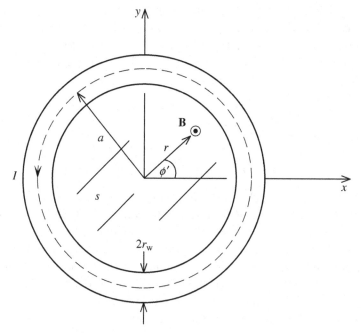

FIGURE 4.4. Circular loop.

Note that the integral with respect to r is from $r = 0$ out to the inner edge of the wires at $r = a - r_w$ as with the rectangular loop. Once this is completed, the self inductance of the circular current loop is again determined from

$$L = \frac{\psi}{I}$$

In Section 2.6 three methods for determining the $\mathbf{B} = B_z \mathbf{a}_z$ field over the loop surface were evaluated. First the Biot–Savart law was the simplest method and gave the result in (2.73):

$$B_z(r) = 2\frac{\mu_0 I a}{4\pi} \int_{\phi=0}^{\pi} \frac{a - r\cos\phi}{\left(a^2 + r^2 - 2ar\cos\phi\right)^{3/2}} \, d\phi$$

$$= \frac{\mu_0 I a}{2\pi} \int_{\phi=0}^{\pi} \frac{a - r\cos\phi}{\left(a^2 + r^2 - 2ar\cos\phi\right)^{3/2}} \, d\phi \qquad (2.73)$$

Next, we obtained the $\mathbf{B}$ field over the loop surface from the vector magnetic potential of a current loop given in (2.59). That general result in (2.59) specialized for the problem of Fig. 4.4 for the field in the plane of the loop ($z = 0$) is

$$A_\phi = \frac{\mu_0 I a}{2\pi} \int_{\phi=0}^{\pi} \frac{\cos\phi}{\sqrt{a^2 + r^2 - 2ar\cos\phi}} \, d\phi \qquad (4.21)$$

We obtained the magnetic flux density over the loop surface contained by the loop from $\mathbf{B} = \nabla \times \mathbf{A} = 1/r[\partial\,(rA_\phi)/\partial r]\,\mathbf{a}_z$ using the result in (4.21). The magnetic flux density in the plane of the loop ($z = 0$) is totally z directed (out of the page within the interior of the loop and into the page outside the loop) according to the right-hand rule and is

$$
\begin{aligned}
B_z &= \frac{1}{r}\frac{\partial\,(rA_\phi)}{\partial r} \\
&= \frac{\mu_0 I}{2\pi r}\int_{\phi=0}^{\pi}\frac{a^2\,\cos\phi\,(a - r\cos\phi)}{\left(a^2 + r^2 - 2ar\cos\phi\right)^{3/2}}d\phi
\end{aligned}
\tag{4.22}
$$

The third method for obtaining the $\mathbf{B}$ field over the loop surface is to use directly the result obtained from (2.59) by Smythe [10] and Weber [11] and given in (2.67c). We showed in Section 2.6 that all three results give the same value for the $\mathbf{B}$ field over the surface of the loop. So the choice of which result to use is whichever one provides the simplest integral for obtaining the total flux through the loop. It is for this reason that we choose to use the result obtained from differentiating $\mathbf{A}$ and given in (4.22).

The total flux through the surface of the loop is

$$
\begin{aligned}
\psi_{\text{loop}} &= \int_{\phi'=0}^{2\pi}\int_{r=0}^{a-r_w}B_z\,r\,dr\,d\phi' \\
&= \frac{\mu_0 I}{2\pi}\int_{\phi'=0}^{2\pi}\int_{r=0}^{a-r_w}\frac{1}{r}\left[\int_{\phi=0}^{\pi}\frac{a^2\,\cos\phi\,(a - r\cos\phi)}{\left(a^2 + r^2 - 2ar\cos\phi\right)^{3/2}}d\phi\right]r\,dr\,d\phi' \\
&= \mu_0 I\int_{\phi=0}^{\pi}\left[\int_{r=0}^{a-r_w}\frac{a^2\,\cos\phi\,(a - r\cos\phi)}{\left(a^2 + r^2 - 2ar\cos\phi\right)^{3/2}}dr\right]d\phi
\end{aligned}
$$

and we have interchanged the order of integration. The interior integral can be evaluated using integrals 380.003 and 380.013 in Dwight [7]:

$$
\int\frac{dx}{\left[ax^2 + bx + c\right]^{3/2}} = \frac{4ax + 2b}{(4ac - b^2)\left[ax^2 + bx + c\right]^{1/2}}
\tag{D380.003}
$$

$$
\int\frac{x\,dx}{\left[ax^2 + bx + c\right]^{3/2}} = -\frac{2bx + 4c}{(4ac - b^2)\left[ax^2 + bx + c\right]^{1/2}}
\tag{D380.013}
$$

to yield

$$
\psi_{\text{loop}} = \mu_0 I a\,(a - r_w)\int_{\phi=0}^{\pi}\frac{\cos\phi}{\sqrt{a^2 + (a - r_w)^2 - 2a\,(a - r_w)\cos\phi}}\,d\phi
$$

$$
\tag{4.23}
$$

This integral cannot be evaluated in closed form, but the result can be given in terms of complete elliptic integrals of the first and second kind [7]:

$$K = \int_{\theta=0}^{\pi/2} \frac{d\theta}{\sqrt{1 - k^2 \sin^2 \theta}} \qquad (D773.1)$$

and

$$E = \int_{\theta=0}^{\pi/2} \sqrt{1 - k^2 \sin^2 \theta}\, d\theta \qquad (D774.1)$$

Making a change of variables in (4.23) as $\phi = \pi - 2\theta$, $d\phi = -2d\theta$ gives $\cos \phi = 2 \sin^2 \theta - 1$ and

$$\psi_{\text{loop}} = 2\mu_0 I a (a - r_w) \int_{\theta=0}^{\pi/2} \frac{2 \sin^2 \theta - 1}{(2a - r_w)\sqrt{1 - k^2 \sin^2 \theta}}\, d\theta \qquad (4.24)$$

where k^2 is defined here as

$$k^2 = \frac{4a(a - r_w)}{(2a - r_w)^2} \qquad (4.25)$$

This can be written in terms of the complete elliptic integrals as

$$\psi_{\text{loop}} = \mu_0 I \sqrt{a(a - r_w)} \left[\left(\frac{2}{k} - k\right) K(k) - \frac{2}{k} E(k) \right] \qquad (4.26)$$

Hence, the loop inductance is

$$\boxed{\begin{aligned} L_{\text{loop}} &= \frac{\psi_{\text{loop}}}{I} \\ &= \mu_0 \sqrt{a(a - r_w)} \left[\left(\frac{2}{k} - k\right) K(k) - \frac{2}{k} E(k) \right] \end{aligned}} \qquad (4.27)$$

This result can be simplified by assuming that the loop radius is much larger than the wire radius, $a \gg r_w$. For this reasonable approximation we obtain $\sqrt{a(a - r_w)} \cong a$ and $k^2 \cong 1$. From series expansions of the complete elliptic integrals given by Dwight [7], we obtain

$$K(k) \cong \ln\left(\frac{8a}{r_w} - 4\right) \qquad a \gg r_w$$

$$E(k) \cong 1 \qquad a \gg r_w$$

Hence, the loop inductance of the circular loop approximates to

$$L_{\text{loop}} \cong \mu_0 a \left(\ln \frac{8a}{r_{\text{w}}} - 2 \right) \qquad a \gg r_{\text{w}} \qquad (4.28)$$

The self inductance of coils consisting of a thin wire of radius r_{w} and the same total length, denoted as Len, are approximately independent of their shape. For example, the circular loop of radius a has a total circumference of Len $= 2\pi a$ and an inductance in (4.28) of

$$L_{\text{circular loop}} = \frac{\mu_0 \, \text{Len}}{2\pi} \left(\ln \frac{4 \, \text{Len}}{r_{\text{w}}} - 3.145 \right)$$

whereas the square loop of equal side lengths of l has a total circumference of Len $= 4l$ and an inductance in (4.20) of

$$L_{\text{square loop}} = \frac{\mu_0 \, \text{Len}}{2\pi} \left(\ln \frac{4 \, \text{Len}}{r_{\text{w}}} - 3.547 \right)$$

4.1.3   Coaxial Cable

In this section we determine the inductance of a coaxial cable shown in Fig. 4.5(a). The cable is assumed to be infinite in length (or very long compared with the cable radius) in order to avoid having to deal with fringing of the fields at the ends of a finite-length section. The magnetic flux density for this cable was determined in Chapter 2. Consider a section of length 1 m. Because of the infinite length and symmetry, the magnetic field between the inner wire and the inside of the shield is circumferentially directed in the ϕ direction as shown in Fig. 4.5(b) and is determined in Chapter 2 as

$$B_\phi = \frac{\mu_0 I}{2\pi r} \qquad r_{\text{w}} < r < r_s \qquad (2.35a)$$

We determine the flux through a flat surface that extends from the outer edge of the inner wire, $r = r_{\text{w}}$, to the inner edge of the outer shield, $r = r_s$, and is of length along the cable of 1 m. We have shown two choices for this surface. One (which we will choose) is perpendicular the inner wire surface, and the other extends at an angle from the inner wire surface to the inner surface of the shield as shown in Fig. 4.5(b). The best choice is the first surface that is perpendicular to the inner wire surface and extends directly across perpendicular to the inner surface of the shield. The reason that this is preferred is that the magnetic field,

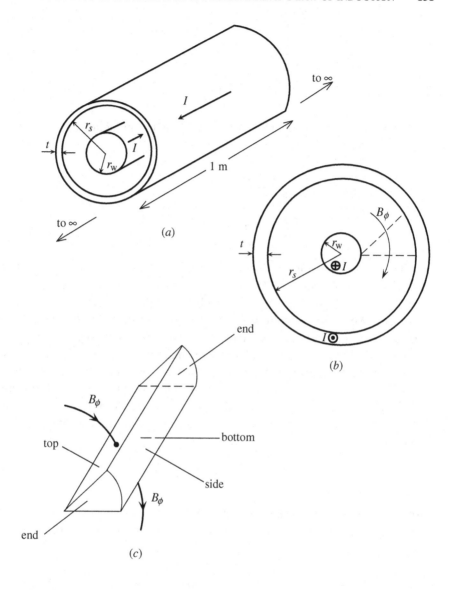

FIGURE 4.5. Coaxial cable.

B_ϕ, is perpendicular to that surface and hence we easily obtain the flux through this surface as

$$\psi = \int_s \mathbf{B} \cdot d\mathbf{s}$$

$$= \int_{z=0}^{1m} \int_{r=r_w}^{r_s} B_\phi \underbrace{dr\, dz}_{ds}$$

$$= \int_{z=0}^{1m} \int_{r=r_w}^{r_s} \frac{\mu_0 I}{2\pi r} \underbrace{dr\,dz}_{ds}$$

$$= \frac{\mu_0 I}{2\pi} \ln \frac{r_s}{r_w}$$

The *per-unit-length* inductance of the cable is the inductance of this section and is denoted as l:

$$
\boxed{
\begin{aligned}
l &= \frac{\psi}{I} \\
&= \frac{\mu_0}{2\pi} \ln \frac{r_s}{r_w} \qquad \text{H/m}
\end{aligned}
}
$$

(4.29)

There were two choices for the flat surface through which we were to determine the flux. Figure 4.5(c) shows this situation. Consider this as a closed, "wedge-shaped" surface. The top and bottom sides were the two choices for surfaces. We chose the bottom surface because the magnetic flux density vector is perpendicular to that surface, thus allowing us to remove the dot product in the flux integral and deal only with the magnitude of the field over the surface. Would computing the flux through the other surface, the top surface, have given a different answer? Certainly that computation would be more difficult since the magnetic flux density vector would not be perpendicular to it and the dot product could not be removed from the flux integral. Recall Gauss's law for the magnetic field:

$$\oint_s \mathbf{B} \cdot d\mathbf{s} = 0$$

In other words, the net magnetic flux leaving a *closed surface* s is zero for the magnetic field. Consider the *closed* wedge-shaped surface in Fig. 4.5(c). Applying Gauss's law gives

$$\oint_s \mathbf{B} \cdot d\mathbf{s} = \int_{top} \mathbf{B} \cdot d\mathbf{s} + \int_{bottom} \mathbf{B} \cdot d\mathbf{s} + \underbrace{\int_{side} \mathbf{B} \cdot d\mathbf{s}}_{0} + \underbrace{\int_{left\ end} \mathbf{B} \cdot d\mathbf{s}}_{0}$$

$$+ \underbrace{\int_{right\ end} \mathbf{B} \cdot d\mathbf{s}}_{0} = 0$$

The flux through the side of constant radius r_s is zero because on that surface B_ϕ is parallel to the surface. Similarly, the flux through the left and right ends of the surface are also zero because on the surfaces B_ϕ is also parallel to the surfaces. Hence, we see that

$$\int_{\text{top}} \mathbf{B} \cdot d\mathbf{s} = -\int_{\text{bottom}} \mathbf{B} \cdot d\mathbf{s}$$

But obtaining the flux through the bottom surface is much easier than obtaining the flux through the top surface since the magnetic field is perpendicular to the bottom surface.

4.2   THE CONCEPT OF FLUX LINKAGES FOR MULTITURN LOOPS

Consider a single, circular current loop carrying a current I. Denote the magnetic flux through the surface of the loop due to this current I as $\psi_{\text{one loop}}$. The emf voltage induced in that single loop is

$$V_{\text{one loop}} = \frac{d\psi_{\text{one loop}}}{dt}$$

$$= L_{\text{one loop}} \frac{dI}{dt}$$

The magnetic flux $\psi_{\text{one loop}}$ is said to *link* current I.

Now consider a *multiturn loop* where we add N such identical loops that are in very close proximity (virtually on top of each other) so that all of the magnetic flux that passes through one of the loops that is due to the current of that loop, $\psi_{\text{one loop}}$, also passes through all the other loop surfaces. The total are concentrically located and are tightly wound together such that they resemble one loop carrying a current of NI amperes as shown in Fig. 4.6. The *total* flux through *each loop* is therefore the *sum of the fluxes* from all the

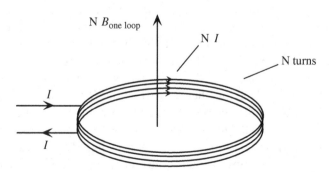

FIGURE 4.6. A multiturn loop consisting of N loops close together and connected in series.

other N loops or $N\psi_{\text{one loop}}$. Hence, we say that each loop has N *flux linkages* linking its current I. The emf voltage induced in *each loop* is therefore

$$V_{\text{one loop}} = N\frac{d\psi_{\text{one loop}}}{dt}$$

The loops are connected in *series* so that each carry current I in the same direction around the loops. Since all N loops are connected in series, the total emf voltage at the terminals of the N loops is

$$\boxed{\begin{aligned} V_{\text{total}} &= N\left(V_{\text{one loop}}\right) \\ &= N^2\frac{d\psi_{\text{one loop}}}{dt} \end{aligned}}$$

This is an important property of N identical loops that surround a common core; the inductance is proportional to N^2 times the inductance of one of the loops:

$$\boxed{L_{N\,\text{loops}} \propto N^2 L_{\text{one loop}}}$$

4.2.1 Solenoid

For example, consider the *solenoid* shown in Fig. 4.7(a), consisting of N turns of wire wound in one layer on a ferromagnetic core that has a relative permeability of μ_r and a radius r. The purpose of a ferromagnetic core having a large μ_r is to *concentrate the flux in that core, thereby minimizing the flux that leaks out into the air*, which has $\mu_r = 1$ [3]. Hence, if the turns of wire are closely wound on the core, there will be very little leakage of the magnetic field between the adjacent turns of wire. In fact, if the solenoid is infinite in length, $l \to \infty$, and the turns of wire are tightly wound, the magnetic field in the core will be (1) in the z direction parallel to the axis of the core, (2) constant along that axis, (3) uniformly distributed across the core cross section, and (4) the magnetic field outside the solenoid will be zero. These properties of an infinite-length solenoid are also approximate properties of a solenoid of finite length on which the wires are tightly wound.

To determine the H field in the core, assume that the solenoid is infinite in length ($l \to \infty$). Thinking of this as an infinite number of current loops that are infinitesimally close together shows that the H field in the core will be in the z direction and independent of z and r. Draw a rectangular closed contour c whose sides are parallel to the core axis and whose ends are perpendicular to it and which encloses N turns (wires) within a length l as shown in Fig. 4.7(b). If this rectangle were moved outside the core, it would enclose no current and the line integral of H around it must, by Ampere's law, be zero. But this would

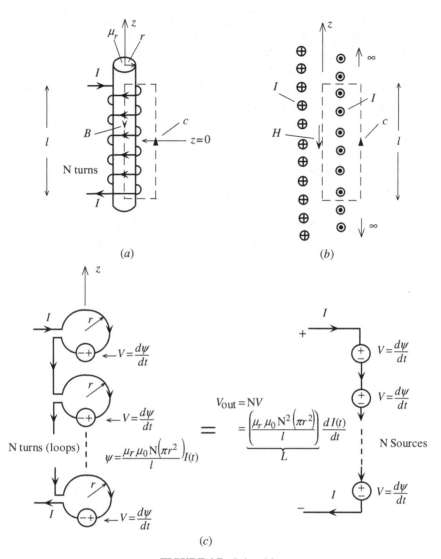

FIGURE 4.7. Solenoid.

imply that the H field along the sides would be constant. Hence, we conclude that the H field outside the infinite-length core is zero since the magnetic field must go to zero as $r \rightarrow \infty$. Therefore, the magnetic field along the right-hand part of contour c (that passes along the outside of the coil of wire) is zero. From Ampère's law and Fig. 4.7(b), we obtain

$$\oint_c \mathbf{H} \cdot d\mathbf{l} = Hl = NI \qquad (4.30)$$

since the closed contour c encloses N currents. From this result for a coil of infinte length the magnetic field intensity is $H = NI/l$. For a core of finite length, this result is approximately the same and relies on our assumption that (1) the turns are tightly wound, (2) the relative permeability of the core is very large, $\mu_r \gg 1$, and (3) the coil length, l, is long, $l \gg r$. Hence, the magnetic flux density in the core and parallel to the core axis is

$$B = \mu_r \mu_0 H = \frac{\mu_r \mu_0 NI}{l} \tag{4.31}$$

This result can be derived in a different fashion by using the result for the magnetic field on the axis of a single current loop derived in Chapter 2 and given in (2.24). Since we assume that the coil of wires is tightly wound, think of the coil of wires as being a cylindrical sheet of current with a surface current distribution of $K = NI/l$ A/m uniformly distributed along the core surface and directed in the circumferential direction about the core. Hence, we may think of a section of the coil of differential length dz as being a single turn carrying a current of $K\,dz = NI/l\,dz$ amperes. Using (2.24) and summing the fields of these turns of differential lengths dz gives the magnetic flux density on the axis of the core and midway between the two ends of the coil of wire at $z = 0$ as

$$\begin{aligned}
B &= \frac{\mu_r \mu_0}{2} \frac{NI}{l} \int_{z=-l/2}^{l/2} \frac{r^2}{\left(r^2 + z^2\right)^{3/2}} dz \\
&= \frac{\mu_r \mu_0}{2} \frac{NI}{l} \left[\frac{z}{\sqrt{r^2 + z^2}} \right]_{z=-l/2}^{l/2} \\
&= \frac{\mu_r \mu_0}{2} \frac{NI}{l} \left[\frac{l/2}{\sqrt{r^2 + (l/2)^2}} + \frac{l/2}{\sqrt{r^2 + (l/2)^2}} \right] \\
&= \frac{\mu_r \mu_0 NI}{\sqrt{4r^2 + l^2}} \tag{4.32}
\end{aligned}$$

and we have used integral 200.03 from Dwight [7]:

$$\int \frac{1}{\left(x^2 + a^2\right)^{3/2}} dx = \frac{x}{a^2 \sqrt{x^2 + a^2}} \tag{D200.03}$$

For a very long coil length with respect to the radius, $l \gg r$, (4.32) reduces to (4.31) derived by the previous method using Ampère's law.

Since the field for a very long coil, $l \gg r$, is (approximately) uniformly distributed over the core cross section, which has an area of πr^2, the magnetic flux through each turn of the solenoid is

$$\psi_{\text{each loop}} = \mu_r \mu_0 \frac{NI}{l} \pi r^2 \tag{4.33}$$

Hence, the emf voltage induced in each loop is

$$V_{\text{each loop}} = \text{N} \underbrace{\frac{\mu_r \mu_0 \pi r^2}{l}}_{L_{\text{each loop}}} \frac{dI(t)}{dt} \tag{4.34}$$

If we visualize the entire coil of wire as being the N loops connected in series as shown in Fig. 4.7(c), the emf voltage sources of each loop are connected in series so that the voltage across the terminals of the entire coil is $V = NV_{\text{each loop}}$. Hence, the total inductance of the solenoid is

$$\boxed{L = \text{N}^2 \frac{\mu_r \mu_0 \pi \, r^2}{l}} \tag{4.35}$$

4.2.2 Toroid

Next, consider the *toroid* shown in Fig. 4.8(a). The toroid consists of N turns of wire wound tightly around a toroidal core of ferromagnetic material having relative permeability of μ_r, an inner radius a, and an outer radius b. The cross section of the toroid is usually rectangular with thickness t and width $w = b - a$, as shown in Fig. 4.8(b). If we assume that the turns are tightly wound on the core and $\mu_r \gg 1$ so that there is no significant leakage of the

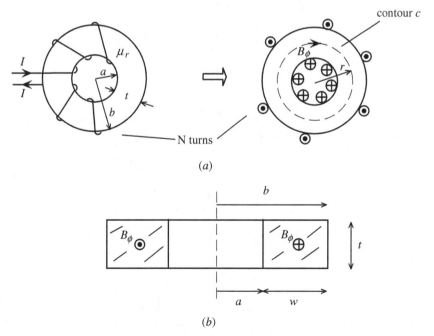

FIGURE 4.8. Toroid.

magnetic field outside the core, we may assume as an approximation that the magnetic field is contained within the core and is in the circumferential or ϕ direction. Alternatively, we can view the toroid as a finite-length solenoid that is formed into a circle.

To determine that magnetic field, we choose a circular contour c of radius r in the core as shown in Fig. 4.8(a) and write Ampère's law as

$$\oint_c \mathbf{H} \cdot d\mathbf{l} = H_\phi (2\pi r) = NI \tag{4.36a}$$

since the contour c surrounds NI currents. Hence, the magnetic field intensity in the core is $H_\phi = NI/2\pi r$, and the flux density in the core is

$$B_\phi = \mu_r \mu_0 \, H_\phi$$
$$= N \frac{\mu_r \mu_0}{2\pi r} I \qquad a < r < b \tag{4.36b}$$

Expanding the contour to a radius $r > b$ encloses zero net current, and hence the H field outside the toroid is zero, as is the field for $r < a$. If the core cross section is rectangular with width w and thickness t as shown in Fig. 4.8(b), we can determine the total magnetic flux through each loop as

$$\psi_{\text{each loop}} = \int_s \mathbf{B} \cdot d\mathbf{s}$$
$$= \int_{z=0}^t \int_{r=a}^b NI \frac{\mu_r \mu_0}{2\pi r} \, dr \, dz \tag{4.37}$$
$$= NI \frac{\mu_r \mu_0}{2\pi} t \ln \frac{b}{a}$$

where surface s is the rectangular flat surface of a cross section of the core. The emf voltage induced in each turn is

$$V_{\text{each loop}} = N \underbrace{\frac{\mu_r \mu_0}{2\pi} t \ln \frac{b}{a}}_{L_{\text{each loop}}} \frac{dI(t)}{dt} \tag{4.38}$$

Since the loops are connected in series, the total inductance is

$$\boxed{\begin{aligned} L &= N L_{\text{each loop}} \\ &= N^2 \frac{\mu_r \mu_0}{2\pi} t \ln \frac{b}{a} \end{aligned}} \tag{4.39}$$

This expression can be simplified for cores of rectangular cross section where the width, $w = b - a$, is much less than the inner radius, $w \ll a$, by using the approximation of the natural logarithm:

$$\ln\frac{b}{a} = \ln\left(\frac{w}{a} + 1\right)$$

$$\cong \frac{w}{a} \qquad w \ll a \tag{D601}$$

Evaluating (4.39) gives

$$L \cong N^2 \frac{\mu_r \mu_0}{2\pi a} tw \tag{4.40a}$$

Since the cross-sectional area of the core is $A = tw$, we can write a general relation for the inductance of a toroid as

$$\boxed{L \cong \frac{\mu_r \mu_0 N^2 A}{2\pi a}} \tag{4.40b}$$

4.3 LOOP INDUCTANCE USING THE VECTOR MAGNETIC POTENTIAL

The inductance of a current loop is defined fundamentally by Faraday's law as the ratio of the magnetic flux penetrating the open surface s that is surrounded by the current and the current I as illustrated in Fig. 4.9:

$$L = \frac{\psi = \int_s \mathbf{B} \cdot d\mathbf{s}}{I} \tag{4.41}$$

In the previous examples we evaluated this by first computing the magnetic flux density $\mathbf{B}$ and then evaluating (4.41) by computing the flux ψ through the surface that is surrounded by the current loop. This required the evaluation of two integrals: one to obtain $\mathbf{B}$ (by the Biot–Savart law or Ampère's law) and *two* [since (4.41) is a surface integral] to obtain ψ.

There is another way of obtaining this result by using the *vector magnetic potential* $\mathbf{A}$ rather than using $\mathbf{B}$. To obtain this alternative result, recall from Chapter 2 that $\mathbf{A}$ is defined by

$$\mathbf{B} = \nabla \times \mathbf{A} \tag{4.42}$$

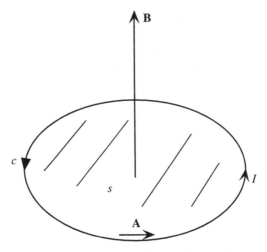

FIGURE 4.9. Using the vector magnetic potential **A** to obtain the magnetic flux through an open surface s.

Hence, the magnetic flux through surface s can alternatively be obtained in terms of **A** as

$$\psi = \int_s \mathbf{B} \cdot d\mathbf{s}$$

$$= \int_s (\nabla \times \mathbf{A}) \cdot d\mathbf{s}$$

$$= \oint_c \mathbf{A} \cdot d\mathbf{l} \tag{4.43}$$

where we have used Stokes's theorem (see the Appendix) and c is the *closed contour* that surrounds the *open surface s*. Hence, to obtain the total magnetic flux penetrating the open surface s we only need to obtain A (which is usually easier to obtain than B) and then integrate with only *one* integral, a line integral, the component of **A** that is tangent to the contour c around the perimeter of that open surface, as illustrated in Fig. 4.9. The inductance calculation becomes

$$L = \frac{\psi = \displaystyle\int_s \mathbf{B} \cdot d\mathbf{s}}{I}$$

$$= \frac{\displaystyle\oint_c \mathbf{A} \cdot d\mathbf{l}}{I} \tag{4.44}$$

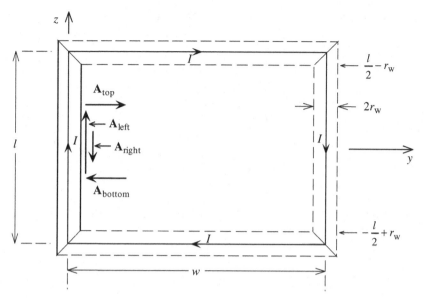

FIGURE 4.10. Determining the inductance of a rectangular loop by using the vector magnetic potential **A**.

4.3.1   Rectangular Loop

We now apply this to the calculation of the self inductance of a rectangular loop composed of four wires of radii r_w having lengths w and l as shown in Fig. 4.3. The basic idea is to integrate the line integral of **A** along the interior edge of the wire of one side as illustrated in Fig. 4.10 and then repeat this for the other three sides. The total magnetic flux threading the loop, according to (4.43), is

$$\psi_{\text{loop}} = 2\int_{\text{left side}} \mathbf{A}\cdot d\mathbf{l} + 2\int_{\text{top side}} \mathbf{A}\cdot d\mathbf{l} \qquad (4.45)$$

It is very important to realize that the total vector magnetic potential tangent to *each* side has contributions from the current of that side *and* the currents of the other three sides. This is illustrated for the left side in Fig. 4.10. Again we assume that the currents are dc and are uniformly distributed over the wire cross sections so that they can be represented by filaments on the axes of the wires. Two of these contributions, $\mathbf{A}_{\text{left}}$ (that is due to the current in the left side) and $\mathbf{A}_{\text{right}}$ (that is due to the current in the right side), are parallel to the left side and are oppositely directed. $\mathbf{A}_{\text{left}}$ is larger in magnitude than $\mathbf{A}_{\text{right}}$ since the current of the right side is further away. The other contributions along the left side, $\mathbf{A}_{\text{top}}$ and $\mathbf{A}_{\text{bottom}}$, are due to the currents in the top and bottom sides and are perpendicular to the left side since the vector magnetic potential is in the direction of the current producing it. Hence, the line integral of the *total* vector magnetic field along the left side is

$$\psi_{\text{left side}} = \int_{\text{left side}} \mathbf{A} \cdot d\mathbf{l}$$

$$= \int_{z=r_w-l/2}^{l/2-r_w} \mathbf{A}_{\text{left}} \cdot d\mathbf{l} + \int_{z=r_w-l/2}^{l/2-r_w} \mathbf{A}_{\text{right}} \cdot d\mathbf{l}$$

$$+ \underbrace{\int_{z=r_w-l/2}^{l/2-r_w} \mathbf{A}_{\text{top}} \cdot d\mathbf{l}}_{0} + \underbrace{\int_{z=r_w-l/2}^{l/2-r_w} \mathbf{A}_{\text{bottom}} \cdot d\mathbf{l}}_{0} \qquad (4.46)$$

The contributions to $\mathbf{A}_{\text{left}}$ and $\mathbf{A}_{\text{right}}$ are derived in Chapter 2 and given in (2.57) with respect to Fig. 2.24:

$$A_z = \frac{\mu_0 I}{4\pi} \left(\sinh^{-1} \frac{Z+L/2}{r} + \sinh^{-1} \frac{L/2-Z}{r} \right) \qquad (2.57)$$

Hence, the contribution to the magnetic flux through the loop surface integrated along the left side is the same as obtained in Section 4.1.1 and given in (4.13):

$$\psi_{\text{left side}} = \frac{\mu_0 I}{4\pi} \int_{Z=r_w-l/2}^{l/2-r_w} \left(\sinh^{-1} \frac{Z+l/2}{r_w} + \sinh^{-1} \frac{l/2-Z}{r_w} \right.$$

$$\left. - \sinh^{-1} \frac{Z+l/2}{w-r_w} - \sinh^{-1} \frac{l/2-Z}{w-r_w} \right) dZ$$

$$(4.13)$$

Notice that the vector magnetic potential in (2.57) is evaluated over the left wire surface, giving

$$\mathbf{A}_{\text{left}} = \mathbf{A}_z|_{r=r_w} \qquad (4.47a)$$

and

$$\mathbf{A}_{\text{right}} = -\mathbf{A}_z|_{r=w-r_w} \qquad (4.47b)$$

since $\mathbf{A}_{\text{left}}$ along the left side is at a distance $r = r_w$ from the current of that side, and $\mathbf{A}_{\text{right}}$ along the left side is at a distance $r = w - r_w$ from the current of the right side that produces it. The integral of (4.13) was evaluated in Section 4.1.1, giving

$$
\psi_{\text{left side}} = \frac{\mu_0 I}{2\pi} \left[-(l - r_w) \sinh^{-1} \frac{l - r_w}{w - r_w} + \sqrt{(l - r_w)^2 + (w - r_w)^2} \right.
$$

$$
+ (l - r_w) \sinh^{-1} \frac{l - r_w}{r_w} - \sqrt{(l - r_w)^2 + (r_w)^2}
$$

$$
+ r_w \sinh^{-1} \frac{r_w}{w - r_w} - \sqrt{(r_w)^2 + (w - r_w)^2}
$$

$$
\left. - r_w \underbrace{\sinh^{-1} \left(\frac{r_w}{r_w} \right)}_{\ln\left(1 + \sqrt{2}\right)} + \underbrace{\sqrt{(r_w)^2 + (r_w)^2}}_{\sqrt{2}\, r_w} \right] \tag{4.15}
$$

Similarly, we obtain the contributions to the magnetic flux through the surface from the right side and the top and bottom sides by integrating **A** along those remaining three sides of the loop ($\psi_{\text{left side}} = \psi_{\text{right side}}$ and $\psi_{\text{top side}} = \psi_{\text{bottom side}}$).

The total flux through the loop is

$$
\psi_{\text{loop}} = 2\psi_{\text{left side}}(l, w, r_w) + 2\psi_{\text{top side}}(w, l, r_w) \tag{4.48}
$$

where we simply interchange l and w in $\psi_{\text{left side}}(l, w, r_w)$ to obtain $\psi_{\text{top side}}(w, l, r_w)$. Since the result is identical to that obtained in Section 4.1.1 using **B**, the inductance of the loop is identical to that obtained in Section 4.1.1:

$$
L_{\text{loop}} = 2\frac{\psi_{\text{left side}}(l, w, r_w) + \psi_{\text{top side}}(w, l, r_w)}{I}
$$

$$
= \frac{\mu_0}{\pi} \left[-(l - r_w) \sinh^{-1} \frac{l - r_w}{w - r_w} - (w - r_w) \sinh^{-1} \frac{w - r_w}{l - r_w} \right.
$$

$$
+ (l - r_w) \sinh^{-1} \frac{l - r_w}{r_w} + (w - r_w) \sinh^{-1} \frac{w - r_w}{r_w}
$$

$$
+ r_w \sinh^{-1} \frac{r_w}{w - r_w} + r_w \sinh^{-1} \frac{r_w}{l - r_w}
$$

$$+ 2\sqrt{(l - r_w)^2 + (w - r_w)^2} - 2\sqrt{(w - r_w)^2 + (r_w)^2}$$
$$- 2\sqrt{(l - r_w)^2 + (r_w)^2} - 2r_w \ln\left(1 + \sqrt{2}\right) + 2\sqrt{2}\, r_w\Bigg]$$

$$(4.17)$$

The remaining results in (4.18), (4.19), and (4.20) for a square loop and for loop side lengths greater than the wire radius are obtained from (4.17) and are identical to those obtained with this method. But the method of this section is much simpler since it avoids having to integrate **B** over the surface of the loop, thereby eliminating one integration.

4.3.2 Circular Loop

The circular loop of radius a is composed of a wire having radius r_w and shown in Fig. 4.4 and is illustrated for this problem in Fig. 4.11. Again we assume that the current is dc and is uniformly distributed over the wire cross section so that it can be represented by a filament on the axis of the wire. To obtain the magnetic flux through the loop enclosed by the wire surface using

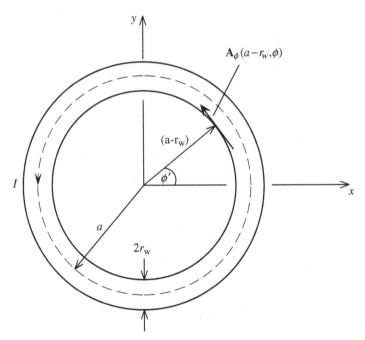

FIGURE 4.11. Determining the inductance of a circular loop by using the vector magnetic potential **A**.

the vector magnetic potential method in (4.44), we first obtain the vector magnetic potential along the inner surface of the wire at $r = a - r_w$. The vector magnetic potential for a circular current loop was obtained in Chapter 2 and given in (2.59) with reference to Fig. 2.25. That result is used to give the magnetic vector potential over the loop surface given in (4.21):

$$A_\phi = \frac{\mu_0 I a}{2\pi} \int_{\phi=0}^{\pi} \frac{\cos\phi}{\sqrt{a^2 + r^2 - 2ar\cos\phi}}\, d\phi \qquad (4.21)$$

Evaluating (4.21) at $r = a - r_w$ gives the vector magnetic potential along the inner wire surface as

$$A_\phi(a - r_w, \phi) = \frac{\mu_0 I a}{2\pi} \int_{\phi=0}^{\pi} \frac{\cos\phi}{\sqrt{a^2 + (a - r_w)^2 - 2a(a - r_w)\cos\phi}}\, d\phi$$
$$(4.49)$$

Then we obtain the result for the total flux through the loop as

$$\psi_{\text{loop}} = \oint_c \mathbf{A} \cdot d\mathbf{l}$$

$$= \int_{\phi'=0}^{2\pi} A_\phi\big|_{r=a-r_w} \underbrace{(a - r_w)\, d\phi'}_{r\, d\phi'}$$

$$= \frac{\mu_0 I a}{2\pi} \int_{\phi'=0}^{2\pi} \int_{\phi=0}^{\pi} \frac{(a - r_w)\cos\phi}{\sqrt{a^2 + (a - r_w)^2 - 2a(a - r_w)\cos\phi}}\, d\phi\, d\phi'$$

$$= \mu_0 I a (a - r_w) \int_{\phi=0}^{\pi} \frac{\cos\phi}{\sqrt{a^2 + (a - r_w)^2 - 2a(a - r_w)\cos\phi}}\, d\phi$$
$$(4.50)$$

But this is identical to the result obtained by integrating $\mathbf{B}$ in Section 4.1.2 and given in (4.23). Hence, the remaining results in Section 4.1.2 and the inductance of the loop obtained in (4.27) and (4.28) are identical to those obtained by this method. But the method of this section is much simpler since it avoids having to integrate $\mathbf{B}$ over the surface of the loop, thereby eliminating one (difficult) integration.

4.4 NEUMANN INTEGRAL FOR SELF AND MUTUAL INDUCTANCES BETWEEN CURRENT LOOPS

Mutual inductance between two current loops was discussed at the beginning of this chapter with reference to Fig. 4.2. With the first loop carrying a current

I_1, the mutual inductance between the two loops is

$$M_{12} = \frac{\psi_2}{I_1} \tag{4.6}$$

where ψ_2 is the flux penetrating the surface of the second loop, s_2, that is caused by the current of the first loop:

$$\psi_2 = \int_{s_2} \mathbf{B}_{12} \cdot d\mathbf{s} \tag{4.7}$$

and $\mathbf{B}_{12}$ is the magnetic flux density through loop 2 that is due to the current I_1 of loop 1. This result can be put into a more compact form by recalling that the magnetic flux through the second loop can be written in terms of the vector magnetic potential around the perimeter of that loop (the interior edge of the wire), $\mathbf{A}_{12}$, as

$$\psi_2 = \oint_{c_2} \mathbf{A}_{12} \cdot d\mathbf{l}_2 \tag{4.51}$$

and c_2 is the contour surrounding the surface of the second loop, s_2. But $\mathbf{A}_{12}$ is the magnetic vector potential around contour c_2 of loop 2 that is due to the current of loop 1 as

$$\mathbf{A}_{12} = \frac{\mu_0}{4\pi} \oint_{c_1} \frac{I_1}{R_{12}} d\mathbf{l}_1 \tag{4.52}$$

and c_1 is the contour of the current of loop 1. The distance R_{12} is the distance from a "chunk" of current $I_1 d\mathbf{l}_1$ of loop 1 to the point on the contour of loop 2, c_2, where we are evaluating the integral in (4.51). Substituting (4.52) into (4.51) yields

$$\psi_2 = \frac{\mu_0 I_1}{4\pi} \oint_{c_1} \oint_{c_2} \frac{d\mathbf{l}_1 \cdot d\mathbf{l}_2}{R_{12}} \tag{4.53}$$

Hence, the mutual inductance between the two loops is

$$\boxed{\begin{aligned} M_{12} &= \frac{\psi_2}{I_1} \\ &= \frac{\mu_0}{4\pi} \oint_{c_1} \oint_{c_2} \frac{d\mathbf{l}_1 \cdot d\mathbf{l}_2}{R_{12}} \end{aligned}} \tag{4.54}$$

This result is called the *Neumann integral*. It shows that the mutual inductance between two loops is only a function of the shapes of the two loops and their orientation with respect to each other. It is also important to remember that if the currents are not filamentary but are uniformly distributed over the cross sections of wires of radii r_{w1} in loop 1 and r_{w2} in loop 2, contour c_1 is along the filamentary current I_1 but contour c_2 is along the interior surface of the second wire, which bounds the surface s_2 that is enclosed by that wire. The order of integration is immaterial. This important result shows that $M_{12} = M_{21}$ simply by interchanging the roles of the two loops in (4.54).

The Neumann integral for mutual inductance between two current loops in (4.54) can also be used to determine the self inductance of a loop by letting the two loops be coincident:

$$
\begin{aligned}
L &= \frac{\psi}{I} \\
&= \frac{\mu_0}{4\pi} \oint_{c'} \oint_c \frac{d\mathbf{l} \cdot d\mathbf{l}'}{R}
\end{aligned}
$$
(4.55)

Contour c' is along the filamentary current bearing current I at the center of the wire, and contour c is along the interior edge of that wire that bounds the surface of the loop through which we desire to compute the flux through that loop.

4.4.1 Mutual Inductance Between Two Circular Loops

Consider two coaxial loops having N_1 and N_2 turns, respectively, that are tightly wound, as shown in Fig. 4.12. The two loops are parallel, have radii a and b, and are separated by distance d. First we fix the point on the second loop

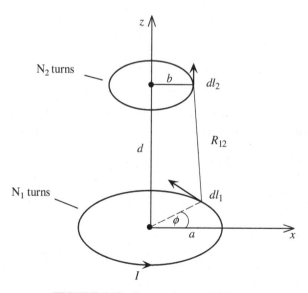

FIGURE 4.12. Concentric, coaxial loops.

and vary the angle ϕ of the first loop. Using the law of cosines, the distance between the two differential arc lengths is

$$R_{12} = \sqrt{a^2 + b^2 + d^2 - 2ab \cos \phi} \tag{4.56}$$

First, we perform the calculation for one turn in each loop and then we multiply the result by the square of the number of turns, N_1^2 and N_2^2, as discussed previously. The dot product in the Neumann integral depends on the dot product $d\mathbf{l}_1 \cdot d\mathbf{l}_2 = \cos \phi \, dl_1 dl_2$ and $dl_1 = a \, d\phi$ and $dl_2 = b \, d\phi'$. Once we integrate with respect to ϕ from $\phi = 0$ to $\phi = 2\pi$, we finally integrate with respect to the angle of loop 2: $\phi' = 0$ to $\phi' = 2\pi$, giving the Neumann integral as

$$\begin{aligned}
M_{12} &= \frac{\mu_0 ab}{4\pi} \int_{\phi'=0}^{2\pi} \int_{\phi=0}^{2\pi} \frac{\cos \phi}{\sqrt{a^2 + b^2 + d^2 - 2ab \cos \phi}} \, d\phi \, d\phi' \\
&= \frac{\mu_0 ab}{2} \int_{\phi=0}^{2\pi} \frac{\cos \phi}{\sqrt{a^2 + b^2 + d^2 - 2ab \cos \phi}} \, d\phi \tag{4.57}
\end{aligned}$$

Making a change of variables to $\phi = 2\theta$ so that $\cos \phi = \cos 2\theta = 2\cos^2 \theta - 1$ and $d\phi = 2d\theta$ gives

$$\begin{aligned}
M_{12} &= \frac{\mu_0 ab}{2} \int_{\theta=0}^{\pi} \frac{2 \cos 2\theta}{\sqrt{(a+b)^2 + d^2 - 4ab \cos^2 \theta}} \, d\theta \\
&= \frac{\mu_0 \sqrt{ab}}{2} \int_{\theta=0}^{\pi} \frac{k \cos 2\theta}{\sqrt{1 - k^2 \cos^2 \theta}} \, d\theta \tag{4.58a}
\end{aligned}$$

where

$$\begin{aligned}
k^2 &= \frac{4ab}{(a+b)^2 + d^2} \\
&= 4 \frac{(a/d)(b/d)}{(a/d + b/d)^2 + 1} \tag{4.58b}
\end{aligned}$$

and the factor k depends on the ratios of the circle radii to their separation. But the numerator of the integrand of (4.58a) can be written as

$$\begin{aligned}
k \cos 2\theta &= k \left(2\cos^2 \theta - 1 \right) \\
&= \left(\frac{2}{k} - k \right) - \frac{2}{k} \left(1 - k^2 \cos^2 \theta \right)
\end{aligned}$$

Note that

$$\begin{aligned}
\int_{\theta=0}^{\pi} \sqrt{1 - k^2 \cos^2 \theta} \, d\theta &= 2 \int_{\theta=0}^{\pi/2} \sqrt{1 - k^2 \cos^2 \theta} \, d\theta \\
&= 2 \int_{\theta=0}^{\pi/2} \sqrt{1 - k^2 \sin^2 \theta} \, d\theta
\end{aligned}$$

$$\int_{\theta=0}^{\pi} \frac{1}{\sqrt{1-k^2\cos\theta}}\,d\theta = 2\int_{\theta=0}^{\pi/2} \frac{1}{\sqrt{1-k^2\cos\theta}}\,d\theta$$

$$= 2\int_{\theta=0}^{\pi/2} \frac{1}{\sqrt{1-k^2\sin\theta}}\,d\theta$$

Hence, (4.58) can be written as

$$M_{12} = \frac{\mu_0\sqrt{ab}}{2}\int_{\theta=0}^{\pi} \frac{k\cos 2\theta}{\sqrt{1-k^2\cos^2\theta}}\,d\theta$$

$$= \mu_0\sqrt{ab}\int_{\theta=0}^{\pi/2}\left[\left(\frac{2}{k}-k\right)\frac{1}{\sqrt{1-k^2\cos^2\theta}} - \frac{2}{k}\sqrt{1-k^2\cos^2\theta}\right]d\theta$$

$$= \mu_0\sqrt{ab}\int_{\theta=0}^{\pi/2}\left[\left(\frac{2}{k}-k\right)\frac{1}{\sqrt{1-k^2\sin^2\theta}} - \frac{2}{k}\sqrt{1-k^2\sin^2\theta}\right]d\theta$$

$$= \mu_0\sqrt{ab}\left[\left(\frac{2}{k}-k\right)K(k) - \frac{2}{k}E(k)\right] \tag{4.59}$$

where $K(k)$ and $E(k)$ are the complete elliptic integrals of the first and second kind that are tabulated by Dwight [7]:

$$K(k) = \int_{\theta=0}^{\pi/2} \frac{d\theta}{\sqrt{1-k^2\sin^2\theta}} \tag{D773.1}$$

$$E(k) = \int_{\theta=0}^{\pi/2} \sqrt{1-k^2\sin^2\theta}\,d\theta \tag{D774.1}$$

This was first obtained by Maxwell [23]. (In [23] 4π denotes μ_0, the units of length taken to be 10^7m.) Setting $d=0$ and $b=a-r_\mathrm{w}$ gives the self inductance of a loop of radius a composed of a wire of radius r_w given in (4.27).

If the separation between the two coils, d, is much larger than the radii (i.e., $d \gg a,b$), R_{12} in (4.56) approximates to

$$\frac{1}{R_{12}} = \frac{1}{\sqrt{a^2+b^2+d^2-2ab\cos\phi}}$$

$$= \left(a^2+b^2+d^2-2ab\cos\phi\right)^{-1/2}$$

$$\cong \frac{1}{d}\left(1-\frac{2ab\cos\phi}{d^2}\right)^{-1/2}$$

$$\cong \frac{1}{d}\left(1+\frac{ab\cos\phi}{d^2}\right) \qquad d \gg a,b \tag{4.56}$$

and we used the binomial theorem:

$$(1 - x)^{-1/2} \cong 1 + \frac{1}{2}x + \cdots \tag{D1}$$

Using this result, the integral in (4.57) approximates to

$$
\begin{aligned}
M_{12} &= \mathrm{N}_1^2\,\mathrm{N}_2^2\,\frac{\mu_0 ab}{4\pi}\int_{\phi'=0}^{2\pi}\int_{\phi=0}^{2\pi}\frac{\cos\phi}{\sqrt{a^2 + b^2 + d^2 - 2ab\cos\phi}}\,d\phi\,d\phi' \\
&= \mathrm{N}_1^2\,\mathrm{N}_2^2\,\frac{\mu_0 ab}{2}\int_{\phi=0}^{2\pi}\frac{\cos\phi}{\sqrt{a^2 + b^2 + d^2 - 2ab\cos\phi}}\,d\phi \\
&\cong \mathrm{N}_1^2\,\mathrm{N}_2^2\,\frac{\mu_0 ab}{2d}\int_{\phi=0}^{2\pi}\left(1 + \frac{ab\cos\phi}{d^2}\right)\cos\phi\,d\phi \\
&= \mathrm{N}_1^2\,\mathrm{N}_2^2\,\frac{\mu_0\pi\,a^2 b^2}{2d^3} \qquad d \gg a, b
\end{aligned}
\tag{4.60}
$$

and we have multiplied by the squares of the number of turns in each coil for multiturn coils.

4.4.2   Self Inductance of the Rectangular Loop

Figure 4.13 shows a rectangular loop for computing the self inductance using the Neumann integral in (4.55):

$$
\begin{aligned}
L &= \frac{\psi}{I} \\
&= \frac{\mu_0}{4\pi}\oint_{c'}\oint_c \frac{d\mathbf{l}\cdot d\mathbf{l}'}{R}
\end{aligned}
\tag{4.55}
$$

The differential element along the filament of current at the center of the wires is denoted as $d\mathbf{l}'$, and the differential element along the inside of the wire (bounding the surface of the loop) is denoted as $d\mathbf{l}$. For the left and right sides of the loop these become $d\mathbf{l}'_{\text{left}} = dz'\,\mathbf{a}_z$, $d\mathbf{l}_{\text{left}} = dz\,\mathbf{a}_z$ and $d\mathbf{l}'_{\text{right}} = -dz'\,\mathbf{a}_z$, $d\mathbf{l}_{\text{right}} = -dz\,\mathbf{a}_z$. For the top and bottom sides of the loop these become $d\mathbf{l}'_{\text{top}} = dy'\,\mathbf{a}_y$, $d\mathbf{l}_{\text{top}} = dy\,\mathbf{a}_y$ and $d\mathbf{l}'_{\text{bottom}} = -dy'\,\mathbf{a}_y$, $d\mathbf{l}_{\text{bottom}} = -dy\,\mathbf{a}_y$.

We first compute the integral along the inside surface of the left wire (at $y = r_{\text{w}}$), $d\mathbf{l}_{\text{left}}$, which is due to the currents of each of the four sides, $d\mathbf{l}'$, according to (4.55) to give the contribution of that side to the total loop inductance of the loop, L_{left}. Then repeat this for the contributions to the inductance of the loop for each of the other three sides that are due to the currents of each of the four sides according to (4.55): L_{top}, L_{right}, and L_{bottom}. The dot product $d\mathbf{l}' \cdot d\mathbf{l}$ equals $dz'dz$ along the left side and $-dz'dz$ along the right side but is zero along the top and bottom sides since $d\mathbf{l}'$ and $d\mathbf{l}$ are orthogonal to each other

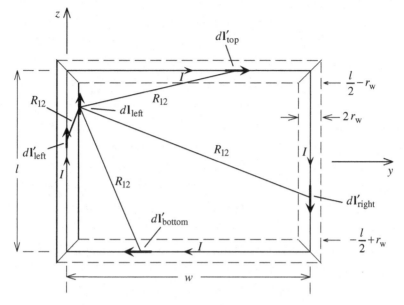

FIGURE 4.13. Neumann integral and the rectangular loop.

along those sides. So the contributions along the left side due to the currents of each of the other four sides is

$$
L_{\text{left}} = \frac{\mu_0}{4\pi} \int_{z=r_w-l/2}^{l/2-r_w} \left[\underbrace{\int_{z'=-l/2}^{l/2} \frac{1}{\sqrt{(z'-z)^2 + r_w^2}} \, dz'}_{\text{left}} \right.
$$

$$
\left. - \underbrace{\int_{z'=-l/2}^{l/2} \frac{1}{\sqrt{(z'-z)^2 + (w-r_w)^2}} \, dz'}_{\text{right}} \right] \, dz
$$

$$
= \frac{\mu_0}{4\pi} \int_{z=r_w-l/2}^{l/2-r_w} \left[\int_{\lambda=-l/2-z}^{l/2-z} \frac{1}{\sqrt{\lambda^2 + r_w^2}} \, d\lambda \right.
$$

$$
\left. - \int_{\lambda=-l/2-z}^{l/2-z} \frac{1}{\sqrt{\lambda^2 + (w-r_w)^2}} \, d\lambda \right] \, dz
$$

$$
= \frac{\mu_0}{4\pi} \int_{z=r_w-l/2}^{l/2-r_w} \left\{ \left[\ln \left(\lambda + \sqrt{\lambda^2 + r_w^2} \right) \right]_{\lambda=-l/2-z}^{l/2-z} \right.
$$

$$
\left. - \left[\ln \left(\lambda + \sqrt{\lambda^2 + (w - r_w)^2} \right) \right]_{\lambda=-l/2-z}^{l/2-z} \right\} dz
$$

$$
= \frac{\mu_0}{4\pi} \int_{z=r_w-l/2}^{l/2-r_w} \left[\ln \frac{(l/2 - z) + \sqrt{(l/2 - z)^2 + r_w^2}}{(-l/2 - z) + \sqrt{(-l/2 - z)^2 + r_w^2}} \right.
$$

$$
\left. - \ln \frac{(l/2 - z) + \sqrt{(l/2 - z)^2 + (w - r_w)^2}}{(-l/2 - z) + \sqrt{(-l/2 - z)^2 + (w - r_w)^2}} \right] dz \qquad (4.61)
$$

where we used a change of variables $\lambda = z' - z$ and integral 200.01. Using

$$
\sinh^{-1} x = \ln \left(x + \sqrt{x^2 + 1} \right)
$$

$$
= - \sinh^{-1} (-x)
$$

$$
= - \ln \left(-x + \sqrt{x^2 + 1} \right) \qquad (D700.1)
$$

(4.61) reduces to

$$
L_{\text{left}} = \frac{\mu_0}{4\pi} \int_{z=r_w-l/2}^{l/2-r_w} \left(\sinh^{-1} \frac{z + l/2}{r_w} + \sinh^{-1} \frac{l/2 - z}{r_w} \right.
$$

$$
\left. - \sinh^{-1} \frac{z + l/2}{w - r_w} - \sinh^{-1} \frac{l/2 - z}{w - r_w} \right) dz \qquad (4.62)
$$

Essentially, we have rederived the equation for A_z in (2.57) and then repeated the derivation using A_z in Section 4.3.1. The integral of (4.62) was evaluated in (4.13)–(4.15), giving

$$
L_{\text{left}} = \frac{\mu_0}{2\pi} \left[-(l - r_w) \sinh^{-1} \frac{l - r_w}{w - r_w} + \sqrt{(l - r_w)^2 + (w - r_w)^2} \right.
$$

$$
\left. + (l - r_w) \sinh^{-1} \frac{l - r_w}{r_w} - \sqrt{(l - r_w)^2 + (r_w)^2} \right.
$$

$$+ r_w \sinh^{-1} \frac{r_w}{w - r_w} - \sqrt{(r_w)^2 + (w - r_w)^2}$$

$$- r_w \underbrace{\sinh^{-1} \frac{r_w}{r_w}}_{\ln(1+\sqrt{2})} + \underbrace{\sqrt{(r_w)^2 + (r_w)^2}}_{\sqrt{2} r_w} \Bigg] \qquad (4.63)$$

Repeating this for the top, right, and left sides gives

$$L_{\text{loop}} = 2L_{\text{left}}(l, w, r_w) + 2L_{\text{top}}(w, l, r_w) \qquad (4.64)$$

where we simply interchange l and w in $L_{\text{left}}(l, w, r_w)$ to obtain $L_{\text{top}}(w, l, r_w)$. Since the result is identical to that obtained in Section 4.1.1 using **B**, the inductance of the loop is identical to that obtained in Section 4.1.1:

$$L_{\text{loop}} = 2\,L_{\text{left side}}(l, w, r_w) + 2\,L_{\text{top side}}(w, l, r_w)$$

$$= \frac{\mu_0}{\pi} \Bigg[-(l - r_w) \sinh^{-1} \frac{l - r_w}{w - r_w} - (w - r_w) \sinh^{-1} \frac{w - r_w}{l - r_w}$$

$$+ (l - r_w) \sinh^{-1} \frac{l - r_w}{r_w} + (w - r_w) \sinh^{-1} \frac{w - r_w}{r_w}$$

$$+ r_w \sinh^{-1} \frac{r_w}{w - r_w} + r_w \sinh^{-1} \frac{r_w}{l - r_w}$$

$$+ 2\sqrt{(l - r_w)^2 + (w - r_w)^2} - 2\sqrt{(w - r_w)^2 + (r_w)^2}$$

$$- 2\sqrt{(l - r_w)^2 + (r_w)^2} - 2r_w \ln\left(1 + \sqrt{2}\right) + 2\sqrt{2}\, r_w \Bigg]$$

$$(4.65)$$

The remaining results in (4.18), (4.19), and (4.20) for a square loop and for loop side lengths greater than the wire radius are obtained from (4.65) and are identical to those obtained with this method.

4.4.3 Self Inductance of the Circular Loop

Applying the Neumann integral in (4.55) to the circular loop in Fig. 4.14 yields

$$L_{\text{loop}} = \frac{\mu_0}{4\pi} \int_{\phi'=0}^{2\pi} \left[\int_{\phi=0}^{2\pi} \frac{\cos\phi}{R_{12}} (a - r_w)\, d\phi \right] a\, d\phi' \qquad (4.66a)$$

where

$$R_{12} = \sqrt{a^2 + (a - r_w)^2 - 2a(a - r_w)\cos\phi} \qquad (4.66b)$$

and the dot product in (4.55) is

$$d\mathbf{l} \cdot d\mathbf{l}' = (a - r_w)\, d\phi\, a\, d\phi'\, \cos\phi \qquad (4.66c)$$

Substituting gives

$$L_{\text{loop}} = \frac{\mu_0 a\,(a - r_w)}{4\pi} \int_{\phi'=0}^{2\pi} \int_{\phi=0}^{2\pi} \frac{\cos\phi}{\sqrt{a^2 + (a - r_w)^2 - 2a(a - r_w)\cos\phi}}\, d\phi\, d\phi'$$

$$= \frac{\mu_0 a\,(a - r_w)}{2} \int_{\phi=0}^{2\pi} \frac{\cos\phi}{\sqrt{a^2 + (a - r_w)^2 - 2a(a - r_w)\cos\phi}}\, d\phi \quad (4.67)$$

But this is identical to the result in (4.57) for the mutual inductance between two coaxial loops obtained in Section 4.4.1 if we let the two loops in Fig. 4.12 be coincident, $d = 0$, and the radius of the second loop be $b = a - r_w$. Hence, the results for that case given in (4.59) yields the self inductance of the loop in Fig. 4.14 and substitute $d = 0$ and $b = a - r_w$ in that result. But that is identical to (4.27), which, for thin wires, $r_w \ll a$, reduces to (4.28).

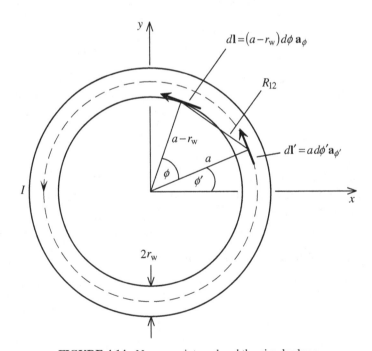

FIGURE 4.14. Neumann integral and the circular loop.

4.5   INTERNAL INDUCTANCE VS. EXTERNAL INDUCTANCE

Thus far, we have determined the *external inductance* of a current loop: that is, the inductance due to the magnetic flux that is external to the wires. In Section 2.4 we determined the magnetic flux density both external and internal to a wire of radius r_w that carries a dc current I that is uniformly distributed over the wire cross section. Those results are given in (2.33a) for $r > r_w$ and in (2.33b) for $r < r_w$. The magnetic flux internal to the wire also links a portion of the current and gives rise to an *internal inductance*. Hence, the internal inductance of the wire should be *added to* the external self-inductances of the current loops that were determined in the previous examples of this chapter as

$$L_{\text{loop}} = L_{\text{external}} + L_{\text{internal}} \qquad (4.68)$$

We next show that the internal inductance of a wire carrying a dc current I that is uniformly distributed over the wire cross section is

$$L_{\text{internal}} = \frac{\mu_0}{8\pi} \times \text{ wire length} \qquad (4.69)$$

Hence the per-unit-length internal inductance is $\mu_0/8\pi = 0.5 \times 10^{-7} \text{ H/m} = 50 \, \text{nH/m} = 1.27 \, \text{nH/in}$. Usually, this is inconsequential compared to the external inductance. For a wire formed into a circular loop of radius $r = a$ such as those shown in Figs. 4.4, 4.11, and 4.14, the total internal inductance is (approximately) $(\mu_0/8\pi)2\pi a$. For currents whose frequency is not zero (dc), the current tends to be concentrated increasingly in an annulus of a skin depth, δ, at the surface, where the skin depth is

$$\delta = \frac{1}{\sqrt{\pi f \mu_0 \sigma}}$$

where σ is the conductivity of the wire material. As $f \to \infty$, the current tends to reside on the surface of the wire and $L_{\text{internal}} \to 0$ since no internal current is linked by the field.

To determine the internal inductance of a wire, consider the cross section shown in Fig. 4.15. The magnetic flux density internal to the wire was determined in Chapter 2, using Ampère's law, to be

$$B_\phi = \frac{\mu_0}{2\pi r} I \left(\frac{\pi r^2}{\pi r_w^2} \right)$$

$$= \frac{\mu_0 I r}{2\pi r_w^2} \qquad r < r_w \qquad (2.33b)$$

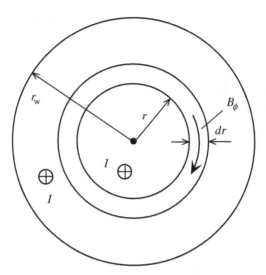

FIGURE 4.15. Determining the internal inductance of a wire.

An annulus of radius r and thickness dr has a total flux through it, for a unit length along the wire axis, of

$$d\psi = \frac{\mu_0 Ir}{2\pi r_w^2}\, dr$$

But this flux links only a portion of the total wire current of

$$I\frac{\pi r^2}{\pi r_w^2}$$

so that the flux linkages for that annulus are

$$d\psi = \frac{\mu_0 Ir}{2\pi\, r_w^2}\frac{r^2}{r_w^2}\, dr$$

$$= \frac{\mu_0 Ir^3}{2\pi r_w^4}\, dr$$

Hence, the total flux linkage per unit length of the wire is

$$\psi = \int_{r=0}^{r_w} \frac{\mu_0 Ir^3}{2\pi\, r_w^4}\, dr$$

$$= \frac{\mu_0 I}{8\pi}$$

and the per-unit-length internal inductance of the wire is

$$
\boxed{
\begin{aligned}
l_{\text{internal}} &= \frac{\psi}{I} \\
&= \frac{\mu_0}{8\pi} \quad \text{H/m}
\end{aligned}
}
$$

(4.70)

EXAMPLE

Consider the coaxial cable shown in Fig. 4.5. The inner wire has radius r_w and an internal inductance per unit length along the cable of

$$
l_{\text{internal, wire}} = \frac{\mu_0}{8\pi} \quad \text{H/m}
$$

(4.71a)

as determined above. The internal inductance of the shield which is of interior radius r_s and thickness t is determined as follows. In Chapter 2 we determined the magnetic flux density in the shield as

$$
B_\phi = \frac{\mu_0 I}{2\pi r} \frac{(r_s + t)^2 - r^2}{(r_s + t)^2 - r_s^2} \qquad r_s < r < r_s + t
$$

(2.35c)

and the dc return current in the shield, $-I$, is distributed uniformly over the cross section of the shield. Again constructing an annulus at radius r and thickness dr within the shield, the magnetic flux through the annulus per unit of cable length is

$$
d\psi = \frac{\mu_0 I}{2\pi r} \frac{(r_s + t)^2 - r^2}{(r_s + t)^2 - r_s^2} dr \qquad r_s < r < r_s + t
$$

But this links only a portion of the total cable current of

$$
I - I \frac{\pi r^2 - \pi r_s^2}{\pi \left[(r_s + t)^2 - r_s^2 \right]} = I \frac{(r_s + t)^2 - r^2}{(r_s + t)^2 - r_s^2} \quad \text{A} \qquad r_s < r < r_s + t
$$

Hence, the total flux linkages for this annulus are

$$
d\psi = \frac{\mu_0 I}{2\pi r} \left[\frac{(r_s + t)^2 - r^2}{(r_s + t)^2 - r_s^2} \right]^2 dr \qquad r_s < r < r_s + t
$$

The total per-unit-length internal inductance of the shield is therefore

$l_{\text{internal, shield}}$

$$= \frac{\psi}{I}$$

$$= \int_{r=r_s}^{r_s+t} \frac{\mu_0}{2\pi r} \left[\frac{(r_s+t)^2 - r^2}{(r_s+t)^2 - r_s^2}\right]^2 dr$$

$$= \frac{\mu_0}{2\pi \left[(r_s+t)^2 - r_s^2\right]^2} \int_{r=r_s}^{r_s+t} \left[(r_s+t)^4 \frac{1}{r} - 2r(r_s+t)^2 + r^3\right]$$

$$= \frac{\mu_0}{2\pi} \left[\frac{(r_s+t)^4 \ln[(r_s+t)/r_s] - (r_s+t)^2 \left[(r_s+t)^2 - r_s^2\right] + \frac{1}{4}\left[(r_s+t)^4 - r_s^4\right]}{\left[(r_s+t)^2 - r_s^2\right]^2}\right] \text{H/m}$$

$$(4.71\text{b})$$

To this is added the per-unit-length external inductance determined in (4.29):

$$l_{\text{external}} = \frac{\psi}{I}$$

$$= \frac{\mu_0 I}{2\pi} \ln \frac{r_s}{r_w} \qquad \text{H/m} \qquad (4.29)$$

Hence the total per-unit-length inductance of the coaxial cable is

$$l = l_{\text{internal, wire}} + l_{\text{external}} + l_{\text{internal, shield}} \qquad \text{H/m} \qquad (4.72)$$

4.6 USE OF FILAMENTARY CURRENTS AND CURRENT REDISTRIBUTION DUE TO THE PROXIMITY EFFECT

Throughout this chapter and in Chapter 2 we have assumed that the (dc) current I in a wire was *uniformly distributed* over the wire cross section. Hence, we were able to represent the wire current as a filament of current on the axis of the wire, thereby simplifying the computations.

If there are no other currents in close proximity, this will be the case. However, if another current is within a few radii of this wire, the current over the wire cross section will not be distributed uniformly but will tend to be concentrated toward the side facing the other wire. This phenomenon is called the *proximity effect*. If that is the case, the previous results for the **B** field in Chapter 2 as well as the inductances associated with the wire (both external and internal inductances) in this chapter will be only approximately correct and will become less so the closer the wires are spaced. Typically, the proximity effect does not substantially alter the results that were obtained by assuming that the current is uniformly distributed over the wire cross section if the separation of the two wires is greater than approximately four wire radii, as we will see. In other words, one wire of the same radius as the other two could be placed exactly between the two wires.

4.6.1   Two-Wire Transmission Line

To demonstrate this dependence, consider a two-wire transmission line consisting of two wires of equal radii, r_w, carrying equal but oppositely directed currents and separated by a distance (center to center) of s as illustrated in Fig. 4.16(a). The wires are considered infinitely long (or at least very long compared with their radii) so that we will not have to deal with fringing of the fields at the endpoints of finite-length wires. For the widely spaced wires

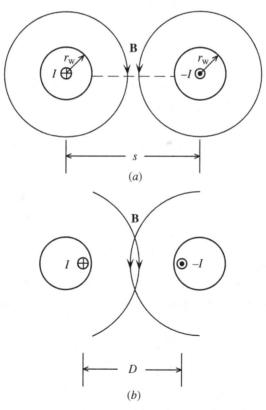

FIGURE 4.16. Proximity effect in a two-wire transmission line.

shown in Fig. 4.16(a), the current is distributed uniformly over the wire cross sections so that the current may be replaced by filaments on the wire axes. Hence, the **B** fields of each wire form circles that are centered on the centers of the respective wires. The magnetic flux density about each wire is

$$B_\phi = \frac{\mu_0 I}{2\pi r}$$

where r is measured from the center of each wire. Constructing a flat surface between the adjacent surfaces, the total flux through the surface per unit of its longitudinal length is

$$\begin{aligned} \psi &= 2 \int_{r_w}^{s-r_w} \frac{\mu_0 I}{2\pi r} dr \\ &= \frac{\mu_0 I}{\pi} \ln \frac{s-r_w}{r_w} \\ &\cong \frac{\mu_0 I}{\pi} \ln \frac{s}{r_w} \end{aligned}$$

We have made the approximation that $s-r_w \cong s$ since the wires are assumed to be widely spaced. Hence, the approximate per-unit-length inductance of the line for widely spaced wires which assumes a uniform current distribution over the wire cross sections is

$$l_{\text{approximate}} = \frac{\mu_0}{\pi} \ln \frac{s}{r_w} \qquad \text{H/m} \qquad (4.73)$$

If the wires are closely spaced as shown in Fig. 4.16(b), the currents will be concentrated toward the facing sides, and the result above for the per-unit-length inductance in (4.73) is an approximation since that relied on the currents being uniformly distributed over the wire cross sections. It can be shown (see [3,8]) that the magnetic fields are as though the total currents are concentrated as filaments but separated by a distance $D \leq s$ as shown in Fig. 4.16(b). The exact per-unit-length inductance for this result can be shown to be [3,8]

$$l_{\text{exact}} = \frac{\mu_0}{\pi} \ln \left[\frac{s}{2r_w} + \sqrt{\left(\frac{s}{2r_w}\right)^2 - 1} \right] \qquad \text{H/m} \qquad (4.74)$$

Observe that (4.74) reduces to (4.73) if $s \gg 2r_w$.

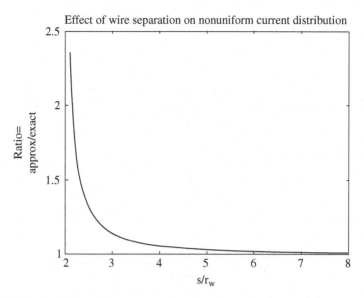

FIGURE 4.17. Ratio of the approximate and exact per-unit-length inductances of a two-wire transmission line as a function of the ratio of separation to wire radius.

Figure 4.17 shows a plot of the ratio

$$R = \frac{l_{\text{approximate}}}{l_{\text{exact}}}$$

$$= \frac{\ln(s/r_{\text{w}})}{\ln\left[(s/2r_{\text{w}}) + \sqrt{(s/2r_{\text{w}})^2 - 1}\right]}$$

for ratios of wire separation to wire radius between 2.1 and 8: $2.1 < s/r_{\text{w}} \leq 8$. (*Note:* For a ratio of $s/r_{\text{w}} = 2$, the two wires would be touching.) For a ratio of $s/r_{\text{w}} = 4$, the error is 5.3%.

4.6.2   One Wire Above a Ground Plane

The results for the two-wire transmission line can be extended rather easily to cover the case of a transmission line consisting of one wire at a height h above an infinite and perfectly conducting ground plane, as shown in Fig. 4.18(a). The basic idea is to use the method of images discussed in Section 2.7 to replace the ground plane with the image of the current as shown in Fig. 4.18(b). The image of the current above the ground plane is the same but with the current direction reversed and at a distance h below the position of the ground plane but with the ground plane removed. All the fields above the position of the ground plane remain the same in the image problem of Fig. 4.18(b). Note

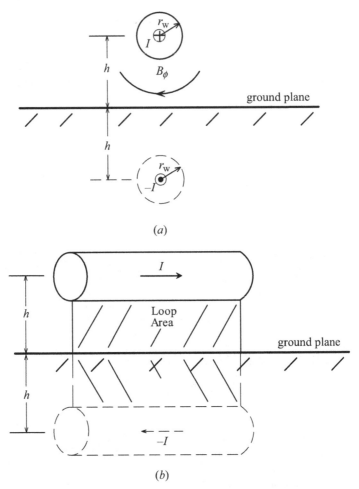

FIGURE 4.18. Transmission line consisting of one wire above and infinite and perfectly conducting ground plane.

that the method of images also applies to the case where the current is not distributed uniformly over the wire cross section.

Now we have an equivalent problem of a two-wire transmission line where the separation between the two wires is $s = 2h$. In the original problem of one wire above a ground plane, the loop area for a 1-m length of the line is between the surface of the wire and the ground plane, whereas the equivalent surface for the image problem is between the surfaces of the wire and its image. Hence, the per-unit-length inductance of the problem of one wire above a ground plane is one-half the value of the per-unit-length inductance of the two-wire but with s replaced by $s = 2h$. Therefore, the per-unit-length inductance of the transmission line consisting of one wire above a ground plane is

$$\boxed{l_{\text{approximate}} = \frac{\mu_0}{2\pi} \ln \frac{2h}{r_{\text{w}}} \qquad \text{H/m}}$$ (4.75)

and

$$\boxed{l_{\text{exact}} = \frac{\mu_0}{2\pi} \ln \left[\frac{h}{r_{\text{w}}} + \sqrt{\left(\frac{h}{r_{\text{w}}}\right)^2 - 1} \right] \qquad \text{H/m}}$$ (4.76)

Figure 4.17, demonstrating the impact of proximity effect and current redistribution, applies to this case, but the horizontal axis is h/r_{w}, which varies from 1 to 4.

The internal inductances of the wires can then be added to all these external inductances to give the total per-unit-length inductances of the lines. In the case of the two-wire line, we add $l_{\text{internal}} = 2(\mu_0/8\pi) = \mu_0/4\pi$ H/m, and in the case of one wire above a ground plane we add $l_{\text{internal}} = \mu_0/8\pi$ H/m.

4.7 ENERGY STORAGE METHOD FOR COMPUTING LOOP INDUCTANCE

In Section 3.6 we obtained the result that the magnetic energy stored in the magnetic field is

$$
\begin{aligned}
W_M &= \frac{1}{2} \int_v \mathbf{B} \cdot \mathbf{H} \, dv \\
&= \frac{\mu_0}{2} \int_v H^2 \, dv \\
&= \frac{1}{2\mu_0} \int_v B^2 \, dv
\end{aligned}
$$ (4.77)

and v is the volume of space containing the magnetic field. From a circuit analysis standpoint, the energy stored in an inductance is

$$W_M = \tfrac{1}{2} L I^2$$ (4.78)

Hence, we can determine the inductance in terms of the stored magnetic field from

$$
\begin{aligned}
L &= \frac{2\,W_M}{I^2} \\
&= \frac{\mu_0}{I^2} \int_v H^2\,dv \\
&= \frac{1}{\mu_0 I^2} \int_v B^2\,dv
\end{aligned}
\tag{4.79}
$$

4.7.1   Internal Inductance of a Wire

In Chapter 2 we determined the magnetic fields both inside and outside a wire of radius r_w that contained a current I that is distributed uniformly over the wire cross section. Hence we assumed that the current is dc and there are no other currents in close proximity to disturb this uniform distribution. Those magnetic fields at a radius r are directed circumferentially in the ϕ direction about the wire axis:

$$
B_\phi =
\begin{cases}
\dfrac{\mu_0\,I\,r}{2\pi\,r_w^2} & 0 < r < r_w \qquad\qquad (2.33b)\\[3mm]
\dfrac{\mu_0\,I}{2\pi\,r} & r_w < r \qquad\qquad\; (2.33a)
\end{cases}
$$

The per-unit-length internal inductance of the wire is obtained from (4.79) using (2.33b) by integrating throughout a cylindrical volume of unit length within the wire as

$$
\begin{aligned}
l_{\text{internal}} &= \frac{1}{\mu_0 I^2} \int_{z=0}^{1m} \int_{\phi=0}^{2\pi} \int_{r=0}^{r_w} \left(\frac{\mu_0 I r}{2\pi\, r_w^2}\right)^2 \underbrace{r\,d\phi\,dr\,dz}_{dv} \\[2mm]
&= \frac{2\pi\mu_0}{4\pi^2 r_w^4} \int_{r=0}^{r_w} r^3\,dr \\[2mm]
&= \frac{\mu_0}{2\pi\, r_w^4} \left[\frac{r^4}{4}\right]_{r=0}^{r=r_w} \\[2mm]
&= \frac{\mu_0}{8\pi} \quad \text{H/m}
\end{aligned}
\tag{4.80}
$$

which was obtained directly by the flux linkage method in Section 4.5.

4.7.2   Two-Wire Transmission Line

The per-unit-length external inductance of a two-wire transmission line consisting of two identical wires of radii r_w with center-to-center separation s as shown in Fig. 4.16(a) was obtained in Section 4.6.1. Again we assume that the wire separation is sufficiently large, $s \gg r_w$, so that the current of each wire remains distributed uniformly (or approximately so) over the wire cross section. We obtain the total per-unit-length inductance by integrating (4.79) throughout a cylindrical volume of unit length using the results for the **B** fields inside and outside the wires given in (2.33a) and (2.33b):

$$
l = 2 \frac{1}{\mu_0 I^2} \underbrace{\int_{z=0}^{1m} \int_{\phi=0}^{2\pi} \int_{r=0}^{r_w} \left(\frac{\mu_0 I r}{2\pi r_w^2} \right)^2 \underbrace{r d\phi \, dr \, dz}_{dv}}_{l_{\text{internal}}}
$$

$$
+ 2 \frac{1}{\mu_0 I^2} \underbrace{\int_{z=0}^{1m} \int_{\phi=0}^{2\pi} \int_{r=r_w}^{s-r_w} \left(\frac{\mu_0 I}{2\pi r} \right)^2 \underbrace{r d\phi \, dr \, dz}_{dv}}_{l_{\text{external}}}
$$

$$
= 2 \underbrace{\frac{2\pi \mu_0}{4\pi^2 r_w^4} \int_{r=0}^{r_w} r^3 \, dr}_{l_{\text{internal}}} + 2 \underbrace{\frac{2\pi \mu_0}{4\pi^2} \int_{r=r_w}^{s-r_w} \frac{1}{r} \, dr}_{l_{\text{external}}}
$$

$$
= 2 \underbrace{\frac{\mu_0}{2\pi r_w^4} \left[\frac{r^4}{4} \right]_{r=0}^{r_w}}_{l_{\text{internal}}} + \underbrace{\frac{\mu_0}{\pi} \left[\ln r \right]_{r=r_w}^{s-r_w}}_{l_{\text{external}}}
$$

$$
= 2 \underbrace{\frac{\mu_0}{8\pi}}_{l_{\text{internal}}} + \underbrace{\frac{\mu_0}{\pi} \ln \frac{s - r_w}{r_w}}_{l_{\text{external}}} \qquad \text{H/m} \tag{4.81}
$$

4.7.3   Coaxial Cable

Finally, we obtain the per-unit-length inductance of a coaxial cable shown in Fig. 2.19 consisting of an inner wire of radius r_w and an overall shield of inner radius r_s and thickness t. The dc current of the inner wire returns in the shield. Observe that because of symmetry there is no proximity effect regardless of the spacing of the conductors. The magnetic fields of the cable were derived in Chapter 2 and are

$$
B_\phi = \begin{cases}
\dfrac{\mu_0 \, Ir}{2\pi \, r_w^2} & r < r_w & (2.35b) \\[12pt]
\dfrac{\mu_0 \, I}{2\pi r} & r_w < r < r_s & (2.35a) \\[12pt]
\dfrac{\mu_0 \, I}{2\pi r} \dfrac{(r_s + t)^2 - r^2}{(r_s + t)^2 - r_s^2} & r_s < r < r_s + t & (2.35c)
\end{cases}
$$

Integrating (4.79) over a differential volume of unit length, $dv = r \, d\phi \, dr \, dz$, throughout the appropriate regions gives

$$
l_{\text{internal wire}} = \frac{\mu_0}{8\pi} \qquad \text{H/m} \tag{4.82a}
$$

$$
l_{\text{external}} = \frac{1}{\mu_0 \, I^2} \int_{z=0}^{1m} \int_{\phi=0}^{2\pi} \int_{r=r_w}^{r_s} \left(\frac{\mu_0 \, I}{2\pi r} \right)^2 \underbrace{r \, d\phi \, dr \, dz}_{dv}
$$

$$
= \frac{2\pi\mu_0}{4\pi^2} \int_{r=r_w}^{r_s} \frac{1}{r} \, dr = \frac{\mu_0}{2\pi} \, [\ln r]_{r=r_w}^{r_s}
$$

$$
= \frac{\mu_0}{2\pi} \ln \frac{r_s}{r_w} \qquad \text{H/m} \tag{4.82b}
$$

$$
l_{\text{internal, shield}} = \frac{1}{\mu_0 \, I^2} \int_{z=0}^{1m} \int_{\phi=0}^{2\pi} \int_{r=r_s}^{r_s+t} \left(\frac{\mu_0 I}{2\pi r} \right)^2 \left[\frac{(r_s + t)^2 - r^2}{(r_s + t)^2 - r_s^2} \right]^2 \underbrace{r \, d\phi \, dr \, dz}_{dv}
$$

$$
= \frac{2\pi \, \mu_0}{4\pi^2} \int_{r=r_s}^{r_s+t} \frac{1}{r} \left[\frac{(r_s + t)^2 - r^2}{(r_s + t)^2 - r_s^2} \right]^2 dr
$$

$$
= \frac{\mu_0}{2\pi} \left[\frac{(r_s + t)^4 \, \ln[(r_s + t)/r_s] - (r_s + t)^2 \left[(r_s + t)^2 - r_s^2 \right] + \frac{1}{4} \left[(r_s + t)^4 - r_s^4 \right]}{\left[(r_s + t)^2 - r_s^2 \right]^2} \right] \qquad \text{H/m}
$$

$$
\tag{4.82c}
$$

But the result in (4.82c) is the same result as obtained in (4.71b) by the method of flux linkages.

4.8   LOOP INDUCTANCE MATRIX FOR COUPLED CURRENT LOOPS

Figure 4.19(a) shows n current-carrying loops which are in close proximity such that their magnetic fields interact with each other so that the loops are said to be *coupled*. This structure can be characterized by self and mutual

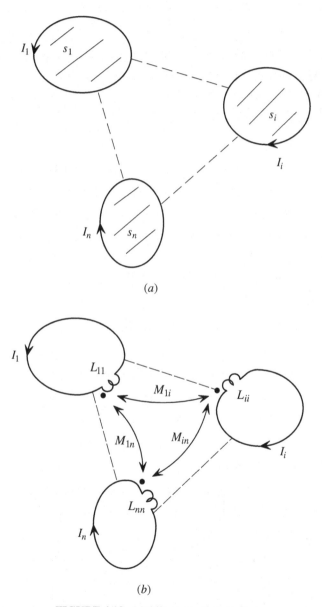

(a)

(b)

FIGURE 4.19. Multiloop coupled structure.

inductances as shown in Fig. 4.19(b). Denote the fluxes through each loop as $\psi_1, \ldots, \psi_i, \ldots, \psi_n$. These fluxes are related to the currents of each loop, $I_1, \ldots, I_i, \ldots, I_n$ as

$$
\begin{aligned}
\psi_1 &= L_{11} I_1 + \cdots + M_{1i} I_i + \cdots + M_{1n} I_n \\
&\vdots \\
\psi_n &= M_{n1} I_1 + \cdots + M_{ni} I_i + \cdots + L_{nn} I_n
\end{aligned}
\tag{4.83}
$$

and $M_{ij} = M_{ji}$. The terms L_{ii} are the *self inductances* of the current loops, and the M_{ij} are the *mutual inductances* between the current loops. These individual inductances can be obtained with the methods of this chapter by setting all but one of the currents in each equation of (4.83) to zero:

$$
L_{ii} = \left. \frac{\psi_i}{I_i} \right|_{I_1 = \cdots = I_{i-1} = I_{i+1} = \cdots = I_n = 0}
\tag{4.84a}
$$

$$
M_{ij} = \left. \frac{\psi_i}{I_j} \right|_{I_1 = \cdots = I_{j-1} = I_{j+1} = \cdots = I_n = 0}
\tag{4.84b}
$$

The equations in (4.83) can be placed in matrix form as

$$
\psi = \mathbf{L} I
\tag{4.85a}
$$

where

$$
\psi = \begin{bmatrix} \psi_1 \\ \vdots \\ \psi_i \\ \vdots \\ \psi_n \end{bmatrix}
\tag{4.85b}
$$

$$
\mathbf{L} = \begin{bmatrix}
L_{11} & \cdots & M_{1i} & \cdots & M_{1n} \\
\vdots & \ddots & \vdots & \ddots & \vdots \\
M_{i1} & \vdots & L_{ii} & \vdots & M_{in} \\
\vdots & \ddots & \vdots & \ddots & \vdots \\
M_{n1} & \cdots & M_{ni} & \cdots & L_{nn}
\end{bmatrix}
\tag{4.85c}
$$

$$I = \begin{bmatrix} I_1 \\ \vdots \\ I_i \\ \vdots \\ I_n \end{bmatrix} \tag{4.85d}$$

The $n \times n$ matrix $\mathbf{L}$ is said to be the *inductance matrix* for this coupled set of current loops. Assuming that the entire structure is electrically small at the frequencies of the currents, the Faraday law voltages induced into each loop (see Fig. 4.1) are obtained by differentiating (4.85) to give

$$\mathbf{V}(t) = \mathbf{L}\frac{d\mathbf{I}(t)}{dt} \tag{4.86a}$$

where the $n \times 1$ vector of induced voltages is

$$\mathbf{V} = \begin{bmatrix} V_1 \\ \vdots \\ V_i \\ \vdots \\ V_n \end{bmatrix} \tag{4.86b}$$

4.8.1 Dot Convention

It is important to review the *dot convention* for computing the contributions to the voltages across each inductance that are due to the mutual inductances between that loop and the other loops [1,2]. Figure 4.20 illustrates this. The dots are placed on the individual inductors in order to give the relative orientations of the loops with respect to each other. For example, consider the two coupled current-carrying loops (shown for simplicity as being rectangular) shown in Fig. 4.20. The currents around each loop, I_1 and I_2, are arbitrarily chosen to be in the clockwise direction around those loops, and the directions of the fluxes through each loop, ψ_1 and ψ_2, are arbitrarily chosen to be into the page for loop 1 and out of the page for loop 2. Using the right-hand rule, we see that

$$\psi_1 = L_{11}I_1 - M_{12}I_2 \tag{4.87a}$$

and

$$\psi_2 = -L_{22}I_2 + M_{12}I_1 \tag{4.87b}$$

(which you should verify using the right-hand rule) where $L_{11}, L_{22,}$ and $M_{12} = M_{21}$ are positive numbers.

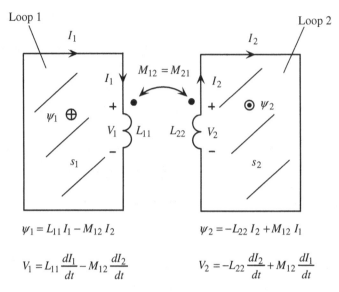

$$\psi_1 = L_{11} I_1 - M_{12} I_2 \qquad\qquad \psi_2 = -L_{22} I_2 + M_{12} I_1$$

$$V_1 = L_{11} \frac{dI_1}{dt} - M_{12} \frac{dI_2}{dt} \qquad\qquad V_2 = -L_{22} \frac{dI_2}{dt} + M_{12} \frac{dI_1}{dt}$$

FIGURE 4.20. Dot convention.

Next, we label the inductors with dots and choose the voltage polarities for the voltages, V_1 and V_2, across those inductances as shown. These voltages are the Faraday law voltage sources that are induced into each loop (see Fig. 4.1). Each current contributes to each voltage, so we initially set up the relation

$$
\begin{aligned}
V_1 &= (?)\, L_{11} I_1 + (?)\, M_{12} I_2 \\
V_2 &= (?)\, M_{12} I_1 + (?)\, L_2 I_2
\end{aligned}
\tag{4.88}
$$

and (?) denotes the *signs* that are to be determined. We determine the *signs* of each of the terms according to the following rules. For each loop we write the equation for the induced voltage in that loop as the sum of a self term and a mutual term, with the signs of each being determined by the following [1,2]:

1. The sign of the self-inductance term is positive if the current of that loop enters the assumed positive or $+$ terminal of the voltage for that loop (the passive sign convention [1,2]). For loop 1, I_1 enters the $+$ terminal of V_1, so the sign of the self inductance contribution, $L_{11} I_1$, is positive. For loop 2, I_2 enters the negative or $-$ terminal of V_2, so the sign of the self inductance contribution, $L_{22} I_2$, is negative.

2. The mutual inductance contribution to the voltage of a loop is positive *at the dotted end* of that loop inductance if the current of the *other loop* enters the dotted end of the inductance of that loop. Otherwise, it is negative. For example, the current of loop 2, I_2, enters the undotted

end of its inductance. Hence, it produces a contribution to the voltage of the first loop, $M_{12}I_2$, that is positive at the undotted end of V_1 and is therefore entered as a negative contribution to the equation for V_1. Hence, the equation for the voltage of loop 1 is

$$V_1 = L_{11}I_1 - M_{12}I_2 \tag{4.89a}$$

The current of loop 1 enters the dotted end of its inductance. Hence, it produces a contribution to the voltage of the second loop, $M_{12}I_1$, that is positive at the dotted end of V_2 and is entered as a positive contribution to the equation for V_2. Hence, the equation for the voltage of loop 2 is

$$V_2 = -L_{22}I_1 + M_{12}I_2 \tag{4.89b}$$

4.8.2 Multiconductor Transmission Lines

These concepts are very useful in constructing transmission-line equations characterizing multiconductor transmission lines (MTLs) [8]. Solving those MTL equations allows the prediction of *crosstalk*, which is the unintended coupling of a signal from one current loop into another current loop [8].

For example, consider the MTL shown in Fig. 4.21(a) consisting of $n + 1$ *parallel* conductors of infinite (or very long) length. The $(n + 1)$st conductor serves as the *return* for all the other currents. The currents of all conductors are directed to the right (in the z direction parallel to the conductor axes). The current of the $(n + 1)$st conductor is therefore $I_{n+1} = -\sum_{i=1}^{n} I_i = -(I_1 + \cdots + I_i + \cdots + I_n)$. Each of the n currents therefore forms a loop between that current and the $(n + 1)$st conductor. Hence, the fluxes of each loop, ψ_i (assumed arbitrarily to be directed into the page) can be related to the currents with a *per-unit-length inductance matrix* as

$$\boxed{\psi = \mathbf{L}\mathbf{I}} \tag{4.90a}$$

where the $n \times n$ matrix of *per-unit-length* inductances (denoted as lowercase) is

$$
\mathbf{L} =
\begin{bmatrix}
l_{11} & \cdots & m_{1i} & \cdots & m_{1n} \\
\vdots & \ddots & \vdots & \ddots & \vdots \\
m_{i1} & \vdots & l_{ii} & \vdots & m_{in} \\
\vdots & \ddots & \vdots & \ddots & \vdots \\
m_{n1} & \cdots & m_{ni} & \cdots & l_{nn}
\end{bmatrix}
\quad \text{H/m} \tag{4.90b}
$$

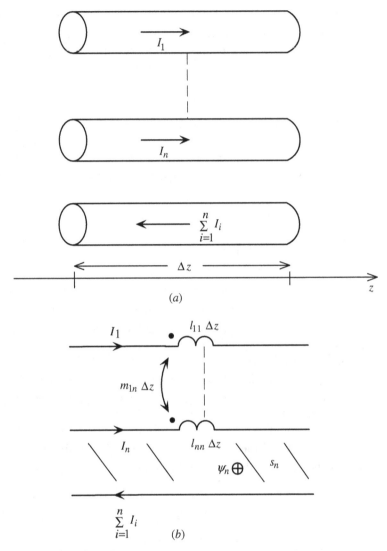

FIGURE 4.21. Multiconductor transmission line.

and $m_{ij} = m_{ji}$. The *per-unit-length* equivalent circuit of a Δz length of line is shown in Fig. 4.21(b). From this we can determine the voltages across each of the n inductors with the $+$ terminal assumed at the dotted end as

$$\mathbf{V}(t) = \mathbf{L}\,\Delta z\,\frac{d\mathbf{I}(t)}{dt} \qquad (4.91)$$

To this circuit are added the per-unit-length self and mutual capacitances between the $n + 1$ conductors from which the MTL equations are derived and whose solution can be used to predict crosstalk between the n circuits (loops) [8].

In the following subsections we determine approximate relations for the per-unit-length self and mutual inductances of MTLs that are composed of n wires. To make our calculations feasible, we assume that all wires are "widely spaced," so that the current of each wire is distributed uniformly over the wire cross section. In other words, the wires are separated sufficiently, so that the proximity effect is not pronounced. As we saw in Section 4.6, this will be a good approximation as long as the ratio of wire separation to wire radius is larger than about $4:1$. This is not an unduly restrictive assumption since it means that one wire can just be placed between two other wires of the same radii, so that wires separated by this ratio are rather "closely spaced." With this assumption of "widely spaced" wires we can replace the current of each wire with a filament on its axis containing the total current of the wire. The n wires are assumed to be infinite in length to avoid having to deal with fringing of the field at the endpoints of a finite-length line. Therefore, the magnetic flux density of each wire is in the circumferential or ϕ direction about the wire, and we obtain the familiar result for the magnetic flux density at a radius r about an infinitely long wire that has a uniform current distribution over its cross section that was obtained in Chapter 2:

$$B_\phi = \frac{\mu_0 I}{2\pi r} \tag{2.14}$$

In determining the per-unit-length inductances of the line, we determine the total magnetic flux through a surface by using superposition to give the total contribution from all wires of the line.

Using the result in (2.14), we can develop a useful wide-separation approximation for calculating the total magnetic flux through a surface. Consider the problem shown in Fig. 4.22(a) of an isolated wire where we wish to calculate the total magnetic flux through a tilted surface s whose edges are at radii R_1 and R_2 from the wire axis with $R_2 > R_1$. The total per-unit-length flux through this surface for $R_2 > R_1$ is in the direction indicated through surface s and is obtained from

$$\psi = \int_s \mathbf{B} \cdot d\mathbf{s}$$

But as shown in the figure, this is a difficult calculation because the magnetic flux density B_ϕ is not perpendicular to the surface s, so that the dot product cannot be removed from the integrand. However, consider the *closed* "wedge-shaped" surface that is 1 m in length into the page and has a side s (the original side through which the flux is desired), a side s_2 that is at a constant radius

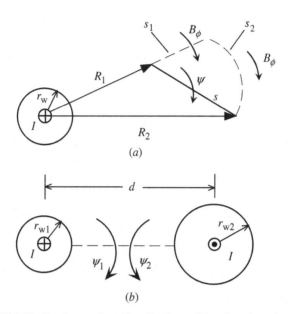

FIGURE 4.22. Fundamental problem for determining flux through a surface.

$r = R_2$ from the wire, a side s_1 that extends *radially* from $r = R_1$ to $r = R_2$, and two "end caps." Gauss's law provides that the total flux *leaving* this *closed* surface is

$$\oint \mathbf{B} \cdot d\mathbf{s} = \int_s \mathbf{B} \cdot d\mathbf{s} + \int_{s_1} \mathbf{B} \cdot d\mathbf{s} + \int_{s_2} \mathbf{B} \cdot d\mathbf{s} + \int_{\text{end caps}} \mathbf{B} \cdot d\mathbf{s}$$
$$= 0$$

From the figure we see that

$$\int_{s_2} \mathbf{B} \cdot d\mathbf{s} = 0$$

because the magnetic flux density B_ϕ is tangent to this side. Similarly, we see that

$$\int_{\text{end caps}} \mathbf{B} \cdot d\mathbf{s} = 0$$

because the magnetic flux density B_ϕ is tangent to these sides. Hence, we obtain the important result that

$$\int_s \mathbf{B} \cdot d\mathbf{s} = -\int_{s_1} \mathbf{B} \cdot d\mathbf{s}$$
$$= \int_{z=0}^{1\text{ m}} \int_{r=R_1}^{R_2} B_\phi \, dr$$

$$= \int_{z=0}^{1\,\text{m}} \int_{r=R_1}^{R_2} \frac{\mu_0 I}{2\pi r} \, dr$$

$$= \frac{\mu_0 I}{2\pi} \ln \frac{R_2}{R_1} \tag{4.92}$$

and the dot product may be removed from the integrand since B_ϕ is perpendicular to surface s_1.

In addition, when the surface is between two wires that are separated center to center by distance d, we would integrate from $r = r_w$ to $r = d - r_w \cong d$, which results from our assumption that $d \gg r_w$. This is illustrated in Fig. 4.22(b). By superposition the total flux through the flat surface between the interior edges of the two wires is the sum of the fluxes due to each current:

$$\psi = \psi_1 + \psi_2$$

$$= \frac{\mu_0 I}{2\pi} \ln \frac{d - r_{w2}}{r_{w1}} + \frac{\mu_0 I}{2\pi} \ln \frac{d - r_{w1}}{r_{w2}}$$

$$\cong \frac{\mu_0 I}{2\pi} \ln \frac{d}{r_{w1}} + \frac{\mu_0 I}{2\pi} \ln \frac{d}{r_{w2}}$$

$$= \frac{\mu_0 I}{2\pi} \ln \frac{d^2}{r_{w1} r_{w2}}$$

since we must assume that $d \gg r_{w1}, r_{w2}$ in order for the current to be uniformly distributed over the wire cross sections and for (2.14) to apply.

Lines Composed of $n + 1$ Wires   Consider the case of $n + 1$ wires of radii r_{wi} that are parallel to each other as shown in cross section in Fig. 4.23(a). The $(n + 1)$st conductor through which the other n currents "return" is denoted as the zeroth conductor. Figures 4.23(b) and (c) show the calculation of the per-unit-length self and mutual inductances of the line:

$$
\boxed{
\begin{aligned}
l_{ii} &= \frac{\psi_i}{I_i} \bigg|_{I_1 = \cdots = I_{i-1} = I_{i+1} = \cdots = I_n = 0} \\
&= \frac{\mu_0}{2\pi} \ln \frac{d_{i0}}{r_{w0}} + \frac{\mu_0}{2\pi} \ln \frac{d_{i0}}{r_{wi}} \\
&= \frac{\mu_0}{2\pi} \ln \frac{d_{i0}^2}{r_{w0} \, r_{wi}}
\end{aligned}
}
$$

$$\tag{4.93a}$$

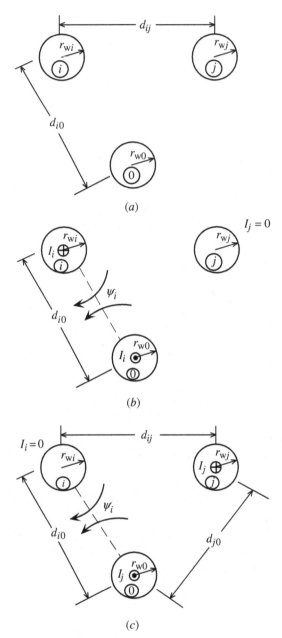

FIGURE 4.23. $(n + 1)$ wires.

and

$$
\begin{aligned}
l_{ij} = l_{ji} &= \left. \frac{\psi_i}{I_j} \right|_{I_1=\cdots=I_{j-1}=I_{j+1}=\cdots=I_n=0} \\
&= \frac{\mu_0}{2\pi} \ln \frac{d_{j0}}{d_{ij}} + \frac{\mu_0}{2\pi} \ln \frac{d_{i0}}{r_{w0}} \\
&= \frac{\mu_0}{2\pi} \ln \frac{d_{i0}d_{j0}}{d_{ij}r_{w0}}
\end{aligned}
$$

(4.93b)

Lines Composed of n Wires Above an Infinite Ground Plane   Figure 4.24 shows the case of n parallel wires of radii r_{wi} situated at heights h_i above an infinite "ground plane" which is designated as the zeroth conductor through which all the other n currents "return." Replacing the ground plane with the images of the currents according to Section 2.8 allows calculation of the

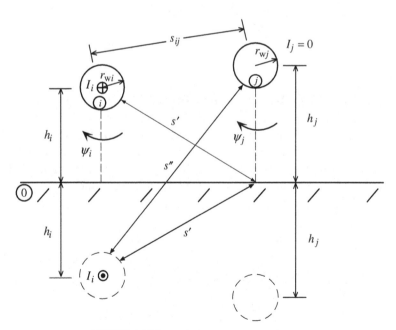

FIGURE 4.24. n wires above a ground plane.

per-unit-length self and mutual inductances as shown:

$$
\begin{aligned}
l_{ii} &= \left. \frac{\psi_i}{I_i} \right|_{I_1=\cdots=I_{i-1}=I_{i+1}=\cdots=I_n=0} \\
&= \frac{\mu_0}{2\pi} \ln \frac{h_i}{r_{wi}} + \frac{\mu_0}{2\pi} \ln \frac{2h_i}{h_i} \\
&= \frac{\mu_0}{2\pi} \ln \frac{2h_i}{r_{wi}}
\end{aligned}
$$

(4.94a)

and

$$
\begin{aligned}
l_{ij} = l_{ji} &= \left. \frac{\psi_j}{I_i} \right|_{I_1=\cdots=I_{i-1}=I_{i+1}=\cdots=I_n=0} \\
&= \frac{\mu_0}{2\pi} \ln \frac{s'}{s_{ij}} + \frac{\mu_0}{2\pi} \ln \frac{s''}{s'} \\
&= \frac{\mu_0}{2\pi} \ln \frac{\sqrt{s_{ij}^2 + 4h_i h_j}}{s_{ij}} \\
&= \frac{\mu_0}{4\pi} \ln \left[1 + \frac{4h_i h_j}{s_{ij}^2} \right]
\end{aligned}
$$

(4.94b)

Lines Composed of n Wires Within an Overall Shield   Figure 4.25(a) shows the case of n parallel wires of radii r_{wi} within an overall circular shield of interior radius r_s which is designated as the zeroth conductor through which all the other n currents "return." We can replace the shield with the wire images that are located at radii r_s^2/d_i from the axis of the shield [10]. This allows calculation of the per-unit-length self and mutual inductances as shown in Fig. 4.25(b) [8]:

$$
\begin{aligned}
l_{ii} &= \left. \frac{\psi_i}{I_i} \right|_{I_1=\cdots=I_{i-1}=I_{i+1}=\cdots=I_n=0} \\
&= \frac{\mu_0}{2\pi} \ln \frac{r_s - d_i}{r_{wi}} + \frac{\mu_0}{2\pi} \ln \frac{r_s^2/d_i - d_i}{r_s^2/d_i - r_s} \\
&= \frac{\mu_0}{2\pi} \ln \frac{r_s^2 - d_i^2}{r_s r_{wi}}
\end{aligned}
$$

(4.95a)

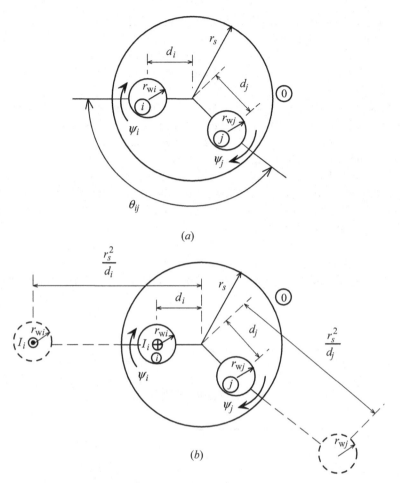

FIGURE 4.25. n wires within an overall shield.

and

$$
\begin{aligned}
l_{ij} = l_{ji} &= \left.\frac{\psi_j}{I_i}\right|_{I_1=\cdots=I_{i-1}=I_{i+1}=\cdots=I_n=0} \\
&= \frac{\mu_0}{2\pi} \ln\left(\frac{d_j}{r_s}\sqrt{\frac{(d_id_j)^2 + r_s^4 - 2d_id_jr_s^2\cos\theta_{ij}}{(d_id_j)^2 + d_j^4 - 2d_id_j^3\cos\theta_{ij}}}\right)
\end{aligned}
\tag{4.95b}
$$

4.9   LOOP INDUCTANCES OF PRINTED CIRCUIT BOARD LANDS

So far we have concentrated on conductors having circular, cylindrical cross sections (i.e., wires). Finally, we turn our attention to transmission lines that are constructed of conductors that have rectangular cross sections. These appear on printed circuit boards (PCBs) and are referred to as *lands*. The calculation of the loop inductances of structures composed of lands is considerably more difficult than for wires, and only approximate relations are generally obtained.

Figure 4.26 shows three common configurations used in constructing PCBs. Figure 4.26(a) shows the *stripline* that appears in PCBs that contain innerplanes. Innerplanes are layers of conductors sandwiched at various levels within the board substrate, which is glass epoxy, a dielectric with $\varepsilon_r \cong 4.7$ and $\mu_r = 1$. Hence the board substrate is not ferromagnetic and does not affect the magnetic fields (but does affect the electric fields). A land of thickness t is situated between two "ground planes." The thickness of the land is commonly that of 1-oz copper, which is $t = 1.4$ mils $= 0.036$ mm (1 mil $= 0.001$ in). However, in the following results it is assumed that $t = 0$. The separation between the two surrounding ground planes is denoted as s, and the land is situated midway between the two ground planes (as is common). The

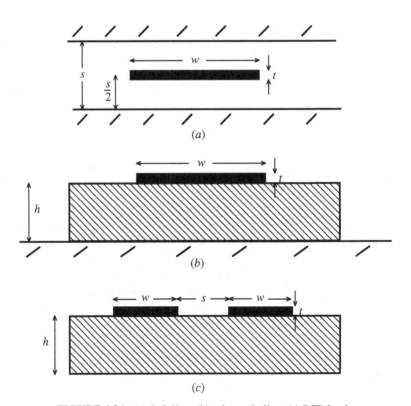

FIGURE 4.26. (a) Stripline; (b) microstrip line; (c) PCB lands.

per-unit-length loop inductance of the stripline is [8]

$$
l = \begin{cases} \dfrac{30}{v_0} \ln \left[2\dfrac{1+\sqrt{k}}{1-\sqrt{k}} \right] & \dfrac{1}{\sqrt{2}} \leq k \leq 1 \\[3ex] \dfrac{30\pi^2}{v_0 \ln \left[2\left(1+\sqrt{k'}\right) / \left(1-\sqrt{k'}\right) \right]} & 0 \leq k \leq \dfrac{1}{\sqrt{2}} \end{cases} \qquad \text{H/m}
$$

$$(4.96a)$$

where

$$
k = \frac{1}{\cosh\left(\pi w / 2s\right)} \tag{4.96b}
$$

and $k' = \sqrt{1 - k^2}$. This can be approximated as [8]

$$
l \cong \frac{30\pi}{v_0} \frac{1}{w_e/s + 0.441} \qquad \text{H/m} \tag{4.96c}
$$

and the effective width of the conductor is

$$
\frac{w_e}{s} = \begin{cases} \dfrac{w}{s} & \dfrac{w}{s} \geq 0.35 \\[3ex] \dfrac{w}{s} - \left(0.35 - \dfrac{w}{s}\right)^2 & \dfrac{w}{s} \leq 0.35 \end{cases} \tag{4.96d}
$$

The speed of light is denoted as $v_0 \cong 3 \times 10^8$ m/s.

The *microstrip line* shown in Fig. 4.26(b) is typical of the outer layers of a PCB that has innerplanes. A land of thickness t lies on top of a dielectric substrate of thickness h, and a ground plane (representing an adjacent inner-plane) is below the substrate. Assuming that the thickness of the land is zero, $t = 0$, approximate relations for the per-unit-length loop inductance are [8]

$$
l = \begin{cases} \dfrac{60}{v_0} \ln \left(\dfrac{8h}{w} + \dfrac{w}{4h} \right) & \dfrac{w}{h} \leq 1 \\[3ex] \dfrac{120\pi}{v_0} \left[\dfrac{w}{h} + 1.393 + 0.667 \ln \left(\dfrac{w}{h} + 1.444 \right) \right]^{-1} & \dfrac{w}{h} \geq 1 \end{cases} \qquad \text{H/m}
$$

$$(4.97)$$

Finally, the case of two lands on the surface of a PCB is shown in Fig. 4.26(c). The approximate per-unit-length loop inductance is [8]

$$
l = \begin{cases} \dfrac{120}{v_0} \ln \left(2 \dfrac{1 + \sqrt{k}}{1 - \sqrt{k}} \right) & \dfrac{1}{\sqrt{2}} \le k \le 1 \\[2em] \dfrac{120\pi^2}{v_0 \ln \left[2(1 + \sqrt{k'})/(1 - \sqrt{k'}) \right]} & 0 \le k \le \dfrac{1}{\sqrt{2}} \end{cases} \quad \text{H/m}
$$

(4.98a)

where

$$
k = \frac{s}{s + 2w} \tag{4.98b}
$$

and

$$
k' = \sqrt{1 - k^2} \tag{4.98c}
$$

4.10 SUMMARY OF METHODS FOR COMPUTING LOOP INDUCTANCE

There are several methods for computing the **B** field of currents: the Biot–Savart law, Ampère's law, the vector magnetic potential **A**, and the method of images for problems with ground planes. There are also several methods for computing the loop inductance of a closed current loop, but all these methods fundamentally require computation of the flux through the open surface s that is enclosed by the closed current loop:

$$
\psi = \int_s \mathbf{B} \cdot d\mathbf{s} \tag{4.99a}
$$

The loop inductance is computed from this result as

$$
L = \frac{\psi}{I} \tag{4.99b}
$$

We investigated several methods in this chapter for calculating L either directly or indirectly. The direct method is to use the Biot-Savart law:

$$
\mathbf{B} = \frac{\mu_0 I}{4\pi} \int_l \frac{d\mathbf{l} \times \mathbf{a}_R}{R^2} \tag{4.100a}
$$

or Ampère's law for problems with symmetry:

$$\oint_c \mathbf{B} \cdot d\mathbf{l} = \mu_0 I \tag{4.100b}$$

to compute **B** over the surface enclosed by the current loop, and then to compute the inductance of the loop via (4.99a) and (4.99b).

The next method is to compute the vector magnetic potential **A** directly from

$$\mathbf{A} = \frac{\mu_0 I}{4\pi} \int_l \frac{1}{R} \, d\mathbf{l} \tag{4.101a}$$

Substituting

$$\mathbf{B} = \nabla \times \mathbf{A} \tag{4.101b}$$

into (4.99) yields

$$L = \frac{\oint_c \mathbf{A} \cdot d\mathbf{l}}{I} \tag{4.101c}$$

where c is the contour around the open surface that the current loop surrounds.

The third method is via the Neumann integral. Substituting (4.101a) into (4.101c) yields

$$L = \frac{\mu_0}{4\pi} \oint_c \oint_{c'} \frac{d\mathbf{l} \cdot d\mathbf{l'}}{R_{12}} \tag{4.102}$$

A fourth indirect method is by computing the energy stored in the magnetic field:

$$W_M = \tfrac{1}{2} L I^2$$
$$= \tfrac{1}{2} \int_v \mathbf{B} \cdot \mathbf{H} \, dv \tag{4.103a}$$

giving

$$L = \frac{1}{\mu_0 I^2} \int_v B^2 \, dv \tag{4.103b}$$

Generally, this energy method works best for *closed structures* where the magnetic field is contained within a finite region of space as with a coaxial cable or in computing the internal inductance of a wire.

For some structures, such as the circular current loop, the vector magnetic potential method in (4.101) and the Neumann integral in (4.102) are easiest, whereas for rectangular current loops, the direct method of computing the magnetic flux through the loop in (4.99) is somewhat simpler. All of these methods can be applied in a similar fashion to the computation of the mutual inductance between two closed current loops:

$$M_{12} = \frac{\psi_2}{I_1} \qquad (4.104)$$

4.10.1 Mutual Inductance Between Two Rectangular Loops

We finally illustrate all three methods by computing the mutual inductance between two rectangular loops that lie in the same plane and whose sides are either parallel or perpendicular as shown in Fig. 4.27. We consider the loops to be composed of wires which can be approximated by filaments on the axes of those wires on the assumption that the currents of the wires are

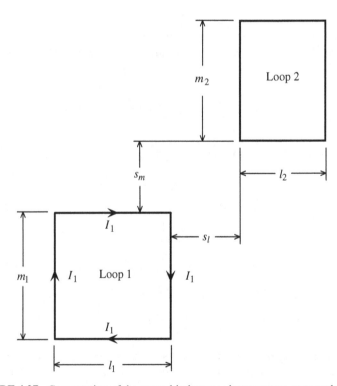

FIGURE 4.27. Computation of the mutual inductance between two rectangular loops.

uniformly distributed over their cross sections (i.e., all of the parallel wires are "widely spaced"). Loop 1 carries a current I_1 that circulates about that loop in the clockwise direction. We first compute the mutual inductance between the two loops from the fundamental definition of mutual inductance in (4.104) by computing the total magnetic flux penetrating the surface enclosed by loop 2:

$$\psi_2 = \int_{s_2} \mathbf{B} \cdot d\mathbf{s} \tag{4.105}$$

where s_2 is the open surface enclosed by loop 2. Using the Biot–Savart law, we see that the magnetic flux density $\mathbf{B}$ is perpendicular to the surface of loop 2, and hence the dot product in (4.105) can be removed:

$$\psi_2 = \int_{s_2} B \, ds \tag{4.106}$$

Note that the total magnetic flux through loop 2 can be obtained as the super-position of the fluxes through that loop, due to each of the four currents of the four sides comprising loop 1. Hence, we essentially need first to solve the fundamental problem shown in Fig. 4.28 of determining the total magnetic flux through a rectangular loop due to a current filament of length L and then use that fundamental result and the right-hand rule to superimpose the fluxes due to the currents of the four sides of loop 1. The magnetic flux density of a line current was determined in (2.15). This was for the origin located at the

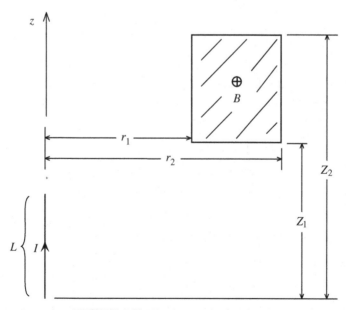

FIGURE 4.28. Fundamental subproblem.

midpoint of the current. We modify that result for the origin at the lower end of the current filament, giving

$$B = \frac{\mu_0 I}{4\pi r} \left[\frac{Z}{\sqrt{Z^2 + r^2}} - \frac{Z - L}{\sqrt{(Z - L)^2 + r^2}} \right] \qquad (4.107)$$

Hence, the magnetic flux through the loop due to this current filament is

$$\psi = \frac{\mu_0 I}{4\pi} \int_{r=r_1}^{r_2} \int_{Z=Z_1}^{Z_2} \frac{1}{r} \left[\frac{Z}{\sqrt{Z^2 + r^2}} - \frac{Z - L}{\sqrt{(Z - L)^2 + r^2}} \right] dZ \, dr \qquad (4.108)$$

The inner integral with respect to Z is evaluated as

$$(\mathrm{I}) = \int_{Z=Z_1}^{Z_2} \left[\frac{Z}{\sqrt{Z^2 + r^2}} - \frac{Z - L}{\sqrt{(Z - L)^2 + r^2}} \right] dZ$$

Using an integral from Dwight [7],

$$\int \frac{x}{\sqrt{x^2 + a^2}} \, dx = \sqrt{x^2 + a^2} \qquad (\mathrm{D}201.01)$$

gives

$$(\mathrm{I}) = \sqrt{Z_2^2 + r^2} - \sqrt{Z_1^2 + r^2} - \sqrt{(Z_2 - L)^2 + r^2} + \sqrt{(Z_1 - L)^2 + r^2}$$

where we have used a change of variables $\lambda = Z - L, \, d\lambda = dZ$ in the second part of the integral. The second integral with respect to r is

$$(\mathrm{II}) = \int_{r=r_1}^{r_2} \left[\frac{\sqrt{Z_2^2 + r^2}}{r} - \frac{\sqrt{Z_1^2 + r^2}}{r} \right.$$

$$\left. - \frac{\sqrt{(Z_2 - L)^2 + r^2}}{r} + \frac{\sqrt{(Z_1 - L)^2 + r^2}}{r} \right] dr$$

This can be evaluated using an integral from Dwight [7]:

$$\int \frac{\sqrt{x^2 + a^2}}{x} \, dx = \sqrt{x^2 + a^2} - a \ln \frac{a + \sqrt{x^2 + a^2}}{x}$$

$$= \sqrt{x^2 + a^2} - a \ln \left(a + \sqrt{x^2 + a^2} \right) + a \ln x \quad (\mathrm{D}241.01)$$

giving the flux through the loop as

$$
\begin{aligned}
\psi = \frac{\mu_0 I}{4\pi} \Bigg[&\sqrt{Z_2^2 + r_2^2} - Z_2 \ln\left(Z_2 + \sqrt{Z_2^2 + r_2^2} \right) - \sqrt{Z_2^2 + r_1^2} \\
&+ Z_2 \ln\left(Z_2 + \sqrt{Z_2^2 + r_1^2} \right) - \sqrt{Z_1^2 + r_2^2} + Z_1 \ln\left(Z_1 + \sqrt{Z_1^2 + r_2^2} \right) \\
&+ \sqrt{Z_1^2 + r_1^2} - Z_1 \ln\left(Z_1 + \sqrt{Z_1^2 + r_1^2} \right) - \sqrt{(Z_2 - L)^2 + r_2^2} \\
&+ (Z_2 - L) \ln\left((Z_2 - L) + \sqrt{(Z_2 - L)^2 + r_2^2} \right) + \sqrt{(Z_2 - L)^2 + r_1^2} \\
&- (Z_2 - L) \ln\left((Z_2 - L) + \sqrt{(Z_2 - L)^2 + r_1^2} \right) + \sqrt{(Z_1 - L)^2 + r_2^2} \\
&- (Z_1 - L) \ln\left((Z_1 - L) + \sqrt{(Z_1 - L)^2 + r_2^2} \right) - \sqrt{(Z_1 - L)^2 + r_1^2} \\
&+ (Z_1 - L) \ln\left((Z_1 - L) + \sqrt{(Z_1 - L)^2 + r_1^2} \right) \Bigg]
\end{aligned}
\tag{4.109a}
$$

This can be written more compactly as

$$
\psi = \frac{\mu_0 I}{4\pi} K\left(Z_1, Z_2, r_1, r_2, L\right)
\tag{4.109b}
$$

where

$$
K\left(Z_1, Z_2, r_1, r_2, L\right) = \sum_{i=1}^{2} \sum_{j=1}^{2} (-1)^{i+j} \left[f\left(Z_i, r_j, 0\right) - f\left(Z_i, r_j, L\right) \right]
$$

$$
\tag{4.109c}
$$

and

$$
f\left(Z, r, L\right) = \sqrt{(Z - L)^2 + r^2} - (Z - L) \ln\left[(Z - L) + \sqrt{((Z - L)^2 + r^2} \right]
$$

$$
\tag{4.109d}
$$

Hence, by superimposing the magnetic fluxes through loop 2 in Fig. 4.27 due to each of the four sides of loop 1 (using the right-hand rule and matching each case to Fig. 4.28), we obtain the mutual inductance between the two rectangular loops in Fig. 4.27 as

$$
\begin{aligned}
M_{12} = \frac{\mu_0}{4\pi} \big[\, & K\,(m_1 + s_m, m_1 + s_m + m_2, l_1 + s_l, l_1 + s_l + l_2, m_1) \\
& -K\,(m_1 + s_m, m_1 + s_m + m_2, s_l, s_l + l_2, m_1) \\
& -K\,(l_1 + s_l, l_1 + s_l + l_2, s_m, s_m + m_2, l_1) \\
& +K\,(l_1 + s_l, l_1 + s_l + l_2, m_1 + s_m, m_1 + s_m + m_2, l_1)\big]
\end{aligned}
$$

$$(4.110)$$

Note that the s_m and s_l may be negative. If both loops are *identical* and *square* (i.e., $l_1 = l_2 = m_1 = m_2 = L$), this result simplifies to

$$
\begin{aligned}
M_{12} = \frac{\mu_0}{4\pi} \big[\, & K\,(L + s_m, 2L + s_m, L + s_l, 2L + s_l, L) \\
& -K\,(L + s_m, 2L + s_m, s_l, L + s_l, L) \\
& -K\,(L + s_l, 2L + s_l, s_m, L + s_m, L) \\
& +K\,(L + s_l, 2L + s_l, L + s_m, 2L + s_m, L)\big]
\end{aligned}
\qquad (4.111)
$$

Next, we obtain the mutual inductance by determining the total magnetic flux through the second loop using the vector magnetic potential, $\mathbf{A}$, as

$$
\begin{aligned}
\psi_2 &= \int_{S_2} \mathbf{B} \cdot ds \\
&= \oint_{c_2} \mathbf{A} \cdot dl
\end{aligned}
\qquad (4.112)
$$

This requires that we solve the fundamental subproblem shown in Fig. 4.29. Once this is done, we superimpose the result around the four sides of loop 2 from each of the four currents in loop 1. Since the current I and the left segment of loop 2 are parallel, the vector magnetic potential $\mathbf{A}$ is tangent to the conductor and the dot product in (4.112) can be removed. The vector magnetic potential for the case in Fig. 4.29 was derived in equation (2.57). That was derived for the origin at the midpoint of the current. Rederiving that for the origin at the bottom of the current as in Fig. 4.29 gives

$$
A = \frac{\mu_0 I}{4\pi} \left[\ln\left(Z + \sqrt{Z^2 + r^2}\right) - \ln\left((Z - L) + \sqrt{(Z - L)^2 + r^2}\right) \right]
$$

$$(4.113)$$

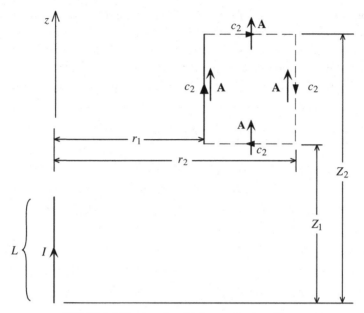

FIGURE 4.29. Another fundamental subproblem.

Integrating this along the left side of the second loop gives

$$
\int_{Z=Z_1}^{Z_2} A\,dZ = \frac{\mu_0 I}{4\pi} \int_{Z=Z_1}^{Z_2} \left[\ln\left(Z + \sqrt{Z^2 + r_1^2} \right) \right.
$$

$$
\left. - \ln\left((Z-L) + \sqrt{(Z-L)^2 + r_1^2} \right) \right] dZ
$$

$$
= \frac{\mu_0 I}{4\pi} \left[Z_2 \ln\left(Z_2 + \sqrt{Z_2^2 + r_1^2} \right) - \sqrt{Z_2^2 + r_1^2} \right.
$$

$$
- Z_1 \ln\left(Z_1 + \sqrt{Z_1^2 + r_1^2} \right) + \sqrt{Z_1^2 + r_1^2}
$$

$$
- (Z_2 - L) \ln\left((Z_2 - L) + \sqrt{(Z_2 - L)^2 + r_1^2} \right) + \sqrt{(Z_2 - L)^2 + r_1^2}
$$

$$
\left. + (Z_1 - L) \ln\left((Z_1 - L) + \sqrt{(Z_1 - L)^2 + r_1^2} \right) - \sqrt{(Z_1 - L)^2 + r_1^2} \right]
$$

$$(4.114)$$

where we have used an integral from Dwight [7]:

$$
\int \ln\left(x + \sqrt{x^2 + a^2} \right) dx = x \ln\left(x + \sqrt{x^2 + a^2} \right) - \sqrt{x^2 + a^2}
$$

$$(D625)$$

and have made a change of variables in the second half of the integral of $\lambda = Z - L, d\lambda = dZ$. Realizing that the vector magnetic potential is perpendicular to the top and bottom sides of the loop in Fig. 4.29 and contribute nothing to the line integral around the loop, the contribution to integration around loop 2 in (4.112) due to the current in the left side of loop 1 as in Fig. 4.28 is obtained as

$$
\begin{aligned}
\psi = \frac{\mu_0 I}{4\pi} \Bigg[& Z_2 \ln\left(Z_2 + \sqrt{Z_2^2 + r_1^2}\right) - \sqrt{Z_2^2 + r_1^2} \\
& -Z_1 \ln\left(Z_1 + \sqrt{Z_1^2 + r_1^2}\right) + \sqrt{Z_1^2 + r_1^2} \\
& -(Z_2 - L)\ln\left((Z_2 - L) + \sqrt{(Z_2 - L)^2 + r_1^2}\right) + \sqrt{(Z_2 - L)^2 + r_1^2} \\
& +(Z_1 - L)\ln\left((Z_1 - L) + \sqrt{(Z_1 - L)^2 + r_1^2}\right) - \sqrt{(Z_1 - L)^2 + r_1^2} \\
& -Z_2 \ln\left(Z_2 + \sqrt{Z_2^2 + r_2^2}\right) + \sqrt{Z_2^2 + r_2^2} \\
& +Z_1 \ln\left(Z_1 + \sqrt{Z_1^2 + r_2^2}\right) - \sqrt{Z_1^2 + r_2^2} \\
& +(Z_2 - L)\ln\left((Z_2 - L) + \sqrt{(Z_2 - L)^2 + r_2^2}\right) - \sqrt{(Z_2 - L)^2 + r_2^2} \\
& -(Z_1 - L)\ln\left((Z_1 - L) + \sqrt{(Z_1 - L)^2 + r_2^2}\right) + \sqrt{(Z_1 - L)^2 + r_2^2} \Bigg]
\end{aligned}
$$

(4.115)

giving the same result as in (4.109). Using this result and superimposing the fluxes through loop 2 due to the the top, right and bottom currents of loop 1 gives the same result as the previous direct computation of the flux through loop 2 and given in (4.110).

Using the Neumann integral we compute the mutual inductance between loops 1 and 2 directly from

$$
M_{12} = \frac{\mu_0}{4\pi} \oint_{c_2} \oint_{c_1} \frac{d\mathbf{l} \cdot d\mathbf{l}_2}{R_{12}}
\tag{4.116}
$$

where c_1 and c_2 are the contours of loops 1 and 2, respectively, and R_{12} is the distance between a point on loop 1 and a point on loop 2. This was derived by substituting the expression for the vector magnetic potential produced along the contour of loop 2 by the current of loop 1:

$$
\mathbf{A}_{12} = \frac{\mu_0 I_1}{4\pi} \oint_{c_1} \frac{1}{R_{12}} d\mathbf{l}_1
\tag{4.117a}
$$

into the basic expression for the magnetic flux through loop 2:

$$\psi_{12} = \oint_{s_2} \mathbf{B}_{12} \cdot d\mathbf{s}$$

$$= \oint_{c_2} \mathbf{A}_{12} \cdot d\mathbf{l} \tag{4.117b}$$

The mutual inductance is obtained by dividing the flux by the current I_1 according to the basic definition in (4.104). Once again we need to solve a basic subproblem shown in Fig. 4.30. This represents the contribution to the Neumann integral along the left side of loop 1 and the left side of loop 2. The portion of the Neumann integral due to the left side of loop 1 and the left side of loop 2 represented in Fig. 4.30 beomes

$$\text{Int}_{\text{left-left}} = \frac{\mu_0}{4\pi} \int_{c_2} \int_{c_1} \frac{1}{R_{12}} dl_1 \, dl_2$$

$$= \frac{\mu_0}{4\pi} \int_{z_2=Z_1}^{Z_2} \int_{z_1=0}^{L} \frac{1}{\sqrt{(z_2 - z_1)^2 + r_1^2}} dz_1 \, dz_2 \tag{4.118}$$

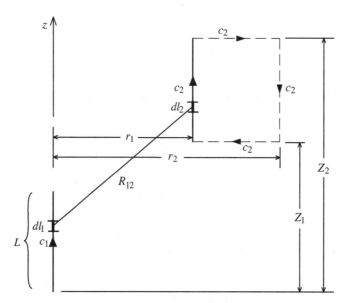

FIGURE 4.30. Basic subproblem for the Neumann integral.

The inner integral with respect to z_1 is integrated as

$$(\mathrm{I}) = \int_{z_1=0}^{L} \frac{1}{\sqrt{(z_2 - z_1)^2 + r_1^2}} dz_1$$

$$= \int_{\lambda=z_2-L}^{z_2} \frac{1}{\sqrt{\lambda^2 + r_1^2}} d\lambda$$

$$= \ln\left(z_2 + \sqrt{z_2^2 + r_1^2}\right) - \ln\left[(z_2 - L) + \sqrt{(z_2 - L)^2 + r_1^2}\right]$$

and we have used an integral from Dwight [7]:

$$\int \frac{dx}{\sqrt{x^2 + a^2}} = \ln\left(x + \sqrt{x^2 + a^2}\right) \qquad \text{(D200.01)}$$

and a change of variables $\lambda = z_2 - z_1$, $d\lambda = -dz_1$. Integrating with respect to z_2 gives

$$\mathrm{Int}_{\text{left-left}} = \frac{\mu_0}{4\pi} \int_{z_2=Z_1}^{Z_2} \left[\ln\left(z_2 + \sqrt{z_2^2 + r_1^2}\right) \right.$$

$$\left. - \ln\left((z_2 - L) + \sqrt{(z_2 - L)^2 + r_1^2}\right) \right] dz_2$$

Using a change of variables $\lambda = z_2 - L$, $d\lambda = dz_2$ gives

$$\mathrm{Int}_{\text{left-left}} = \frac{\mu_0}{4\pi} \int_{z_2=Z_1}^{Z_2} \left[\ln\left(z_2 + \sqrt{z_2^2 + r_1^2}\right) \right] dz_2$$

$$- \frac{\mu_0}{4\pi} \int_{\lambda=Z_1-L}^{Z_2-L} \left[\ln\left(\lambda + \sqrt{\lambda^2 + r_1^2}\right) \right] d\lambda$$

Integrating this using an integral from Dwight [7],

$$\int \ln\left(x + \sqrt{x^2 + a^2}\right) dx = x \ln\left(x + \sqrt{x^2 + a^2}\right) - \sqrt{x^2 + a^2} \qquad \text{(D625)}$$

yields

$$
\begin{aligned}
\text{Int}_{\text{left-left}} = \frac{\mu_0}{4\pi} \Bigg[&Z_2 \ln\left(Z_2 + \sqrt{Z_2^2 + r_1^2}\right) - \sqrt{Z_2^2 + r_1^2} \\
&- Z_1 \ln\left(Z_1 + \sqrt{Z_1^2 + r_1^2}\right) + \sqrt{Z_1^2 + r_1^2} \\
&- (Z_2 - L)\ln\left((Z_2 - L) + \sqrt{(Z_2 - L)^2 + r_1^2}\right) \\
&+ \sqrt{(Z_2 - L)^2 + r_1^2} \\
&+ (Z_1 - L)\ln\left((Z_1 - L) + \sqrt{(Z_1 - L)^2 + r_1^2}\right) \\
&- \sqrt{(Z_1 - L)^2 + r_1^2} \Bigg]
\end{aligned}
$$

The contribution from the left side of loop 1 to the right side of loop 2 is the same but negated and with r_1 replaced with r_2. Realizing that $d\mathbf{l}_1 \cdot d\mathbf{l}_2 = 0$ along the top and bottom sides of loop 2 gives the total contribution to the Neumann integral due to the left side of loop 1 as

$$
\begin{aligned}
\text{Int}_{\text{left}} = \frac{\mu_0}{4\pi} \Bigg[&Z_2 \ln\left(Z_2 + \sqrt{Z_2^2 + r_1^2}\right) - \sqrt{Z_2^2 + r_1^2} \\
&- Z_1 \ln\left(Z_1 + \sqrt{Z_1^2 + r_1^2}\right) + \sqrt{Z_1^2 + r_1^2} \\
&- (Z_2 - L)\ln\left((Z_2 - L) + \sqrt{(Z_2 - L)^2 + r_1^2}\right) + \sqrt{(Z_2 - L)^2 + r_1^2} \\
&+ (Z_1 - L)\ln\left((Z_2 - L) + \sqrt{(Z_1 - L)^2 + r_1^2}\right) - \sqrt{(Z_1 - L)^2 + r_1^2} \\
&- Z_2 \ln\left(Z_2 + \sqrt{Z_2^2 + r_2^2}\right) + \sqrt{Z_1^2 + r_2^2} \\
&+ Z_1 \ln\left(Z_1 + \sqrt{Z_1^2 + r_2^2}\right) - \sqrt{Z_1^2 + r_2^2} \\
&+ (Z_2 - L)\ln\left((Z_2 - L) + \sqrt{(Z_2 - L)^2 + r_2^2}\right) - \sqrt{(Z_2 - L)^2 + r_2^2} \\
&- (Z_1 - L)\ln\left((Z_1 - L) + \sqrt{(Z_1 - L)^2 + r_2^2}\right) + \sqrt{(Z_1 - L)^2 + r_2^2} \Bigg]
\end{aligned}
$$

$$\tag{4.119}$$

which is the same as (4.115) with the current I removed from that expression. Using this result and superimposing the contributions to the Neumann integral around loop 2 due to the top, right, and bottom segments of loop 1 gives the same result as the previous direct computation of the flux through loop 2 and given in (4.110).

5

THE CONCEPT OF "PARTIAL" INDUCTANCE

In the preceding chapters we discussed the meaning and calculation of the "loop" inductance of various conducting structures that support a closed loop of current. This "loop" inductance is calculated fundamentally for steady (dc) currents which we showed in Section 2.9 must form closed loops. If we open the loop at a point with a small gap, the loop inductance of that current loop is seen as an inductance L at these input terminals. When we pass a time-varying current around the loop via these terminals a voltage, $V(t) = L dI(t)/dt$, is developed across the terminals. This voltage is essentially the Faraday's law voltage induced into the loop. For electrically small loop dimensions, this lumped inductance and the voltage across its terminals can be represented as a lumped voltage source and placed *anywhere* in the loop perimeter (see Fig. 4.1). It is important, however, to remember that *neither this lumped inductance nor the equivalent voltage source it represents can be placed in a unique position in the loop*! This loop inductance is a property of the entire loop and its use is valid only at the input terminals of the loop. Hence, it is not possible to associate the loop inductance with any particular segment of the loop.

However, there are numerous situations, some of which were described in Chapter 1, where it is useful to develop a lumped-circuit model of a closed current loop wherein the segments of the perimeter of the loop are represented

Inductance: Loop and Partial, By Clayton R. Paul
Copyright © 2010 John Wiley & Sons, Inc.

with a self inductance as well as mutual inductances between that segment and other segments of this and other *adjacent current loops*. The concept of "partial" inductance allows us to do that in a unique way.

It has been said that "you cannot ascribe the properties of inductance to an isolated piece of wire." Of course you can't because an isolated piece of wire is not capable of supporting a dc current, which must form a closed loop (i.e., it must return to its source). This is therefore a misleading statement. The proper question is: Can you ascribe the properties of inductance *uniquely* to a segment of a *closed loop of current*? The answer to this question is yes, and the method for doing so is with "partial" inductances.

There are three significant references regarding partial inductance. Those by Grover [14] and Ruehli [15] are excellent general references, and the paper by Hoer and Love [16] gives results for the partial inductances of conductors of rectangular cross section [e.g., printed circuit board (PCB) lands].

5.1   GENERAL MEANING OF PARTIAL INDUCTANCE

Consider a closed physical loop constructed of a conductor such as a wire, PCB land, and so on, that supports a dc current I. The "loop" inductance of this current loop is defined fundamentally in previous chapters as

$$L = \frac{\psi}{I} \tag{5.1a}$$

where

$$\psi = \int_s \mathbf{B} \cdot d\mathbf{s} \tag{5.1b}$$

is the total magnetic flux that penetrates the open surface s that is surrounded by the closed contour of the loop, c, and $\mathbf{B}$ is the magnetic flux density (caused by current I) through the surface s. In Chapter 2 we calculated $\mathbf{B}$ for various configurations of loop shapes. In Chapter 4 we calculated the flux ψ and hence the inductance according to (5.1) for various loop shapes. Faraday's fundamental law of induction (Chapter 3) gives the induced voltage appearing at the terminals of the loop as

$$
\begin{aligned}
V &= \frac{d\psi}{dt} \\
&= L\frac{dI}{dt}
\end{aligned}
\tag{5.2}
$$

where the current I is now allowed to be "slowly varying with time," as demonstrated in Section 3.4. Essentially, the condition "slowly varying with time" is satisfied approximately as long as the physical dimensions of the loop are much less than a wavelength (e.g., $< \lambda/10$, where the wavelength is $\lambda = v/f$,

f is the highest significant frequency in the waveform of the current I, and v is the velocity of propagation of the current.

In Chapter 4 we developed an alternative means of calculating the inductance by using the vector magnetic potential $\mathbf{A}$, which is defined by

$$\mathbf{B} = \nabla \times \mathbf{A} \tag{5.3}$$

Hence, the total magnetic flux through the surface s is

$$\psi = \int_s \mathbf{B} \cdot d\mathbf{s}$$

$$= \int_s (\nabla \times \mathbf{A}) \cdot d\mathbf{s}$$

$$= \oint_c \mathbf{A} \cdot d\mathbf{l} \tag{5.4}$$

where we have used Stokes's theorem (see the Appendix) to convert the surface integral over surface s to a line integral around contour c that encloses the surface. This gives an alternative way of calculating the flux through the loop, ψ, in terms of $\mathbf{A}$. Hence, an alternative way of calculating the inductance of the current loop is

$$L = \frac{\oint_c \mathbf{A} \cdot d\mathbf{l}}{I} \tag{5.5}$$

where c is the closed contour that bounds the open surface s. Hence, we can compute the inductance of a loop by integrating, with a line integral, the product of the differential path lengths around the contour c that surrounds the open surface s and the components of the vector magnetic potential $\mathbf{A}$ that are tangent to that closed path. But (5.5) can be decomposed into the line integral *along unique segments of the closed loop* as

$$L = \frac{\oint_c \mathbf{A} \cdot d\mathbf{l}}{I}$$

$$= \frac{\int_{c_1} \mathbf{A}_1 \cdot d\mathbf{l}}{I} + \frac{\int_{c_2} \mathbf{A}_2 \cdot d\mathbf{l}}{I} + \cdots + \frac{\int_{c_n} \mathbf{A}_n \cdot d\mathbf{l}}{I} \tag{5.6}$$

where the closed path c is segmented into n *contiguous segments* c_i so that $c = c_1 + c_2 + \cdots + c_n$ and $\mathbf{A}_i$ is the *total* $\mathbf{A}$ along contour c_i that is due to the current of that segment as well as the currents of the other segments of c or of some other current loop. This allows us to *uniquely associate an inductance contribution to each segment of the closed loop* as

$$L_i = \frac{\int_{c_i} \mathbf{A}_i \cdot d\mathbf{l}}{I} \tag{5.7a}$$

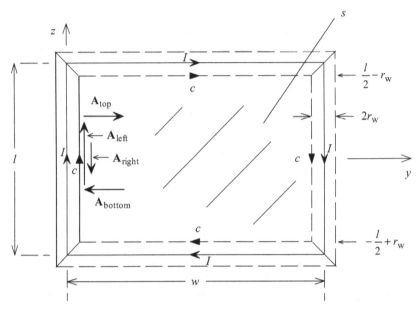

FIGURE 5.1. Rectangular loop.

so that the *total* loop inductance is the sum of these parts:

$$L = L_1 + L_2 + \cdots + L_n \tag{5.7b}$$

For example, in Fig. 5.1 we have shown Fig. 4.10, where in Section 4.3.1 we detailed the calculation of the inductance of a rectangular loop using the vector magnetic potential according to (5.5) Essentially, we are indirectly computing the total magnetic flux threading the loop, which is the region surrounded by the interior surfaces of the wires whose radii are r_w. This contour surrounding the open surface s is denoted as contour c in Fig. 5.1. As discussed in Sections 4.5 and 4.6, two important assumptions in computing the **B** field (and the subsequent calculation of the **A** field) are that (1) the current I is *distributed uniformly over the wire cross section so that the current I can be represented as a filament on the wire axis* (as it is for dc currents), and (2) there are no other currents in close enough proximity to this wire to upset this uniform current distribution over its cross section (i.e., the "proximity effect" is not pronounced). The total vector magnetic potential along the left side of the loop, $\mathbf{A}_1$, is the sum of the vector magnetic potentials along that side that are due to the current of that side, $\mathbf{A}_{\text{left}}$, and those that are due to the currents of the other three sides of the loop, $\mathbf{A}_{\text{right}}$, $\mathbf{A}_{\text{top}}$, and $\mathbf{A}_{\text{bottom}}$:

$$\mathbf{A}_1 = \mathbf{A}_{\text{left}} + \mathbf{A}_{\text{right}} + \mathbf{A}_{\text{top}} + \mathbf{A}_{\text{bottom}} \tag{5.8a}$$

Hence, the portion of the loop inductance *uniquely attributable to the left side* is

$$L_1 = \frac{\int_{\substack{\text{left}\\\text{side}}} \mathbf{A}_1 \cdot d\mathbf{l}}{I}$$

$$= \frac{\int_{\substack{\text{left}\\\text{side}}} \mathbf{A}_{\text{left}} \cdot d\mathbf{l}}{I} + \frac{\int_{\substack{\text{left}\\\text{side}}} \mathbf{A}_{\text{right}} \cdot d\mathbf{l}}{I} + \frac{\int_{\substack{\text{left}\\\text{side}}} \mathbf{A}_{\text{top}} \cdot d\mathbf{l}}{I}$$

$$+ \frac{\int_{\substack{\text{left}\\\text{side}}} \mathbf{A}_{\text{bottom}} \cdot d\mathbf{l}}{I} \tag{5.8b}$$

Observe that $\mathbf{A}_{\text{top}}$ and $\mathbf{A}_{\text{bottom}}$ are in the directions of the currents of those sides and hence are both orthogonal to the left side and do not contribute to the line integral for L_1 along the left side. In a similar fashion we obtain the inductances attributable to the other three sides, L_2, L_3, and L_4. Hence, the rectangular loop can be represented *uniquely* by the lumped equivalent circuit shown in Fig. 5.2.

Observe that the *total* vector magnetic potential *along the left side*, $\mathbf{A}_1$, in (5.8a) has contributions due to its own current as well as the currents of the other three sides. So this leads us to break the inductance of the left side, L_1, into four distinct pieces according to (5.8b):

$$L_1 = L_{p1} + M_{p12} + M_{p13} + M_{p14} \tag{5.9}$$

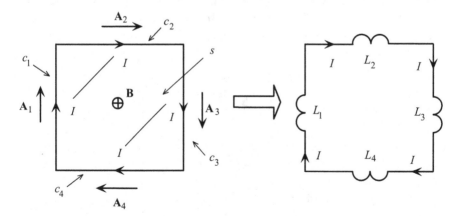

FIGURE 5.2. Uniquely attributing inductances to the sides of the rectangular loop of Fig. 5.1.

The first contribution is the *self partial inductance* of the left side:

$$L_{p1} = \frac{\int_{\substack{\text{left}\\\text{side}}} \mathbf{A}_{\text{left}} \cdot d\mathbf{l}}{I} \tag{5.10a}$$

which is due to the current of the left side. The other three contributions are due to the currents of the other three sides and are referred to as the *mutual partial inductances* between the other three sides and the left side:

$$M_{p12} = \frac{\int_{\substack{\text{left}\\\text{side}}} \mathbf{A}_{\text{top}} \cdot d\mathbf{l}}{I} \tag{5.10b}$$

$$M_{p13} = \frac{\int_{\substack{\text{left}\\\text{side}}} \mathbf{A}_{\text{right}} \cdot d\mathbf{l}}{I} \tag{5.10c}$$

$$M_{p14} = \frac{\int_{\substack{\text{left}\\\text{side}}} \mathbf{A}_{\text{bottom}} \cdot d\mathbf{l}}{I} \tag{5.10d}$$

Hence, the more complete equivalent circuit of the rectangular loop in terms of the *partial inductances* is shown in Fig. 5.3.

According to the dot convention described in Section 4.8.1, the total voltage across the left conductor is

$$\begin{aligned}
V_1 &= L_{p1}\frac{dI}{dt} + M_{p12}\frac{dI}{dt} + M_{p13}\frac{dI}{dt} + M_{p14}\frac{dI}{dt}\\
&= \underbrace{\left(L_{p1} + M_{p12} + M_{p13} + M_{p14}\right)}_{L_1} \frac{dI}{dt} \tag{5.11}
\end{aligned}$$

Observe that because sides 2 and 4 are orthogonal to side 1, $\mathbf{A}_{\text{top}}$ and $\mathbf{A}_{\text{bottom}}$ are orthogonal to the left side, so that $M_{p12} = M_{p14} = 0$. Also, because the direction of $\mathbf{A}_{\text{right}}$ is opposite the direction of the contour c along the left side,

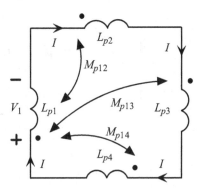

FIGURE 5.3. Rectangular loop equivalent circuit in terms of the partial inductances.

M_{p13} in (5.10c) is *negative*. The effective inductance of the left side of the loop, L_1, in (5.9) is referred to as the *net partial inductance* but has little value or use. Separating the net partial inductance into its constituent parts as in (5.9) gives more information about the contributions of all the other side currents.

5.2 PHYSICAL MEANING OF PARTIAL INDUCTANCE

The self partial inductance of the ith segment of a current loop is

$$L_{pi} = \frac{\int_{c_i} \mathbf{A}_i \cdot d\mathbf{l}}{I_i} \tag{5.12a}$$

and $\mathbf{A}_i$ is the portion of $\mathbf{A}$ along c_i that is produced by the current I_i of that segment. The voltage developed across that self partial inductance is

$$V_i = L_{pi} \frac{dI_i}{dt} \tag{5.12b}$$

as shown in Fig. 5.4.

Although (5.12a) gives the mathematical definition of self partial inductance, we now investigate the physical meaning of self partial inductance. Consider a segment c_i of a current loop carrying current I_i as shown in Fig. 5.5(a). Draw a surface extending from the segment to infinity with sides that are perpendicular to the current segment. Now determine the magnetic flux through that surface:

$$\frac{\psi_\infty}{I_i} = \frac{\int_s \mathbf{B} \cdot d\mathbf{s}}{I_i} = \frac{\oint_c \mathbf{A} \cdot d\mathbf{l}}{I_i}$$

$$= \underbrace{\frac{\int_{c_i} \mathbf{A}_i \cdot d\mathbf{l}}{I_i}}_{c_i} + \underbrace{\frac{\int_c \mathbf{A} \cdot d\mathbf{l}}{I_i}}_{\text{left side}} + \underbrace{\frac{\int_c \mathbf{A} \cdot d\mathbf{l}}{I_i}}_{\text{right side}} + \underbrace{\frac{\int_c \mathbf{A} \cdot d\mathbf{l}}{I_i}}_{\infty}$$

$$= \underbrace{\frac{\int_{c_i} \mathbf{A}_i \cdot d\mathbf{l}}{I_i}}_{c_i}$$

$$= L_{pi} \tag{5.13}$$

FIGURE 5.4. Self partial inductance of the ith segment of a current loop.

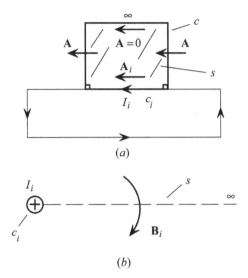

FIGURE 5.5. Physical meaning of self partial inductance.

The line integrals along the left and right sides are zero since the vector magnetic potential $\mathbf{A}$ is parallel to the current I_i that produces it and is therefore perpendicular to the left and right sides of the closed contour. The vector magnetic potential from a line current goes to zero at infinity [see (2.57)], so that the line integral along this portion of the closed contour at infinity is also zero. Hence, we are left with the partial inductance given in (5.12a) and the observation that:

The self partial inductance of a segment of a current loop is the ratio of the magnetic flux between the current segment and infinity and the current of that segment.

This is illustrated in cross section in Fig. 5.5(b).

The mutual partial inductance between two segments c_i and c_j (which may be parts of the same current loop or different current loops) is defined by

$$M_{pij} = \frac{\int_{c_i} \mathbf{A}_{ij} \cdot d\mathbf{l}}{I_j} \qquad (5.14a)$$

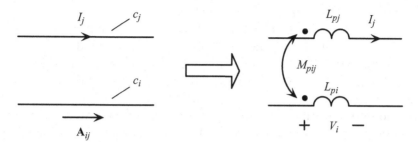

FIGURE 5.6. Mutual partial inductance between two current loop segments c_i and c_j.

where $\mathbf{A}_{ij}$ is along contour c_i and is due to the current of another segment, I_j. The voltage developed across that self partial inductance is

$$V_i = M_{pij} \frac{dI_j}{dt} \qquad (5.14b)$$

as shown in Fig. 5.6.

The physical meaning of mutual partial inductance is illustrated in Fig. 5.7. Consider a current loop and two segments of that loop, c_i and c_j, as shown in Fig. 5.7(a). Again draw a surface s extending from the jth segment (carrying the current) to infinity with sides that are perpendicular to that current

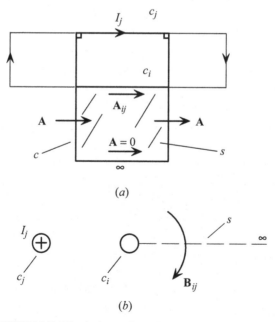

FIGURE 5.7. Physical meaning of mutual partial inductance.

segment. Now determine the magnetic flux through the surface s between the ith segment and infinity. Carrying through a development similar to that in (5.13) we see that the line integrals along the left and right sides are zero since the vector magnetic potential $\mathbf{A}$ is parallel to the current I_j that produces it and is therefore perpendicular to the left and right sides of the contour c that surrounds surface s. Also, the vector magnetic potential from a line current goes to zero at infinity so that the line integral along the portion of the contour at infinity is also zero. Hence, we are left with the mutual partial inductance given in (5.14a) and the observation that

> *The mutual partial inductance between two segments of the same or different current loops is the ratio of the magnetic flux (produced by the current of the first segment) that penetrates the surface between the second segment and infinity and the current of the first segment.*

This is illustrated in cross section in Fig. 5.7(b).

Although Fig. 5.7 shows the result for the mutual partial inductance of two *parallel* conductors, the result also obtains for two conductors at any angle to each other as shown in Fig. 5.8. Again draw two lines to infinity

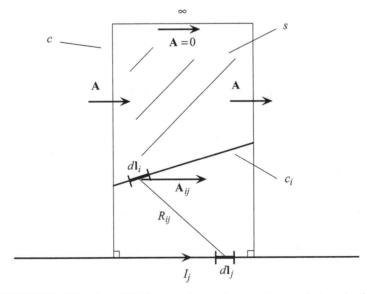

FIGURE 5.8. Mutual partial inductance for conductors at any angle to each other.

that are perpendicular (shown as small rectangles) to current I_j and which enclose the open surface s that lies between those parallel lines and between the skewed conductor and infinity. Integrating the line integral of the vector magnetic potential around the closed contour c surrounding this surface to infinity again gives

$$
\begin{aligned}
M_{pij} &= \frac{\psi_\infty}{I_j} \\
&= \frac{\oint_c \mathbf{A} \cdot d\mathbf{l}}{I_j} \\
&= \frac{\int_{c_i} \mathbf{A}_{ij} \cdot d\mathbf{l}}{I_j}
\end{aligned}
\tag{5.15}
$$

This is obtained again since $\mathbf{A}$ is parallel to I_j at all points in space, so that $\mathbf{A}$ is perpendicular to the left and right sides of s and contribute nothing to the line integral along those sides, and $\mathbf{A}$ goes to zero at infinity. Observe that *the same result is obtained even if the two conductors do not lie in the same plane*, since $\mathbf{A}$ will still be orthogonal to the two sides of the open surface because they were constructed perpendicular to the current I_j and will also go to zero at infinity. (Again draw two lines for the sides of s that are perpendicular to conductor c_j.)

The mutual partial inductance can also be obtained from the Neumann integral by substituting the explicit equation for $\mathbf{A}_{ij}$ into (5.15):

$$
\boxed{M_{pij} = \frac{\mu_0}{4\pi} \int_{c_i} \int_{c_j} \frac{1}{R_{ij}} d\mathbf{l}_i \cdot d\mathbf{l}_j}
\tag{5.16}
$$

where c_j is the contour along the conductor carrying current I_j, and R_{ij} is the distance between differential segments $d\mathbf{l}_i$ along contour c_i and $d\mathbf{l}_j$ along contour c_j, as shown in Fig. 5.8.

5.3 SELF PARTIAL INDUCTANCE OF WIRES

In this section we derive some fundamental results for the self partial inductance of wires having radii r_w. Again we assume that the current of the wire, I, is distributed uniformly over the wire cross section so that for the purpose of computing the $\mathbf{B}$ and $\mathbf{A}$ fields, we can concentrate the current I as a filament on the axis of the wire.

The fundamental problem for computing the self partial inductance of a wire is a wire of length l carrying a current I as shown in Fig. 5.9. We determine the self partial inductance of this segment of wire by integrating the magnetic flux density through the surface s between the wire surface, $y = r_w$,

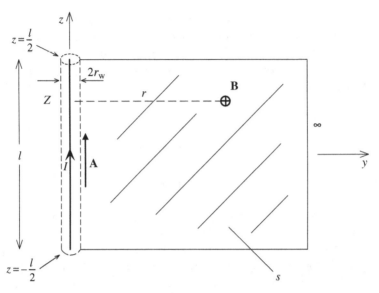

FIGURE 5.9. Determination of the self partial inductance of a wire.

and infinity, $y \to \infty$. The magnetic flux density was derived in Chapter 2 and given in (2.15):

$$\mathbf{B} = \frac{\mu_0 I}{4\pi r} \left[\frac{Z + l/2}{\sqrt{(Z + l/2)^2 + r^2}} - \frac{Z - l/2}{\sqrt{(Z - l/2)^2 + r^2}} \right] \mathbf{a}_\phi \qquad (2.15)$$

The total flux through the surface s is

$$
\begin{aligned}
\psi_\infty &= \int_{r=r_w}^{\infty} \int_{Z=-l/2}^{l/2} B_\phi \, dZ \, dr \\
&= \frac{\mu_0 I}{4\pi} \int_{r=r_w}^{\infty} \frac{1}{r} \int_{Z=-l/2}^{l/2} \left[\frac{Z + l/2}{\sqrt{(Z + l/2)^2 + r^2}} \right. \\
&\qquad \left. - \frac{Z - l/2}{\sqrt{(Z - l/2)^2 + r^2}} \right] dZ \, dr \\
&= 2\frac{\mu_0 I}{4\pi} \int_{r=r_w}^{\infty} \frac{1}{r} \int_{\lambda=0}^{l} \frac{\lambda}{\sqrt{\lambda^2 + r^2}} \, d\lambda \, dr \\
&= \frac{\mu_0 I}{2\pi} \int_{r=r_w}^{\infty} \frac{1}{r} \left[\sqrt{\lambda^2 + r^2} \right]_{\lambda=0}^{l} dr \\
&= \frac{\mu_0 I}{2\pi} \int_{r=r_w}^{\infty} \frac{1}{r} (\sqrt{l^2 + r^2} - r) \, dr \qquad (5.17a)
\end{aligned}
$$

and we have used a change of variables, $\lambda = Z \pm l/2$, $d\lambda = dZ$, and integral 201.01 of Dwight [7]:

$$\int \frac{x}{\sqrt{x^2 + a^2}}\, dx = \sqrt{x^2 + a^2} \qquad \text{(D201.01)}$$

Further integration with respect to r yields

$$\psi_\infty = \frac{\mu_0 I}{2\pi} \int_{r=r_w}^{\infty} \left[\frac{\sqrt{l^2 + r^2}}{r} - 1 \right]\, dr$$

$$= \frac{\mu_0 I}{2\pi} \left[\sqrt{l^2 + r^2} - l \ln \frac{l + \sqrt{l^2 + r^2}}{r} - r \right]_{r=r_w}^{r\to\infty}$$

$$= -\frac{\mu_0 I}{2\pi} l \left[\ln\left(\frac{l}{r} + \sqrt{\left(\frac{l}{r}\right)^2 + 1} \right) - \sqrt{1 + \left(\frac{r}{l}\right)^2} + \frac{r}{l} \right]_{r=r_w}^{r\to\infty}$$

$$= \frac{\mu_0 I}{2\pi} l \left[\ln\left(\frac{l}{r_w} + \sqrt{\left(\frac{l}{r_w}\right)^2 + 1} \right) - \sqrt{1 + \left(\frac{r_w}{l}\right)^2} + \frac{r_w}{l} \right]$$

$$\text{(5.17b)}$$

and we have used integral 241.01 of Dwight [7]:

$$\int \frac{\sqrt{x^2 + a^2}}{x}\, dx = \sqrt{x^2 + a^2} - a \ln \frac{a + \sqrt{x^2 + a^2}}{x} \qquad \text{(D241.01)}$$

Hence, the self partial inductance is

$$
\boxed{
\begin{aligned}
L_p &= \frac{\psi_\infty}{I} \\
&= \frac{\mu_0}{2\pi} l \left[\ln\left(\frac{l}{r_w} + \sqrt{\left(\frac{l}{r_w}\right)^2 + 1} \right) - \sqrt{1 + \left(\frac{r_w}{l}\right)^2} + \frac{r_w}{l} \right] \\
&= 2 \times 10^{-7} l \left[\ln\left(\frac{l}{r_w} + \sqrt{\left(\frac{l}{r_w}\right)^2 + 1} \right) - \sqrt{1 + \left(\frac{r_w}{l}\right)^2} + \frac{r_w}{l} \right]
\end{aligned}
}
$$

$$\text{(5.18a)}$$

and we have substituted $\mu_0/2\pi = 2 \times 10^{-7}$. Using the inverse hyperbolic sine,

$$\sinh^{-1} \frac{x}{a} = \ln\left(\frac{x}{a} + \sqrt{\left(\frac{x}{a}\right)^2 + 1} \right)$$

$$= -\sinh^{-1}\left(-\frac{x}{a} \right) \qquad \text{(D700.1)}$$

gives an alternative form of the result:

$$L_p = \frac{\mu_0}{2\pi} l \left[\sinh^{-1} \frac{l}{r_w} - \sqrt{1 + \left(\frac{r_w}{l}\right)^2} + \frac{r_w}{l} \right] \qquad (5.18b)$$

In a practical case, the length of the segment is usually much larger than the wire radius, $l \gg r_w$, so we have the following approximations:

$$\ln \left[\frac{l}{r_w} + \sqrt{\left(\frac{l}{r_w}\right)^2 + 1} \right] = \ln \frac{2l}{r_w} + \frac{1}{4} \left(\frac{r_w}{l}\right)^2$$

$$-\frac{3}{32} \left(\frac{r_w}{l}\right)^4 + \cdots \qquad \frac{l}{r_w} \gg 1 \quad (D602.1)$$

and

$$\sqrt{1 + \left(\frac{r_w}{l}\right)^2} = 1 + \frac{1}{2} \left(\frac{r_w}{l}\right)^2 - \frac{1}{8} \left(\frac{r_w}{l}\right)^4 + \cdots \qquad \frac{r_w}{l} \le 1 \quad (D5.3)$$

so that (5.18a) approximates to

$$L_p = \frac{\mu_0}{2\pi} l \left[\ln \frac{2l}{r_w} - 1 + \frac{r_w}{l} - \frac{1}{4} \left(\frac{r_w}{l}\right)^2 + \cdots \right]$$

$$\cong 2 \times 10^{-7} l \left(\ln \frac{2l}{r_w} - 1 \right) \qquad l \gg r_w$$

$$(5.18c)$$

Alternatively, we can determine the self partial inductance by integrating the vector magnetic potential along the wire surface also shown in Fig. 5.9. The vector magnetic potential **A** for this case was determined in Chapter 2 and given in (2.57):

$$A_z = \frac{\mu_0 I}{4\pi} \left(\sinh^{-1} \frac{Z + l/2}{r} - \sinh^{-1} \frac{Z - l/2}{r} \right) \qquad (2.57)$$

Hence, we set up the integral

$$L_p = \frac{\int_{Z=-l/2}^{l/2} A_z|_{r=r_w} \, dZ}{I}$$

$$= \frac{\mu_0}{4\pi} \int_{Z=-l/2}^{l/2} \left(\sinh^{-1} \frac{Z + l/2}{r_w} - \sinh^{-1} \frac{Z - l/2}{r_w} \right) dZ$$

$$= 2 \frac{\mu_0}{4\pi} \int_{\lambda=0}^{l} \left(\sinh^{-1} \frac{\lambda}{r_w} \right) d\lambda$$

$$= 2\frac{\mu_0}{4\pi}\left[\lambda\,\sinh^{-1}\frac{\lambda}{r_w} - \sqrt{\lambda^2 + r_w^2}\right]_{\lambda=0}^{l}$$

$$= \frac{\mu_0}{2\pi}\left(l\,\sinh^{-1}\frac{l}{r_w} - \sqrt{l^2 + r_w^2} + r_w\right)$$

$$= \frac{\mu_0}{2\pi}l\left[\sinh^{-1}\frac{l}{r_w} - \sqrt{1 + \left(\frac{r_w}{l}\right)^2} + \frac{r_w}{l}\right]$$

$$= \frac{\mu_0}{2\pi}l\left[\ln\left(\frac{l}{r_w} + \sqrt{\left(\frac{l}{r_w}\right)^2 + 1}\right) - \sqrt{1 + \left(\frac{r_w}{l}\right)^2} + \frac{r_w}{l}\right]$$

$$(5.19)$$

which is the same as (5.18a). We have used a change of variables, $\lambda = Z \pm l/2$, $d\lambda = dZ$, integral 730 of Dwight [7],

$$\int \sinh^{-1}\frac{x}{a}\,dx = x\,\sinh^{-1}\frac{x}{a} - \sqrt{x^2 + a^2} \qquad (D730)$$

and the identity for inverse hyperbolic sine,

$$\sinh^{-1}\frac{x}{a} = \ln\left(\frac{x}{a} + \sqrt{\left(\frac{x}{a}\right)^2 + 1}\right)$$

$$= -\sinh^{-1}\left(-\frac{x}{a}\right) \qquad (D700.1)$$

5.4   MUTUAL PARTIAL INDUCTANCE BETWEEN PARALLEL WIRES

Next, we determine another fundamental result: the mutual partial inductance between two parallel wires shown in Fig. 5.10. We first assume that both wires are of the same length and their endpoints are aligned. In the next section we derive the result for this situation but with the wires offset and their lengths different. The only difference between this computation and those for the self partial inductance of Section 5.3 is that here we integrate from $y = d + r_w$ to $y \to \infty$ rather than from the surface of the first wire. Hence, the integral in (5.17a) becomes

$$\psi_\infty = \int_{r=d+r_w}^{\infty}\int_{Z=-l/2}^{l/2} B_\phi\,dZ\,dr \qquad (5.20)$$

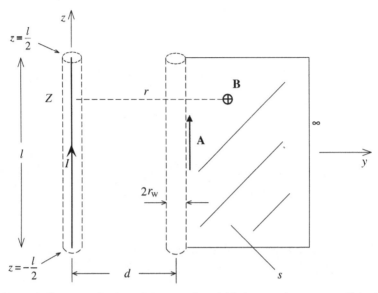

FIGURE 5.10. Determination of the mutual partial inductance between parallel wires.

It is easy to see that we only need to replace r_w with $d + r_w$ in the previous derivation for the self partial inductance in (5.17a)–(5.17b) and obtain

$$
\begin{aligned}
M_p &= \frac{\psi_\infty}{I} \\
&= \frac{\mu_0}{2\pi} l \left[\ln\left(\frac{l}{d + r_w} + \sqrt{\left(\frac{l}{d + r_w}\right)^2 + 1} \right) \right. \\
&\quad \left. - \sqrt{1 + \left(\frac{d + r_w}{l}\right)^2} + \frac{d + r_w}{l} \right] \\
&\cong 2 \times 10^{-7} l \left[\ln\left(\frac{l}{d} + \sqrt{\left(\frac{l}{d}\right)^2 + 1} \right) \right. \\
&\quad \left. - \sqrt{1 + \left(\frac{d}{l}\right)^2} + \frac{d}{l} \right] \qquad d \gg r_w
\end{aligned}
$$

(5.21a)

Using the inverse hyperbolic sine,

$$
\begin{aligned}
\sinh^{-1} \frac{x}{a} &= \ln\left(\frac{x}{a} + \sqrt{\left(\frac{x}{a}\right)^2 + 1} \right) \\
&= -\sinh^{-1}\left(-\frac{x}{a}\right)
\end{aligned}
$$

(D700.1)

gives an alternative form of the result:

$$M_p = \frac{\mu_0}{2\pi} l \left[\sinh^{-1} \frac{l}{d} - \sqrt{1 + \left(\frac{d}{l}\right)^2} + \frac{d}{l} \right] \qquad d \gg r_w \qquad (5.21b)$$

For wires that are very long compared to their separation, $l/d \gg 1$, or, equivalently, separations much smaller than their length, $d/l \ll 1$, the result in (5.21a) can be approximated by using

$$\ln\left[\frac{l}{d} + \sqrt{\left(\frac{l}{d}\right)^2 + 1}\right] = \ln\frac{2l}{d} + \frac{1}{4}\left(\frac{d}{l}\right)^2 - \frac{3}{32}\left(\frac{d}{l}\right)^4 + \cdots \qquad \frac{l}{d} > 1$$

$$\text{(D602.1)}$$

$$\sqrt{1 + \left(\frac{d}{l}\right)^2} = 1 + \frac{1}{2}\left(\frac{d}{l}\right)^2 - \frac{1}{8}\left(\frac{d}{l}\right)^4 + \cdots \qquad \frac{d}{l} \leq 1$$

$$\text{(D5.3)}$$

giving

$$M_p = \frac{\mu_0}{2\pi} l \left[\ln\frac{2l}{d} - 1 + \frac{d}{l} - \frac{1}{4}\left(\frac{d}{l}\right)^2 + \frac{1}{32}\left(\frac{d}{l}\right)^4 - \cdots \right]$$

$$\cong \frac{\mu_0}{2\pi} l \left(\ln\frac{2l}{d} - 1 \right) \qquad l \gg d$$

$$\text{(5.21c)}$$

For wires that are very short compared to their separation, $l/d \ll 1$, or, equivalently, separations much greater than their length, $d/l \gg 1$, the result in (5.21a) can be approximated by using

$$\ln\left[\frac{l}{d} + \sqrt{\left(\frac{l}{d}\right)^2 + 1}\right] = \frac{l}{d} - \frac{1}{6}\left(\frac{l}{d}\right)^3 + \frac{3}{40}\left(\frac{l}{d}\right)^5 - \cdots \qquad \frac{l}{d} < 1$$

$$\text{(D602.1)}$$

$$\sqrt{1 + \left(\frac{d}{l}\right)^2} = \frac{d}{l}\sqrt{\left(\frac{l}{d}\right)^2 + 1}$$

$$= \frac{d}{l} + \frac{1}{2}\left(\frac{l}{d}\right) - \frac{1}{8}\left(\frac{l}{d}\right)^3 + \frac{1}{16}\left(\frac{l}{d}\right)^5 - \cdots \qquad \frac{l}{d} \leq 1$$

$$\text{(D5.3)}$$

giving

$$M_p = \frac{\mu_0}{2\pi} \frac{l}{2d} \left[1 - \frac{1}{12} \left(\frac{l}{d}\right)^2 + \frac{1}{40} \left(\frac{l}{d}\right)^4 - \cdots \right] \qquad l \ll d$$

(5.21d)

We can obtain the same result as in (5.21a) from the vector magnetic potential **A**:

$$M_p = \frac{\int_{Z=-l/2}^{l/2} A_z|_{r=d+r_w} \, dZ}{I}$$

(5.22)

and evaluating A_z along the second wire at $y = d + r_w$. Carrying through the same integration in (5.19) but with $r = r_w$ replaced by $r = d + r_w$ again gives (5.21a).

Finally, we show that the mutual partial inductance in (5.21a) can also be derived from the Neumann integral in (5.16):

$$
\begin{aligned}
M_p &= \frac{\mu_0}{4\pi} \int_{c_1} \int_{c_2} \frac{d\mathbf{l}_1 \cdot d\mathbf{l}_2}{R_{12}} \\
&= \frac{\mu_0}{4\pi} \int_{z_2=-l/2}^{l/2} dz_2 \int_{z_1=-l/2}^{l/2} \frac{1}{\sqrt{(d+r_w)^2 + (z_1 - z_2)^2}} dz_1 \\
&= \frac{\mu_0}{4\pi} \int_{z_2=-l/2}^{l/2} dz_2 \int_{\lambda=-l/2-z_2}^{l/2-z_2} \frac{1}{\sqrt{(d+r_w)^2 + \lambda^2}} d\lambda \\
&= \frac{\mu_0}{4\pi} \int_{z_2=-l/2}^{l/2} \left[\ln \left(\lambda + \sqrt{(d+r_w)^2 + \lambda^2} \right) \right]_{\lambda=-l/2-z_2}^{l/2-z_2} dz_2 \\
&= \frac{\mu_0}{4\pi} \int_{z_2=-l/2}^{l/2} \left(\sinh^{-1} \frac{l/2 - z_2}{d + r_w} + \sinh^{-1} \frac{l/2 + z_2}{d + r_w} \right) dz_2 \\
&= \frac{\mu_0}{4\pi} \int_{\zeta=0}^{l} \left(2 \sinh^{-1} \frac{\zeta}{d + r_w} \right) d\zeta \\
&= 2 \frac{\mu_0}{4\pi} \left[\zeta \sinh^{-1} \frac{\zeta}{d + r_w} - \sqrt{\zeta^2 + (d+r_w)^2} \right]_{\zeta=0}^{l} \\
&= \frac{\mu_0}{2\pi} \left[l \sinh^{-1} \frac{l}{d + r_w} - \sqrt{l^2 + (d+r_w)^2} + (d+r_w) \right] \\
&= \frac{\mu_0}{2\pi} l \left[\sinh^{-1} \frac{l}{d + r_w} - \sqrt{1 + \left(\frac{d+r_w}{l}\right)^2} + \frac{d+r_w}{l} \right]
\end{aligned}
$$

(5.23)

and the differential lengths $d\mathbf{l}_1$ and $d\mathbf{l}_2$ are parallel so that the dot product goes away: $d\mathbf{l}_1 \cdot d\mathbf{l}_2 = dl_1\,dl_2 = dz_1\,dz_2$. But (5.23) is the same as (5.21a). We have substituted a change of variables in the inner integral: $\lambda = z_1 - z_2, d\lambda = dz_1$, and a change of variables in the outer integral: $\zeta = l/2 \pm z_2, d\zeta = \pm dz_2$, and have again used the integrals

$$\int \frac{1}{\sqrt{x^2 + a^2}} dx = \ln\left(x + \sqrt{x^2 + a^2}\right) \qquad \text{(D200.01)}$$

and

$$\int \sinh^{-1}\frac{x}{a}\,dx = x\sinh^{-1}\frac{x}{a} - \sqrt{x^2 + a^2} \qquad \text{(D730)}$$

We also used the important identity

$$\ln\frac{a + \sqrt{x^2 + a^2}}{-b + \sqrt{x^2 + b^2}} = \sinh^{-1}\frac{a}{x} - \sinh^{-1}\left(-\frac{b}{x}\right)$$

$$= \sinh^{-1}\frac{a}{x} + \sinh^{-1}\frac{b}{x}$$

From these results we see that the self partial inductance L_p can be obtained from the mutual partial inductance simply by replacing $d + r_w$ in M_p, with r_w, and vice versa. In other words,

$$L_p = M_p\big|_{d+r_w \to r_w} \qquad (5.24)$$

Using M_p to get L_p in this way presupposes that both wires are of the same length and radii, and their endpoints are aligned.

5.5 MUTUAL PARTIAL INDUCTANCE BETWEEN PARALLEL WIRES THAT ARE OFFSET

Consider the case of two offset, parallel wires whose lengths are l and m shown in Fig. 5.11. The two wires are parallel to the z axis, have a center-to-center separation of d, and their endpoints are offset by a distance s. The radius of the second wire of length l is r_w. The radius of the first wire of length m carrying the current I which produces the magnetic field is immaterial since we assume that the current I is distributed uniformly over the cross section of that wire so that this current can be concentrated as a filament on the axis of the wire. The first wire carrying the current I has its lower end at the origin of the coordinate system, $z = 0$. The two ends of the other wire of length l are at positions $z = z_1$ and $z = z_2$. In all such problems of determining the mutual partial inductance between two parallel but offset wires using the result derived in this section, it is important to determine these wire lengths and positions, $z = 0, z_1$, and z_2, for each particular problem.

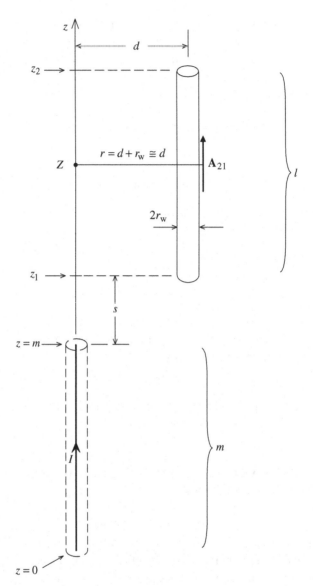

FIGURE 5.11. Mutual partial inductance between offset wires.

Again we have three methods for calculating the mutual partial inductance between the two wire segments: the magnetic flux linkage method using **B**, the vector magnetic potential method using **A**, and the Neumann integral. For this problem we choose to use the vector magnetic potential method using **A**. We must integrate the vector magnetic potential due to the current I of the first wire of length m along the surface of the second wire of length l with a line integral:

$$M_p = \frac{\int_l \mathbf{A}_{21}|_{r=d+r_w \cong d} \cdot d\mathbf{l}}{I}$$

$$= \frac{\int_{Z=z_1}^{z_2} A_{21}(Z,r)|_{r=d} \, dZ}{I} \tag{5.25}$$

where $\mathbf{A}_{21}$ is the vector magnetic potential along the surface of the second wire that is produced by the current I of the first wire. Hence, we need the result for the vector magnetic potential from a wire of length m carrying a current I. This was derived in Chapter 2 from Fig. 2.24 and given in (2.57). Note in Fig. 2.24 that the origin of the coordinate system at $z = 0$ was located at the midpoint of the wire. We must modify that result to fit Fig. 5.11 by rederiving the result for the case where the lower end of the wire is at $z = 0$. Carrying through the development that led to (2.57) yields for this case

$$A_{21}(Z,r) = \frac{\mu_0 I}{4\pi} \left\{ \ln \left(Z + \sqrt{Z^2 + r^2} \right) \right.$$

$$\left. - \ln \left[(Z - m) + \sqrt{(Z - m)^2 + r^2} \right] \right\}$$

$$= \frac{\mu_0 I}{4\pi} \left(\sinh^{-1} \frac{Z}{r} - \sinh^{-1} \frac{Z - m}{r} \right) \tag{5.26}$$

and we have again used the identity

$$\sinh^{-1} \frac{x}{a} = -\sinh^{-1} \left(-\frac{x}{a} \right)$$

$$= \ln \left[\frac{x}{a} + \sqrt{\left(\frac{x}{a} \right)^2 + 1} \right]$$

$$= \ln \left(x + \sqrt{x^2 + a^2} \right) - \ln a \tag{D700.1}$$

Hence, (5.25) becomes

$$M_p = \frac{\int_{Z=z_1}^{z_2} A_{21}|_{r=d} \, dZ}{I}$$

$$= \frac{\mu_0}{4\pi} \int_{Z=z_1}^{z_2} \left(\sinh^{-1} \frac{Z}{d} - \sinh^{-1} \frac{Z - m}{d} \right) dZ \tag{5.27}$$

Carrying through with the integration of (5.27) gives

$$
\begin{aligned}
M_p &= \frac{\mu_0}{4\pi} \int_{Z=z_1}^{z_2} \left(\sinh^{-1}\frac{Z}{d} - \sinh^{-1}\frac{Z-m}{d} \right) dZ \\
&= \frac{\mu_0}{4\pi} \left(\int_{Z=z_1}^{z_2} \sinh^{-1}\frac{Z}{d}\, dZ - \int_{\lambda=z_1-m}^{z_2-m} \sinh^{-1}\frac{\lambda}{d}\, d\lambda \right) \\
&= \frac{\mu_0}{4\pi} \left[z_2 \sinh^{-1}\frac{z_2}{d} - z_1 \sinh^{-1}\frac{z_1}{d} - (z_2-m)\sinh^{-1}\frac{z_2-m}{d} \right. \\
&\quad + (z_1-m)\sinh^{-1}\frac{z_1-m}{d} - \sqrt{z_2^2+d^2} + \sqrt{z_1^2+d^2} \\
&\quad \left. + \sqrt{(z_2-m)^2+d^2} - \sqrt{(z_1-m)^2+d^2} \right]
\end{aligned}
$$

(5.28)

where we have used a change of variables, $\lambda = Z - m$, $d\lambda = dZ$, in the second integral and have used integral 730 of Dwight [7]:

$$
\int \sinh^{-1}\frac{x}{a}\, dx = x\sinh^{-1}\frac{x}{a} - \sqrt{x^2+a^2}
$$

(D730)

In the case where the two wires lie on the z axis, $d = 0$, as shown in Fig. 5.12, we could reintegrate (5.27) for $r = r_w$ or simply substitute $d = r_w$ into (5.28) to give

$$
\begin{aligned}
M_{p(d=r_w)} &= \frac{\mu_0}{4\pi} \left[z_2 \sinh^{-1}\frac{z_2}{r_w} - z_1 \sinh^{-1}\frac{z_1}{r_w} - (z_2-m)\sinh^{-1}\frac{z_2-m}{r_w} \right. \\
&\quad + (z_1-m)\sinh^{-1}\left(\frac{z_1-m}{r_w}\right) - \sqrt{z_2^2+r_w^2} + \sqrt{z_1^2+r_w^2} \\
&\quad \left. + \sqrt{(z_2-m)^2+r_w^2} - \sqrt{(z_1-m)^2+r_w^2} \right]
\end{aligned}
$$

(5.29a)

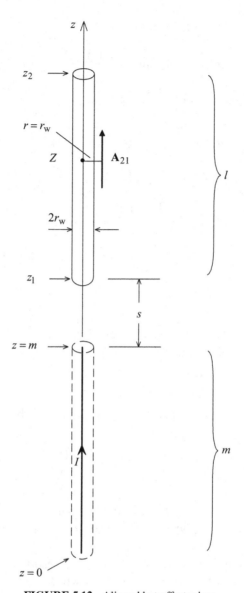

FIGURE 5.12. Aligned but offset wires.

Substituting the dimensions gives

$$
\begin{aligned}
M_{p(d=r_w)} &= \frac{\mu_0}{4\pi} \left[(l+s+m) \sinh^{-1} \frac{l+s+m}{r_w} - (m+s) \sinh^{-1} \frac{m+s}{r_w} \right. \\
&\quad - (l+s) \sinh^{-1} \frac{l+s}{r_w} + s \sinh^{-1} \frac{s}{r_w} - \sqrt{(l+s+m)^2 + r_w^2} \\
&\quad \left. + \sqrt{(m+s)^2 + r_w^2} + \sqrt{(l+s)^2 + r_w^2} - \sqrt{s^2 + r_w^2} \right] \\
&\cong \frac{\mu_0}{4\pi} \left\{ (l+s+m) \left[\ln\left(\frac{2(l+s+m)}{r_w} \right) - 1 \right] \right. \\
&\quad - (m+s) \left[\ln\left(\frac{2(m+s)}{r_w} \right) - 1 \right] \\
&\quad \left. - (l+s) \left[\ln\left(\frac{2(l+s)}{r_w} \right) - 1 \right] + s \left[\ln\left(\frac{2s}{r_w} \right) - 1 \right] \right\}
\end{aligned}
$$

(5.29b)

In terms of the self partial inductances of a wire of radius r_w and length l obtained in (5.18b),

$$
L_l = \frac{\mu_0}{2\pi} \left(l \sinh^{-1} \frac{l}{r_w} - \sqrt{l^2 + r_w^2} + r_w \right)
$$ (5.18b)

the result for aligned but offset wires in (5.29a,b) can be written as

$$
\begin{aligned}
2M_{p(d=r_w)} &= \left(L_{z_2} + L_{z_1-m} \right) - \left(L_{z_2-m} + L_{z_1} \right) \\
&= \left(L_{l+s+m} + L_s \right) - \left(L_{l+s} + L_{m+s} \right)
\end{aligned}
$$

(5.29c)

Notice that (5.29c) gives $2M_p$ since the self partial inductance L_l in (5.18b) is multiplied by $\mu_0/2\pi$, whereas the result for M_p in (5.29a,b) is multiplied by $\mu_0/4\pi$.

There is a simple explanation for why the result for the mutual partial inductance between two aligned but offset wires can be written in terms of the self partial inductances of wires of various lengths obtained previously, as in (5.29c). Recall that the self partial inductance of a wire is the ratio of the magnetic flux between that wire and infinity, ψ_l, and the current of that wire:

$$
L_l = \frac{\psi_l}{I}
$$

Figure 5.13 shows that a current on each wire segment produces not only flux between that segment and infinity but also between each of the other segments and infinity. For example, observe from Fig. 5.13 that superimposing the fluxes

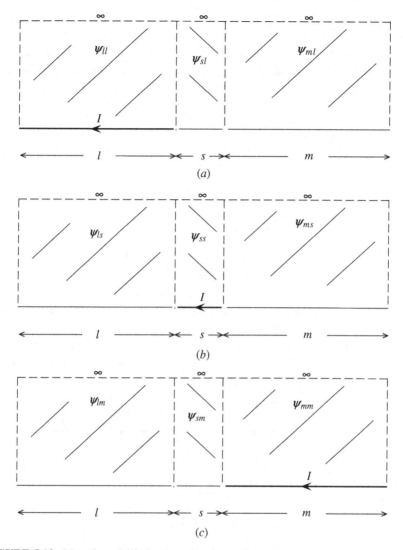

FIGURE 5.13. Mutual partial inductance for aligned but offset wires in terms of fluxes to infinity.

opposite each segment that are due to currents on the other three segments gives

$$\psi_l = \psi_{ll} + \psi_{ls} + \psi_{lm} \tag{5.30a}$$

$$\psi_s = \psi_{sl} + \psi_{ss} + \psi_{sm} \tag{5.30b}$$

$$\psi_m = \psi_{ml} + \psi_{ms} + \psi_{mm} \tag{5.30c}$$

where the notation ψ_{ij} denotes the flux to infinity opposite segment i due to a current I *only* on segment j. Keep in mind that these mutual inductances are reciprocal (i.e., $M_{ij} = M_{ji}$). But if all three segments have current I on them, they produce the total flux ψ_{l+s+m}. Hence, the self inductance of a wire of total length $l + s + m$ can be written as

$$
\begin{aligned}
L_{l+s+m} &= \frac{\psi_{l+s+m}}{I} \\
&= \frac{\psi_l}{I} + \frac{\psi_s}{I} + \frac{\psi_m}{I} \\
&= \frac{\psi_{ll}}{I} + \frac{\psi_{ls}}{I} + \frac{\psi_{lm}}{I} \\
&\quad + \frac{\psi_{sl}}{I} + \frac{\psi_{ss}}{I} + \frac{\psi_{sm}}{I} \\
&\quad + \frac{\psi_{ml}}{I} + \frac{\psi_{ms}}{I} + \frac{\psi_{mm}}{I}
\end{aligned}
\tag{5.31}
$$

The key to simplifying this and writing it in the form of (5.29c) is to *write the result in terms of the self partial inductances of segments of a single length so that we can use the result derived in (5.18a,b) without having to rederive a new result* [which we have already done in (5.29)]. To do this, note that the total fluxes given by (5.31) can be written as

$$
L_{l+s+m} = \underbrace{\frac{\psi_{ll}}{I} + \frac{\psi_{ls}}{I} + \frac{\psi_{sl}}{I} + \frac{\psi_{ss}}{I}}_{L_{l+s}} + \underbrace{\frac{\psi_{mm}}{I} + \frac{\psi_{ms}}{I} + \frac{\psi_{sm}}{I} + \frac{\psi_{ss}}{I}}_{L_{m+s}}
$$

$$
- \underbrace{\frac{\psi_{ss}}{I}}_{L_s} + \underbrace{\frac{\psi_{lm}}{I} + \frac{\psi_{ml}}{I}}_{2M_p}
\tag{5.32a}
$$

Solving this gives the result in (5.29c) since

$$
\begin{aligned}
2M_p &= \frac{M_{lm} + M_{ml}}{I} \\
&= (L_{l+s+m} + L_s) - (L_{l+s} + L_{m+s})
\end{aligned}
\tag{5.32b}
$$

This gives a very basic principle for adding inductors in series where the inductors have not only their self inductance but also mutual inductances between each other:

$$
\boxed{
\begin{aligned}
L_{1+2+3} &= L_1 + M_{12} + M_{13} + L_2 + M_{12} + M_{23} + L_3 + M_{13} + M_{23} \\
&= L_1 + L_2 + L_3 + 2M_{12} + 2M_{13} + 2M_{23}
\end{aligned}
}
$$

$$
\tag{5.33}
$$

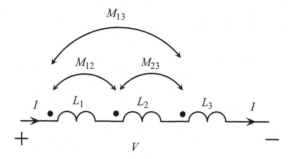

FIGURE 5.14. Adding inductors in series.

This can be verified from the electric circuit diagram in Fig. 5.14 by determining the total voltage across the series combination using the dot convention [1,2].

The basic result in (5.28) for parallel, offset wires with $d \neq 0$ can be written similarly in terms of the result in Section 5.4 for the mutual partial inductance between two identical wires of lengths l and separation d whose endpoints coincide as shown in Fig. 5.10 and given in (5.21b):

$$M_l = \frac{\mu_0}{2\pi} \left(l \sinh^{-1} \frac{l}{d} - \sqrt{l^2 + d^2} + d \right) \qquad d \gg r_w \qquad (5.21b)$$

Hence, the result in (5.28) can be written in terms of (5.21b) as

$$\boxed{\begin{aligned} 2M_p &= \left(M_{z_2} + M_{z_1-m} \right) - \left(M_{z_1} + M_{z_2-m} \right) \\ &= \left(M_{l+s+m} + M_s \right) - \left(M_{m+s} + M_{l+s} \right) \end{aligned}}$$

(5.34)

Notice again that (5.34) gives $2M_p$ since M_l in (5.21b) is multiplied by $\mu_0/2\pi$, whereas the result for M_p in (5.28) is multiplied by $\mu_0/4\pi$. Note from (5.21b) that

$$M_0 = 0 \qquad (5.35a)$$

and

$$M_{-l} = M_l \qquad (5.35b)$$

with (5.35b) resulting from the identity $\sinh^{-1}(-x) = -\sinh^{-1} x$. If the wires overlap, replace s with $-s$ in (5.34).

We can easily determine the mutual partial inductance between the various offset structures shown in Fig. 5.15 by using the basic result in (5.34) *and* comparing each of these structures to Fig. 5.11, from which (5.34) was derived in order to (1) determine the location point of $z = 0$ on those structures, and (2) hence to determine the values of z_1 and z_2 in (5.34). For example,

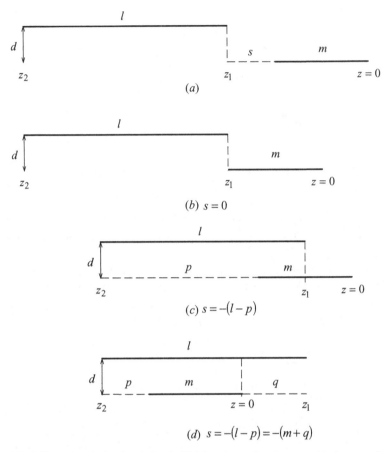

FIGURE 5.15. Using the basic relation in (5.34) to determine the mutual inductance for other offset structures.

in Fig. 5.15(a) we identify $z_2 = l + s + m$ and $z_1 = m + s$. Hence, for the structure in Fig. 5.15(a) we obtain

$$2M_p = (M_{l+s+m} + M_s) - (M_{m+s} + M_{s+l})$$

Similarly, for the case in Fig. 5.15(b) we identify $z_2 = l + m$ and $z_1 = m$ or, equivalently, $s = 0$. Hence, for the structure in Fig. 5.15(b) we obtain

$$2M_p = (M_{l+m} + M_0) - (M_m + M_l)$$
$$= M_{l+m} - M_m - M_l$$

For the case in Fig. 5.15(c) we identify $z_2 = p + m$ and $z_1 = p + m - l$ or, equivalently, $s = -(l - p)$. Hence, for the structure in Fig. 5.15(c) we obtain

$$2M_p = (M_{p+m} + M_{p-l}) - (M_{p+m-l} + M_p)$$
$$= (M_{p+m} + M_{l-p}) - (M_{p+m-l} + M_p)$$

For the case in Fig. 5.15(d) we identify $z_2 = p + m$ and $z_1 = -q = p + m - l$ or, equivalently, $s = -(l - p) = -(m + q)$. Hence, for the structure in Fig. 5.15(d) we obtain

$$2M_p = (M_{p+m} + M_{-q-m}) - (M_{-q} + M_p)$$
$$= (M_{p+m} + M_{q+m}) - (M_q + M_p)$$

Figure 5.16 shows how we could have easily obtained the basic result in (5.34) by using lumped-circuit analysis principles and the dot convention

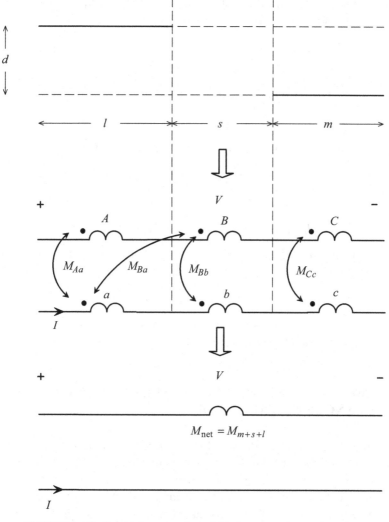

FIGURE 5.16. Combining mutual partial inductances that are in series.

[1,2]. We have shown an equivalent circuit for two parallel conductors of total length $m + s + l$ along with the mutual inductances between the three segments of lengths m, s, and l. Denoting the voltage between the endpoints of the ends of the top conductor as V, and passing a current I through the lower conductor, the total contribution to V due to the mutual inductances between all segments is

$$V = M_{\text{net}} \frac{dI}{dt}$$

Using the dot convention [1,2] and analyzing this circuit for the total voltage contributed to V by the mutual inductances between the segments, the net mutual inductance between the entire lengths is

$$M_{\text{net}} = M_{Aa} + M_{Ab} + M_{Ac} + M_{Ba} + M_{Bb} + M_{Bc} + M_{Ca} + M_{Cb} + M_{Cc}$$

But

$$M_{\text{net}} = M_{m+s+l}$$
$$M_{s+l} = M_{Aa} + M_{Ab} + M_{Ba} + M_{Bb}$$
$$M_{m+s} = M_{Bb} + M_{Bc} + M_{Cb} + M_{Cc}$$
$$M_s = M_{Bb}$$

Hence, we can write

$$M_{m+s+l} = M_{s+l} + M_{m+s} - M_s + M_{Ac} + M_{Ca}$$

However, the mutual inductance we desire between the two conductors of lengths l and m is

$$2M_p = M_{Ac} + M_{Ca}$$

Solving the last two relations gives the basic relation in (5.34), which we derived through a lengthy integration!

5.6 MUTUAL PARTIAL INDUCTANCE BETWEEN WIRES AT AN ANGLE TO EACH OTHER

We first consider a special case of two straight wires of lengths l and m that are inclined with respect to each other at an angle θ and joined at one end (or at least infinitesimally close) as shown in Fig. 5.17. The solution for the mutual partial inductance for this special case can be adapted to give the solution for a large class of similar problems, as we will see. This will be very similar

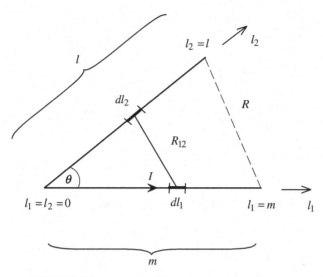

FIGURE 5.17. Wires inclined at an angle to each other.

to our recognizing that the mutual partial inductance for the case for two parallel but offset wires shown in Fig. 5.11 could be obtained in terms of the solution for two equal-length wires whose endpoints are aligned and shown in Fig. 5.10. This adaptation is given in (5.34) in terms of the mutual partial inductance of two equal-length parallel wires whose endpoints are aligned given in (5.21a,b).

We obtain the mutual partial inductance for the configuration in Fig. 5.17 using the Neumann integral:

$$M_p = \frac{\mu_0}{4\pi} \int_{l_2} \int_{l_1} \frac{d\mathbf{l}_1 \cdot d\mathbf{l}_2}{R_{12}}$$

$$= \frac{\mu_0}{4\pi} \cos\theta \int_{l_2} \int_{l_1} \frac{1}{R_{12}} dl_1 \, dl_2 \qquad (5.36a)$$

where l_1 and l_2 are the contours along the axes of the two wires, and R_{12} is the distance between the differential segments dl_1 and dl_2 given by

$$R_{12} = \sqrt{l_1^2 + l_2^2 - 2l_1 l_2 \cos\theta} \qquad (5.36b)$$

and we have used the law of cosines. The dot product of the vector differential segments becomes $d\mathbf{l}_1 \cdot d\mathbf{l}_2 = \cos\theta \, dl_1 \, dl_2$.

We can place this integral in an integrable form using the following technique [17]. We can show that (5.36a) can be written as

$$
\begin{aligned}
M_p &= \frac{\mu_0}{4\pi} \cos\theta \int_{l_2} \int_{l_1} \frac{1}{R_{12}} \, dl_1 \, dl_2 \\
&= \frac{\mu_0}{4\pi} \cos\theta \int_{l_2} \int_{l_1} \left[\frac{d}{dl_1} \left(\frac{l_1}{R_{12}} \right) + \frac{d}{dl_2} \left(\frac{l_2}{R_{12}} \right) \right] dl_1 \, dl_2 \\
&= \frac{\mu_0}{4\pi} \cos\theta \left[l_1 \int_{l_2} \frac{1}{R_{12}} \, dl_2 + l_2 \int_{l_1} \frac{1}{R_{12}} \, dl_1 \right]
\end{aligned}
\tag{5.37}
$$

This first equivalence in (5.37) can be shown, using R_{12} from (5.36b), to give

$$
\begin{aligned}
\frac{d}{dl_1} \left(\frac{l_1}{R_{12}} \right) &= \frac{R_{12} - l_1 R_{12}^{-1} (l_1 - l_2 \cos\theta)}{R_{12}^2} \\
&= \frac{1}{R_{12}} - \frac{l_1 (l_1 - l_2 \cos\theta)}{R_{12}^3}
\end{aligned}
\tag{5.38a}
$$

$$
\begin{aligned}
\frac{d}{dl_2} \left(\frac{l_2}{R_{12}} \right) &= \frac{R_{12} - l_2 R_{12}^{-1} (l_2 - l_1 \cos\theta)}{R_{12}^2} \\
&= \frac{1}{R_{12}} - \frac{l_2 (l_2 - l_1 \cos\theta)}{R_{12}^3}
\end{aligned}
\tag{5.38b}
$$

and we have used

$$
\frac{d\left(\frac{u}{v}\right)}{dx} = \frac{v \dfrac{du}{dx} - u \dfrac{dv}{dx}}{v^2}
\tag{D65}
$$

Hence,

$$
\frac{d}{dl_1} \left(\frac{l_1}{R_{12}} \right) + \frac{d}{dl_2} \left(\frac{l_2}{R_{12}} \right) = \frac{1}{R_{12}}
\tag{5.39}
$$

The second equivalence in (5.37) can easily be shown from

$$
\begin{aligned}
\int_{l_2} \int_{l_1} \left[\frac{d}{dl_1} \left(\frac{l_1}{R_{12}} \right) \right] dl_1 \, dl_2 &= \int_{l_2} \left[\int_{l_1} \frac{d}{dl_1} \left(\frac{l_1}{R_{12}} \right) dl_1 \right] dl_2 \\
&= \int_{l_2} \frac{l_1}{R_{12}} \, dl_2 \\
&= l_1 \int_{l_2} \frac{1}{R_{12}} \, dl_2
\end{aligned}
\tag{5.40a}
$$

$$\int_{l_1} \int_{l_2} \left[\frac{d}{dl_2} \left(\frac{l_2}{R_{12}} \right) \right] dl_2 \, dl_1 = \int_{l_1} \left[\int_{l_2} \frac{d}{dl_2} \left(\frac{l_2}{R_{12}} \right) dl_2 \right] dl_1$$

$$= \int_{l_1} \frac{l_2}{R_{12}} \, dl_1$$

$$= l_2 \int_{l_1} \frac{1}{R_{12}} \, dl_1 \qquad (5.40b)$$

and we have obtained the equivalence in (5.37). But the last result in (5.37) can easily be integrated using integral 380.001 from Dwight [7]:

$$\int \frac{dx}{\sqrt{x^2 + bx + c}} = \ln \left(2\sqrt{x^2 + bx + c} + 2x + b \right) \quad (D380.001)$$

and the equation for R_{12} in (5.36b) to give

$$\int_{l_i} \frac{1}{R_{12}} \, dl_i = \int_{l_i} \frac{1}{\sqrt{l_i^2 + l_j^2 - 2l_i l_j \cos \theta}} \, dl_i$$

$$= \ln \left(2\sqrt{l_i^2 + l_j^2 - 2l_i l_j \cos \theta} + 2l_i - 2l_j \cos \theta \right)$$

$$= \ln \left(\sqrt{l_i^2 + l_j^2 - 2l_i l_j \cos \theta} + l_i - l_j \cos \theta \right) \qquad (5.41)$$

The factor of 2 cancels out when we evaluate at the upper and lower limits of the integral.

Now we apply this result to the problem of Fig. 5.17. For economy of notation we denote the mutual partial inductance between the two segments as

$$M_p = \frac{\mu_0}{4\pi} N \qquad (5.42)$$

and N becomes, in terms of the limits of the integrals,

$$N = \int_{l_2=A}^{B} \int_{l_1=a}^{b} \frac{1}{R_{12}} \, dl_1 dl_2$$

$$= l_1 \int_{l_2} \frac{1}{R_{12}} \, dl_2 + l_2 \int_{l_1} \frac{1}{R_{12}} \, dl_1$$

$$= \left[l_1 \int_{l_2=A}^{B} \frac{1}{R_{12}} dl_2 \right]_{l_1=a}^{b} + \left[l_2 \int_{l_1=a}^{b} \frac{1}{R_{12}} dl_1 \right]_{l_2=A}^{B}$$

$$= b \left\{ \ln \left[\sqrt{b^2 + B^2 - 2bB \cos \theta} + B - b \cos \theta \right] \right.$$

$$- \ln \left[\sqrt{b^2 + A^2 - 2bA \cos \theta} + A - b \cos \theta \right] \Big\}$$

$$-a \left\{ \ln \left[\sqrt{a^2 + B^2 - 2aB \cos \theta} + B - a \cos \theta \right] \right.$$

$$- \ln \left[\sqrt{a^2 + A^2 - 2aA \cos \theta} + A - a \cos \theta \right] \Big\}$$

$$+B \left\{ \ln \left[\sqrt{b^2 + B^2 - 2bB \cos \theta} + b - B \cos \theta \right] \right.$$

$$- \ln \left[\sqrt{a^2 + B^2 - 2aB \cos \theta} + a - B \cos \theta \right] \Big\}$$

$$-A \left\{ \ln \left[\sqrt{b^2 + A^2 - 2bA \cos \theta} + b - A \cos \theta \right] \right.$$

$$- \ln \left[\sqrt{a^2 + A^2 - 2aA \cos \theta} + a - A \cos \theta \right] \Big\}$$

$$= b \ln \frac{R_{bB} + B - b \cos \theta}{R_{bA} + A - b \cos \theta} - a \ln \frac{R_{aB} + B - a \cos \theta}{R_{aA} + A - a \cos \theta}$$

$$+ B \ln \frac{R_{bB} + b - B \cos \theta}{R_{aB} + a - B \cos \theta} - A \ln \frac{R_{Ab} + b - A \cos \theta}{R_{aA} + a - A \cos \theta} \qquad (5.43)$$

The beginning and ending coordinates of the two lines are denoted as a,b for l_1 and A,B for l_2. The distances R_{ij} are the distances between the endpoints of the segments. For the problem in Fig. 5.17, we obtain

$$N = \left[l_1 \int_{l_2} \frac{1}{R_{12}} dl_2 + l_2 \int_{l_1} \frac{1}{R_{12}} dl_1 \right]$$

$$= \left\{ \left[l_1 \int_{l_2=0}^{l} \frac{1}{R_{12}} dl_2 \right]_{l_1=0}^{m} + \left[l_2 \int_{l_1=0}^{m} \frac{1}{R_{12}} dl_1 \right]_{l_2=0}^{l} \right\}$$

$$= m \ln \frac{R_{ml} + l - m \cos \theta}{R_{m0} + 0 - m \cos \theta} - 0 \ln \frac{R_{0l} + l - 0 \cos \theta}{R_{00} + 0 - 0 \cos \theta}$$

$$+ l \ln \frac{R_{ml} + m - l \cos \theta}{R_{0l} + 0 - l \cos \theta} - 0 \ln \frac{R_{0m} + m - 0 \cos \theta}{R_{00} + 0 - 0 \cos \theta}$$

$$= l \ln \frac{R + m - l \cos \theta}{l - l \cos \theta} + m \, \ln \frac{R + l - m \cos \theta}{m - m \cos \theta} \qquad (5.44)$$

and, by using l'Hôpital's rule,

$$\lim_{x \to 0} x \ln (x) = 0 \qquad (D605)$$

The distance between the endpoints is denoted as

$$R = R_{ml}$$
$$= \sqrt{l^2 + m^2 - 2lm\cos\theta} \tag{5.45}$$

and $R_{m0} = m$ and $R_{0l} = l$. Substituting the result in (5.44) into (5.42) gives the mutual partial inductance between the two segments in Fig. 5.17:

$$M_p = \frac{\mu_0}{4\pi}\cos\theta\left(l\ \ln\frac{R+m-l\cos\theta}{l-l\cos\theta} + m\ \ln\frac{R+l-m\cos\theta}{m-m\cos\theta}\right) \tag{5.46a}$$

But this result can be put into an equivalent form as [14]

$$M_p = \frac{\mu_0}{4\pi}\cos\theta\left(l\ \ln\frac{R+m+l}{R+l-m} + m\ \ln\frac{R+l+m}{R+m-l}\right) \tag{5.46b}$$

To demonstrate the equivalence between the two forms of the result in (5.46) we need to show that

$$\frac{R+m-l\cos\theta}{l-l\cos\theta} = \frac{R+m+l}{R+l-m} \tag{5.47}$$

This can be shown directly by multiplying it out to give

$$R^2 + R(l - l\cos\theta) + (m - l\cos\theta)(l - m)$$
$$\stackrel{?}{=} R(l - l\cos\theta) + (m + l)(l - l\cos\theta)$$

or

$$R^2 \stackrel{?}{=} (m + l)(l - l\cos\theta) - (m - l\cos\theta)(l - m)$$
$$= l^2 + m^2 - 2ml\cos\theta$$

which is satisfied.

In fact, a more general result can be proven which will be useful for other situations. Consider the triangles shown in Fig. 5.18. Each triangle is composed of two sides labeled R and R' with included angles θ and θ' with respect to the horizontal axes. These sides R and R' make projections on the horizontal axes of P and P', respectively, where $P = R\cos\theta$ and $P' = R'\cos\theta'$. The total length on the horizontal axis between the intersections of each line with the horizontal axis is denoted as T. We can prove the following important

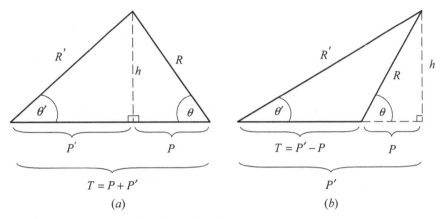

FIGURE 5.18. Important theorem.

equivalences. For the left triangle in Fig. 5.18(a) we have

$$\ln \frac{R+P}{R'-P'} = \ln \frac{R'+P'}{R-P}$$

$$= \ln \frac{R+R'+T}{R+R'-T}$$

$$= 2 \tanh^{-1} \frac{T}{R+R'} \qquad (5.48a)$$

and for the right triangle in Fig. 5.18(b) we have

$$\ln \frac{R-P}{R'-P'} = \ln \frac{R'+P'}{R+P}$$

$$= \ln \frac{R+R'+T}{R+R'-T}$$

$$= 2 \tanh^{-1} \frac{T}{R+R'} \qquad (5.48b)$$

The conversion of (5.48a) to (5.48b) is accomplished simply by replacing P in (5.48a) with $-P$. This is somewhat evident since P in Fig. 5.18(a) adds to P' to give the total length between the endpoints of R and R', which is denoted as $T = P' + P$, whereas in Fig. 5.18(b) P subtracts from P' to give $T = P' - P$. The identity for Fig. 5.17 in (5.47) follows from the identity in (5.48a).

The proofs of (5.48) are fairly simple by comparing the arguments of the log functions. For example, (5.48a) gives

$$\frac{R+P}{R'-P'} \overset{?}{=} \frac{R'+P'}{R-P} \overset{?}{=} \frac{R+R'+T}{R+R'-T}$$

Multiplying these out gives

$$(R + P)(R - P) \overset{?}{=} (R' + P')(R' - P')$$

But $R^2 = P^2 + h^2$ and $R'^2 = P'^2 + h^2$. Substituting $T = P + P'$, we need to show that

$$\frac{R + P}{R' - P'} \overset{?}{=} \frac{R + R' + (P + P')}{R + R' - (P + P')}$$

Multiplying this out and canceling common terms gives

$$R^2 - P^2 \overset{?}{=} R'^2 - P'^2$$

which is satisfied. The results in (5.48b) can be verified similarly. The last results in (5.48) are verified using the identity for the inverse hyperbolic tangent:

$$\tanh^{-1} x = \frac{1}{2} \ln \frac{1 + x}{1 - x} \qquad x^2 < 1 \qquad \text{(D702)}$$

Hence, a further equivalent form for the result in (5.46) for Fig. 5.17 can be obtained in terms of the inverse hyperbolic tangent as

$$\boxed{\begin{aligned} M_p &= \frac{\mu_0}{4\pi} \cos\theta \left(l \ln \frac{R + m + l}{R + l - m} + m \ln \frac{R + l + m}{R + m - l} \right) \\ &= \frac{\mu_0}{2\pi} \cos\theta \left(l \tanh^{-1} \frac{m}{R + l} + m \tanh^{-1} \frac{l}{R + m} \right) \end{aligned}} \qquad \text{(5.46c)}$$

This solution process for the configuration of Fig. 5.17 can readily be adapted to obtain the mutual partial inductance between two segments that do not physically join at a common point but are inclined at an angle θ to each other as shown in Fig. 5.19. Extend the segments of lengths l and m to a point where they join, thereby generating the extension lengths α and β. Adapting

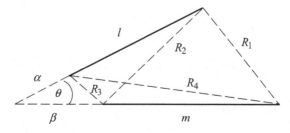

FIGURE 5.19. More general case of Fig. 5.17.

the result in (5.43) gives the result for Fig. 5.19 as

$$N = l_1 \int_{l_2} \frac{1}{R_{12}} dl_2 + l_2 \int_{l_1} \frac{1}{R_{12}} dl_1$$

$$= \left\{ \left[l_1 \int_{l_2=\alpha}^{\alpha+l} \frac{1}{R_{12}} dl_2 \right]_{l_1=\beta}^{\beta+m} + \left[l_2 \int_{l_1=\beta}^{\beta+m} \frac{1}{R_{12}} dl_1 \right]_{l_2=\alpha}^{\alpha+l} \right\}$$

$$= (\beta+m) \ln \frac{R_{(\beta+m)(\alpha+l)} + (\alpha+l) - (\beta+m)\cos\theta}{R_{(\beta+m)\alpha} + \alpha - (\beta+m)\cos\theta}$$

$$-\beta \ln \frac{R_{\beta(\alpha+l)} + (\alpha+l) - \beta\cos\theta}{R_{\beta\alpha} + \alpha - \beta\cos\theta}$$

$$+(\alpha+l) \ln \frac{R_{(\beta+m)(\alpha+l)} + (\beta+m) - (\alpha+l)\cos\theta}{R_{\beta(\alpha+l)} + \beta - (\alpha+l)\cos\theta}$$

$$-\alpha \ln \frac{R_{\alpha(\beta+m)} + (\beta+m) - \alpha\cos\theta}{R_{\beta\alpha} + \beta - \alpha\cos\theta} \qquad (5.49)$$

Denoting the distances between the endpoints of the lines as shown in Fig. 5.19 gives

$$R_1 = R_{(\alpha+l)(\beta+m)}$$
$$= \sqrt{(\alpha+l)^2 + (\beta+m)^2 - 2(\alpha+l)(\beta+m)\cos\theta} \qquad (5.50a)$$

$$R_2 = R_{(\alpha+l)\beta}$$
$$= \sqrt{(\alpha+l)^2 + \beta^2 - 2(\alpha+l)\beta\cos\theta} \qquad (5.50b)$$

$$R_3 = R_{\alpha\beta}$$
$$= \sqrt{\alpha^2 + \beta^2 - 2\alpha\beta\cos\theta} \qquad (5.50c)$$

$$R_4 = R_{\alpha(\beta+m)}$$
$$= \sqrt{\alpha^2 + (\beta+m)^2 - 2\alpha(\beta+m)\cos\theta} \qquad (5.50d)$$

Hence, the mutual partial inductance between the two segments of Fig. 5.19 becomes

$$
\begin{aligned}
M_p &= \frac{\mu_0}{4\pi} N \\
&= \frac{\mu_0}{4\pi} \Bigg[(\beta + m) \ln \frac{R_1 + (\alpha + l) - (\beta + m)\cos\theta}{R_4 + \alpha - (\beta + m)\cos\theta} \\
&\quad - \beta \ln \frac{R_2 + (\alpha + l) - \beta\cos\theta}{R_3 + \alpha - \beta\cos\theta} \\
&\quad + (\alpha + l) \ln \frac{R_1 + (\beta + m) - (\alpha + l)\cos\theta}{R_2 + \beta - (\alpha + l)\cos\theta} \\
&\quad - \alpha \ln \frac{R_4 + (\beta + m) - \alpha\cos\theta}{R_3 + \beta - \alpha\cos\theta} \Bigg]
\end{aligned}
$$
(5.51a)

Using the identities in (5.48) and the inverse hyperbolic tangent identity in (D702) gives equivalent forms as

$$
\begin{aligned}
M_p &= \frac{\mu_0}{4\pi} N \\
&= \frac{\mu_0}{4\pi} \Bigg[(\beta + m) \ln \frac{R_1 + R_4 + l}{R_1 + R_4 - l} - \beta \ln \frac{R_2 + R_3 + l}{R_2 + R_3 - l} \\
&\quad + (\alpha + l) \ln \frac{R_1 + R_2 + m}{R_1 + R_2 - m} - \alpha \ln \frac{R_4 + R_3 + m}{R_4 + R_3 - m} \Bigg] \\
&= \frac{\mu_0}{2\pi} \Bigg[(\beta + m)\tanh^{-1}\frac{l}{R_1 + R_4} - \beta\tanh^{-1}\frac{l}{R_2 + R_3} \\
&\quad + (\alpha + l)\tanh^{-1}\frac{m}{R_1 + R_2} - \alpha\tanh^{-1}\frac{m}{R_4 + R_3} \Bigg]
\end{aligned}
$$
(5.51b)

We can then use the previous result for the mutual partial inductance between two segments of lengths x and y that are joined at a common point that was derived for Fig. 5.17 and given in (5.46c) to obtain the mutual partial inductance for Fig. 5.19 indirectly. Denote the result for Fig. 5.17 as

$$
\begin{aligned}
M_{x,y} &= \frac{\mu_0}{4\pi}\cos\theta \left(x \ln \frac{R+x+y}{R+x-y} + y \ln \frac{R+y+x}{R+y-x} \right) \\
&= \frac{\mu_0}{2\pi}\cos\theta \left(x \tanh^{-1}\frac{y}{R+x} + y \tanh^{-1}\frac{x}{R+y} \right)
\end{aligned}
$$
(5.52a)

where θ is the included angle where they are joined and

$$R = \sqrt{x^2 + y^2 - 2xy\cos\theta} \tag{5.52b}$$

Visualize the structure of Fig. 5.19 as consisting of four such structures, each consisting of the following lengths, with each pair being joined at a common point: (1) $x = \alpha + l$, $y = \beta + m$, (2) $x = \alpha$, $y = \beta$, (3) $x = \alpha + l$, $y = \beta$, and (4) $x = \alpha$, and $y = \beta + m$. We can then obtain the total mutual partial inductance for the structure in Fig. 5.19 of overall lengths $\alpha + l$ and $\beta + m$ to give, in a fashion similar to that of Fig. 5.16,

$$M_{\alpha+l,\beta+m} = M_{\alpha,\beta} + M_{\alpha,m} + M_{l,\beta} + M_{l,m}$$

The desired result is $M_{l,m}$ giving

$$M_{l,m} = M_{\alpha+l,\beta+m} - M_{\alpha,\beta} - M_{\alpha,m} - M_{l,\beta}$$

But the result for Fig. 5.17 does not apply to generating $M_{\alpha,m}$ or $M_{l,\beta}$ since the two lengths in each of these are not joined at a common point. So we write this as

$$M_{l,m} = M_{\alpha+l,\beta+m} - M_{\alpha,\beta} - \underbrace{\left(M_{\alpha,m} + M_{\alpha,\beta}\right)}_{M_{\alpha,\beta+m}} - \underbrace{\left(M_{l,\beta} + M_{\alpha,\beta}\right)}_{M_{\alpha+l,\beta}} + 2M_{\alpha,\beta}$$

giving

$$\boxed{M_p = \left(M_{\alpha+l,\beta+m} + M_{\alpha\beta}\right) - \left(M_{\alpha+l,\beta} + M_{\beta+m,\alpha}\right)} \tag{5.53}$$

Using the result for two segments joined at one end in (5.52) gives the result as

$$
\begin{aligned}
M_p = \frac{\mu_0}{4\pi} \cos\theta \left[(\alpha + l) \ln \frac{R_1 + (\alpha + l) + (\beta + m)}{R_1 + (\alpha + l) - (\beta + m)} \right. \\
+ (\beta + m) \ln \frac{R_1 + (\beta + m) + (\alpha + l)}{R_1 + (\beta + m) - (\alpha + l)} \\
+ \alpha \ln \frac{R_3 + \alpha + \beta}{R_3 + \alpha - \beta} + \beta \ln \frac{R_3 + \beta + \alpha}{R_3 + \beta - \alpha} \\
- (\alpha + l) \ln \frac{R_2 + (\alpha + l) + \beta}{R_2 + (\alpha + l) - \beta} - \beta \ln \frac{R_2 + \beta + (\alpha + l)}{R_2 + \beta - (\alpha + l)} \\
\left. - (\beta + m) \ln \frac{R_4 + (\beta + m) + \alpha}{R_4 + (\beta + m) - \alpha} - \alpha \ln \frac{R_4 + \alpha + (\beta + m)}{R_4 + \alpha - (\beta + m)} \right]
\end{aligned}
$$

$$\tag{5.54a}$$

This result can be simplified to

$$
M_p = \frac{\mu_0}{4\pi} \cos\theta \left[(\alpha + l) \ln \frac{R_1 + R_2 + m}{R_1 + R_2 - m} + (\beta + m) \ln \frac{R_1 + R_4 + l}{R_1 + R_4 - l} \right.
$$
$$
\left. -\alpha \ln \frac{R_3 + R_4 + m}{R_3 + R_4 - m} - \beta \ln \frac{R_2 + R_3 + l}{R_2 + R_3 - l} \right]
$$

(5.54b)

which agrees with (5.51b).

The equivalence of (5.54a) and (5.54b) can be shown with the following important identity for triangles. Consider the three triangles shown in Fig. 5.20. Triangle T_1 has sides of a, b, and R_1. Triangle T_2 has sides of a, c, and R_2. Triangle T_3 has sides of R_1, $b - c$, and R_2 and is formed from triangles T_1 and T_2 as $T_3 = T_1 - T_2$. It is a simple matter to prove the following

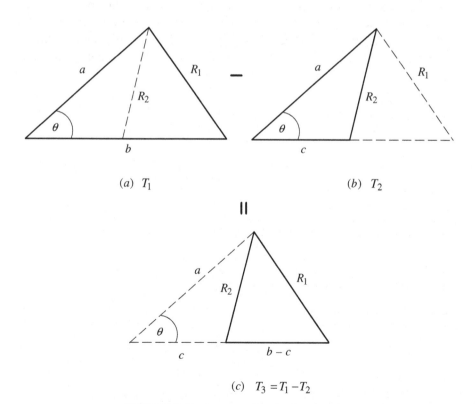

FIGURE 5.20. Important identity for triangles.

identity for these three related triangles:

$$\ln \frac{R_1 + a + b}{R_1 + a - b} - \ln \frac{R_2 + a + c}{R_2 + a - c} = \ln \frac{R_1 + R_2 + (b - c)}{R_1 + R_2 - (b - c)} \tag{5.55a}$$

and R_1 and R_2 are given by the law of cosines:

$$R_1^2 = a^2 + b^2 - 2ab \cos \theta \tag{5.55b}$$

$$R_2^2 = a^2 + c^2 - 2ac \cos \theta \tag{5.55c}$$

The important identity in (5.55a) can easily be verified by multiplying out the arguments of the logarithms as $\ln A - \ln B = \ln C \Rightarrow A/B = C$. Applying (5.55a) to (5.54a) gives the equivalence to (5.54b).

Figure 5.21 shows the general configuration for skewed and displaced conductors. The general result for this was derived by G.A. Campbell in 1915 [17]. This figure is modeled after that of Grover [14], pp. 56, who clearly explained the general result obtained by Campbell. The first conductor is of length l and its endpoints are denoted as A and B. It is shown as lying in a plane. The second conductor is of length m and its endpoints are denoted as a and b. It is shown as lying in another plane. These two planes containing the two conductors are parallel and separated by distance d between the two

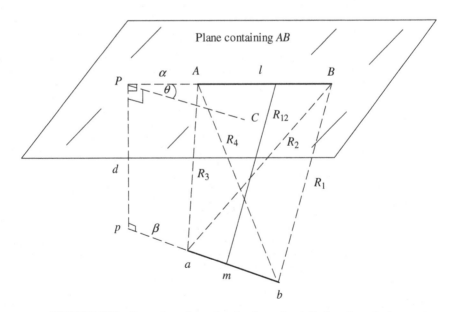

FIGURE 5.21. General configuration for skewed and displaced conductors.

planes. The line Pp between the two planes is of length d and is mutually perpendicular to the two planes containing the two conductors. Hence, Pp is said to be the *common perpendicular* to the two conductors. The endpoints of the conductors, A and a, are displaced from points P and p by distances α and β, respectively. The line PC lying in the plane containing AB is parallel to the line ab representing the second conductor and is at an angle θ to the first conductor AB. This is what is meant by the two conductors AB and ab having an angle of inclination of θ with respect to each other. If the displacement between the planes, d, is zero, $d = 0$, then the angle θ between the two conductors is the same as in the previous results.

The Neumann integral in (5.36) remains the same for this case:

$$M_p = \frac{\mu_0}{4\pi} \cos\theta \int_{l_1} \int_{l_1} \frac{1}{R_{12}} \, dl_1 \, dl_2 \qquad (5.56a)$$

where l_1 and l_2 again denote the contours along the two conductors of lengths m and l, respectively, and

$$R_{12} = \sqrt{d^2 + l_1^2 + l_2^2 - 2l_1 l_2 \cos\theta} \qquad (5.56b)$$

Carrying through with a similar development as before gives

$$M_p = \frac{\mu_0}{4\pi} \cos\theta \int_{l_2} \int_{l_1} \frac{1}{R_{12}} \, dl_1 \, dl_2$$

$$= \frac{\mu_0}{4\pi} \cos\theta \int_{l_2} \int_{l_1} \left[\frac{d}{dl_1}\left(\frac{l_1}{R_{12}}\right) + \frac{d}{dl_2}\left(\frac{l_2}{R_{12}}\right) - \frac{d^2}{R_{12}^3} \right] dl_1 \, dl_2$$

$$= \frac{\mu_0}{4\pi} \cos\theta \left(l_1 \int_{l_2} \frac{1}{R_{12}} \, dl_2 + l_2 \int_{l_1} \frac{1}{R_{12}} \, dl_1 \right.$$

$$\left. - \frac{d}{\sin\theta} \int_{l_2} \int_{l_1} \frac{d \sin\theta}{R_{12}^3} \, dl_1 \, dl_2 \right) \qquad (5.57a)$$

and one can similarly show using R_{12} in (5.56b), as was done previously for $d = 0$, that

$$\frac{d}{dl_1}\left(\frac{l_1}{R_{12}}\right) + \frac{d}{dl_2}\left(\frac{l_2}{R_{12}}\right) - \frac{d^2}{R_{12}^3} = \frac{1}{R_{12}} \qquad (5.57b)$$

Again, the last result in (5.57a) can be integrated, using (D380.001), to yield

$$M_p = \frac{\mu_0}{4\pi} \cos\theta \left(l_1 \int_{l_2} \frac{1}{R_{12}} \, dl_2 + l_2 \int_{l_1} \frac{1}{R_{12}} \, dl_1 \right.$$

$$\left. - \frac{d}{\sin\theta} \int_{l_2} \int_{l_1} \frac{d \, \sin\theta}{R_{12}^3} \, dl_1 \, dl_2 \right)$$

$$= \frac{\mu_0}{4\pi} \cos\theta \left[\left[l_1 \ln\left(R_{12} + l_2 - l_1 \cos\theta\right) + l_2 \ln\left(R_{12} + l_1 - l_2 \cos\theta\right) \right. \right.$$

$$\left. \left. - \frac{\Omega d}{\sin\theta} \right]_{l_2=PA}^{l_2=PB} {}_{\textstyle l_1=pa}^{\textstyle l_1=pb} \right]$$

$$= \frac{\mu_0}{2\pi} \left(pB' \tanh^{-1} \frac{ab}{aB + Bb} - pA' \tanh^{-1} \frac{ab}{aA + Ab} \right.$$

$$\left. + Pb' \tanh^{-1} \frac{AB}{Ab + bB} - Pa' \tanh^{-1} \frac{AB}{Aa + aB} - \frac{\Omega d}{\tan\theta} \right) \qquad (5.58a)$$

where the solid angle Ω is

$$\Omega = \tan^{-1}\left(\frac{Pp}{Bb}\cot\theta + \frac{PB}{Pp}\frac{pb}{Bb}\sin\theta\right) - \tan^{-1}\left(\frac{Pp}{Ba}\cot\theta + \frac{PB}{Pp}\frac{pa}{Ba}\sin\theta\right)$$

$$- \tan^{-1}\left(\frac{Pp}{Ab}\cot\theta + \frac{PA}{Pp}\frac{pb}{Ab}\sin\theta\right)$$

$$+ \tan^{-1}\left(\frac{Pp}{Aa}\cot\theta + \frac{PA}{Pp}\frac{pa}{Aa}\sin\theta\right) \qquad (5.58b)$$

In (5.58a) primes denote the projection of the point on one conductor perpendicular to and onto the other conductor, and the inverse hyperbolic tangent is again defined in terms of the natural logarithm as

$$\tanh^{-1} x = \frac{1}{2} \ln \frac{1+x}{1-x} \qquad x^2 < 1 \qquad (D702)$$

Grover [14] simplified this in terms of the quantities in Fig. 5.21 and the result becomes

$$\boxed{\begin{aligned} M_p = {} & \frac{\mu_0}{2\pi} \cos\theta \left[(\alpha + l) \tanh^{-1} \frac{m}{R_1 + R_2} + (\beta + m) \tanh^{-1} \frac{l}{R_1 + R_4} \right. \\ & \left. - \alpha \tanh^{-1} \frac{m}{R_3 + R_4} - \beta \tanh^{-1} \frac{l}{R_2 + R_3} \right] - \frac{\mu_0}{4\pi} \frac{\Omega d}{\tan\theta} \end{aligned}}$$

$$(5.59a)$$

where the solid angle Ω is

$$\Omega = \tan^{-1} \frac{d^2 \cos\theta + (\alpha + l)(\beta + m)\sin^2\theta}{dR_1 \sin\theta}$$
$$- \tan^{-1} \frac{d^2 \cos\theta + (\alpha + l)\beta \sin^2\theta}{dR_2 \sin\theta}$$
$$+ \tan^{-1} \frac{d^2 \cos\theta + \alpha\beta \sin^2\theta}{dR_3 \sin\theta}$$
$$- \tan^{-1} \frac{d^2 \cos\theta + \alpha(\beta + m)\sin^2\theta}{dR_4 \sin\theta}$$

(5.59b)

The distances between the ends of the two conductors are shown in Fig. 5.21 and are $R_1 = Bb$, $R_2 = Ba$, $R_3 = Aa$, and $R_4 = Ab$. Using the law of cosines, these distances are

$$R_1^2 = d^2 + (\alpha + l)^2 + (\beta + m)^2 - 2(\alpha + l)(\beta + m)\cos\theta \quad \text{(5.60a)}$$

$$R_2^2 = d^2 + (\alpha + l)^2 + \beta^2 - 2\beta(\alpha + l)\cos\theta \quad \text{(5.60b)}$$

$$R_3^2 = d^2 + \alpha^2 + \beta^2 - 2\alpha\beta\cos\theta \quad \text{(5.60c)}$$

$$R_4^2 = d^2 + \alpha^2 + (\beta + m)^2 - 2\alpha(\beta + m)\cos\theta \quad \text{(5.60d)}$$

The only difference between this result and the result for Fig. 5.19 given in (5.51b) is the solid angle Ω, which goes away for $d = 0$.

5.7 NUMERICAL VALUES OF PARTIAL INDUCTANCES AND SIGNIFICANCE OF INTERNAL INDUCTANCE

It is helpful to obtain some representative values of the self and mutual partial inductances for typical configurations. The self partial inductance of a wire of radius r_w and length l is obtained as

$$L_p = 2 \times 10^{-7} l \left[\ln\left(\frac{l}{r_w} + \sqrt{\left(\frac{l}{r_w}\right)^2 + 1}\right) \right.$$
$$\left. - \sqrt{1 + \left(\frac{r_w}{l}\right)^2} + \frac{r_w}{l} \right]$$

(5.18a)

Observe that this depends on the ratio of the wire length and the wire radius: l/r_w. It is typical to specify wire radii r_w in mils (1000 mils=1 in. and 1 in.=2.54 cm). Also observe that the length of the wire, l, also appears outside the equation. Hence, it is not possible to speak of an absolute per-unit-length inductance as is the case for a two-wire transmission line of infinite length. Nevertheless, we can divide both sides of (5.18a) by the wire length and obtain a universal plot of the ratio of self partial inductance per unit length, L_p/l, versus the ratio l/r_w as

$$\frac{L_p}{l} = 5.08 \left[\ln\left(\frac{l}{r_w} + \sqrt{\left(\frac{l}{r_w}\right)^2 + 1}\right) \right.$$
$$\left. - \sqrt{1 + \left(\frac{r_w}{l}\right)^2} + \frac{r_w}{l} \right] \qquad \text{nH/in.} \qquad (5.61)$$

This is shown in Fig. 5.22 for ratios of $10 \le l/r_w \le 500$. For example, a No. 20 gauge (AWG) wire is a common wire size and has a radius of 16 mils. Hence, the last ratio of 500 plotted represents a wire length of 8 in. for a No. 20 gauge wire (30.02 nH/in.), and a ratio of 10 represents a length of 0.16 in., or about 3/16 in. (10.63 nH/in.). Because the wire length l appears outside the result, it is not possible to state a single per-unit-length value of the self partial inductance. But the plot in Fig. 5.22 indicates that a reasonable rule of thumb for practical wire sizes and wire lengths is a value of between 15 and 30 nH/in.

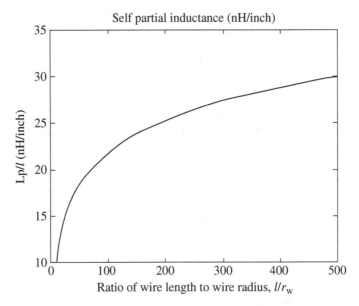

FIGURE 5.22. Plot of L_p/l (nH/in.) vs. the ratio of wire length to wire radius, l/r_w.

We obtained the dc per-unit-length value of the internal inductance of a wire (which is independent of the wire radius) as

$$l_{\text{internal}} = \frac{\mu_0}{8\pi}$$

$$= 0.5 \times 10^{-7} \text{H/m}$$

$$= 1.27 \text{nH/in.} \tag{5.62}$$

Technically, this should be multiplied by the wire length and added to the external self partial inductance in (5.18a) to give the total self partial inductance:

$$L_{p,\text{total}} = L_{p,(5.18a)} + l_{\text{internal}} \times l \tag{5.63}$$

But we see from Fig. 5.22 that for practical situations the internal inductance of the wire can generally be neglected. Furthermore, the value for the internal inductance in (5.62) is its value at dc. As frequency is increased from zero, the current tends to move toward the surface of the wire, and hence the internal inductance goes to zero. This gives further support to the observation that the internal inductance can generally be neglected.

The mutual partial inductance between two wires of common length l and separation d is obtained as

$$M_p = 2 \times 10^{-7} l \left[\ln \left(\frac{l}{d} + \sqrt{\left(\frac{l}{d}\right)^2 + 1} \right) \right.$$
$$\left. - \sqrt{1 + \left(\frac{d}{l}\right)^2} + \frac{d}{l} \right] \tag{5.21a}$$

As was the case for self partial inductance, notice that this depends on the ratio of wire length to wire separation, l/d. But the wire length, l, also appears outside the result, so it is not possible to speak of an absolute value of per-unit-length mutual inductance, as is the case for a transmission line of infinite length. Nevertheless, we can divide both sides of (5.21a) by the wire length and obtain a universal plot of the ratio of the per-unit-length mutual partial inductance, M_p/l, versus the ratio l/d as

$$\frac{M_p}{l} = 5.08 \left[\ln \left(\frac{l}{d} + \sqrt{\left(\frac{l}{d}\right)^2 + 1} \right) - \sqrt{1 + \left(\frac{d}{l}\right)^2} + \frac{d}{l} \right] \quad \text{nH/in.}$$

$$\tag{5.64}$$

This is plotted in Fig. 5.23 for ratios of $1 \leq l/d \leq 100$. For example, a ratio of 80 would apply to two wires of length 5 in. and a separation between them of 0.0625 in., or 1/16 in. (20.77 nH/in.), and a ratio of 10 would apply to

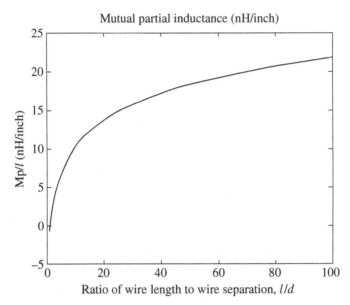

FIGURE 5.23. Plot of M_p/l in (nH/in.) vs. the ratio of wire length to wire separation l/d.

two wires of length 5 in. and a separation between them of 1/2 in. (10.63 nH/in.). Observe that as the wire separation increases without bound (i.e., the ratio goes to zero), the mutual partial inductance goes to zero: an expected result. Similarly, as the wire separation goes to zero (approaches the radii of the wires) (i.e., the ratio increases), the mutual partial inductance approaches the self partial inductance shown in Fig. 5.22: again, an expected result.

5.8 CONSTRUCTING LUMPED EQUIVALENT CIRCUITS WITH PARTIAL INDUCTANCES

Unlike the case of loop inductances, for current loops whose borders are bounded by piecewise-linear segments of wires, there are no further partial inductances to be derived. We simply "put together" the partial inductances (self and mutual) derived previously in this chapter and "turn the crank." We construct an equivalent lumped-circuit model that can be solved with, for example, the SPICE circuit analysis computer program [2]. To do so, we finally need to discuss the allocation of the dots in an inductor equivalent circuit of the segments. The key to doing so is to be able to determine correctly the total voltage developed across the segment, magnitude and polarity, by using the dot convention that replicates the derivation of that inductance in this chapter. The self partial inductance of a segment determines the voltage across

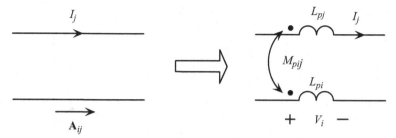

FIGURE 5.24. Mutual partial inductance between pairs of segments.

it that is due to the current through the element according to the passive sign convention: Current entering one end of the inductance produces a voltage across the element that is positive at that end. The dots do not have anything to do with this self voltage.

Let us first address the situation for a pair of segments shown in Fig. 5.24. The dots are placed on the ends of two elements so that the magnitude and polarity of the contribution to the voltage across one of the elements that is due to the current through the other element via the mutual partial inductance between the associated segments will be determined correctly. *The key to doing so is to replicate the situation for which the mutual partial inductance between two segments was as derived in this chapter.* Note that if a current I on one segment enters the dotted end of that segment, a voltage $M_p dI/dt$ will be developed along the other segment that is positive at the dotted end of that segment:

$$V_i = M_{pij} \frac{dI_j}{dt}$$

The key to getting the dots placed correctly *on a pair of segments* is observed to be in the relation between the current in one segment and the direction of the vector magnetic potential $\mathbf{A}$ along the other segment, which was used in the derivations of the mutual partial inductance. The vector magnetic potential $\mathbf{A}$ is everywhere parallel to the current that produced it. Hence, the positive terminal of the induced voltage is on the end of the segment that $\mathbf{A}$ enters, as shown in Fig. 5.24. In other words, $\mathbf{A}$ points *from* the positive terminal of the induced voltage *to* the negative terminal of the induced voltage.

This can be done easily for a pair of elements. For more than two coupled segments we must *arbitrarily* place the dots on the ends of the inductor symbols for each segment. But some of the mutual partial inductances may turn out to be *negative* for that placement. A good example of this is the rectangular loop shown in Fig. 5.1. The inductive equivalent circuit is shown in Fig. 5.3 and the dots are assigned arbitrarily. Observe that a current directed

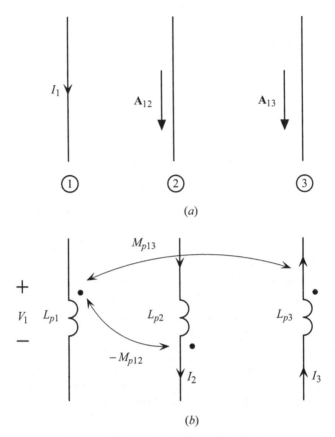

FIGURE 5.25. Assigning dots to the segments.

down through the right segment of the loop produces a vector magnetic potential along the left side of the loop that is also directed downward. But this is opposite the assigned dotted terminal of the left inductor. Hence, M_{13} here is negative.

Figure. 5.25 shows an example of this. Assigning the dots *arbitrarily* gives the inductive equivalent circuit in Fig. 5.25(b). The voltage across the first inductor is assigned the polarity of positive at its dotted end. Directing a current I_1 through the first inductor that enters the assigned dotted end of it as shown in Fig. 5.25(a) generates vector magnetic potentials along the other segments that enters the dotted end of the third segment (assigned arbitrarily) so that the mutual inductance between the first and third inductors is positive as M_{p13}. But current I_1 that enters the dotted terminal of the first inductor (arbitrarily assigned) generates a vector magnetic potential that enters the undotted end of the second segment (assigned arbitrarily) so that the mutual partial inductance between that pair is $-M_{p12}$, where M_{p12} here has a positive value. Hence, using the dot convention, the voltage generated across the first inductor that is due only to the mutual inductances is

$$V_1 = - \left(-M_{p12}\right) \frac{dI_2}{dt} - M_{p13} \frac{dI_3}{dt}$$

$$= +M_{p12} \frac{dI_2}{dt} - M_{p13} \frac{dI_3}{dt}$$

Computer-aided circuit analysis programs such as SPICE require that all self inductances such as L_p here must be positive. However, there are no restrictions on the signs of any of the mutual inductances: Some may have negative values.

6

PARTIAL INDUCTANCES OF CONDUCTORS OF RECTANGULAR CROSS SECTION

In this chapter we obtain the self and mutual partial inductances for conductors of rectangular cross section, referred to here as printed circuit board (PCB) *lands*. Figure 6.1 shows this type of conductor. The width is denoted as w, the length is denoted as l, and the thickness is denoted as t.

In previous chapters we have detailed the computation of inductances for conductors having circular, cylindrical cross sections (i.e., wires). The computation of the partial inductances of and between wires is fairly simple, for an important reason. We consistently made the assumption that *the current carried by a wire is uniformly distributed over the cross section of the wire* which is true for dc and widely spaced wires. In this case, *for the purposes of computing the magnetic fields from that wire, we can replace the wire with a filament containing the total current $I = JA$*, where J is the uniform current density distribution and A is the area of the wire cross section. This is an extraordinarily important simplifying assumption, for a number of reasons. First, we can equate the self partial inductance of a wire having a uniform current distribution over its cross section to the *total magnetic flux threading the surface formed between the surface of the wire and infinity* per unit of that current. No matter what radial direction about the wire we choose to go to infinity, the result is the same since the magnetic field is symmetric about the wire. This provides important alternative methods of directly

Inductance: Loop and Partial, By Clayton R. Paul
Copyright © 2010 John Wiley & Sons, Inc.

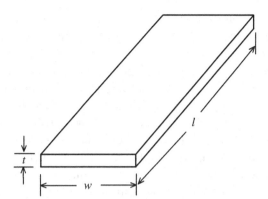

FIGURE 6.1. Printed circuit board land.

computing the self and mutual partial inductances of wires. We can directly compute the magnetic flux through that surface from the surface integral of the magnetic flux density vector **B**, or we could integrate via a line integral the vector magnetic potential **A** along the surface of the wire, which leads to the third method, the Neumann integral.

Now consider the case of the PCB land. Even if we assume that the current is distributed uniformly over the cross section of the land (as we do in this chapter), the magnetic fields about the land do not form concentric circles, and hence we cannot replace the land with an equivalent filament containing the total current for the purposes of computing the magnetic fields due to it. [See Chapter 2 for the fields around an infinitely long, flat conductor of width W given in (2.70).] Hence, for lands, we can no longer relate the partial inductances to the magnetic flux between the land surface and infinity: From which point on the land shall we draw the boundaries of the surface? Although the magnetic fields about a land will appear at very large distances as though they are due to the current from a filament, the computation of the flux at nearer distances is dominant.

We need another way of meaningfully formulating the partial inductances of and between lands. In the next section we derive that formulation in terms of stored energy in the magnetic field. A fundamental assumption in that derivation is again that the current is distributed uniformly over the cross section of the land. This assumption is also true for dc currents and will provide a simplification in the computation, as we will see. When is this assumption of a uniform current distribution over the land cross section invalidated? As in the case of wires, adjacent land currents will cause the currents to migrate toward the facing surfaces of the lands in the same way that closely spaced wires will cause their current distributions to move toward the facing surfaces of the wires. This is again the phenomenon of "proximity effect." For wires, this was not pronounced enough to invalidate the replacement of the wire with a filament as long as the ratio of wire separation to wire radius was larger than about 4 (i.e., for two identical wires, another identical wire would just fit between the two). So this assumption of uniform current distribution over the cross section of the wire and the subsequent replacement of the wire with a filament is not a limiting assumption for wires having typical separations. For two lands this region of separation for the current to be approximately uniformly distributed over the cross section cannot easily be determined, but we will make the assumption that the proximity effect is not pronounced in order to make computation of the self and mutual partial inductances feasible. We provide numerical computations in Section 6.5 to give us some feel for what "too close" means for this problem.

Recall that the fundamental computation of inductance is for dc currents! *If a wire or a PCB land carries a dc current and is isolated from other wires or PCB lands, the current will be distributed uniformly over the wire or land cross section.* For wires, as the frequency of the current increases from dc, the current will crowd to the surface, lying in a region of the surface of thickness on the order of a skin depth, but the current will remain *symmetric about the wire axis* and can still be replaced by a filament as long as the wire is not "too close" to other wires. The internal inductance will go to zero but the external inductances will remain the same. In the case of a PCB land, as the frequency of the current also increases from dc, the current will migrate toward the surface of the land *but it will also peak at the sharp corners of the land*, as numerical computations in Section 6.5 will show. Hence, the advantage of a wire current of increasing frequency remaining symmetric about the wire axis is not shared by the PCB land, even if the proximity effect is not pronounced.

This prior discussion illustrates that although the computation of loop and partial inductances for wires was rather straightforward, *the same computations for PCB lands will be much more difficult.* Even using the assumption that the current is uniformly distributed over the cross section of the land, the computation of its self and mutual partial inductances is considerably more difficult than for wires. Hoer and Love [16] provide general formulas for the self and mutual inductances of PCB lands. These formulas are not derived in [16], but in the following we provide detailed derivations of them. Again the derivations of these formulas for PCB lands are very complicated, as are the resulting formulas themselves.

6.1 FORMULATION FOR THE COMPUTATION OF THE PARTIAL INDUCTANCES OF PCB LANDS

As indicated previously, we need to determine a suitable method for computing the self and mutual partial inductances of PCB lands. The method we use is in terms of the energy stored in the magnetic field. First recall that the magnetic energy stored in the magnetic field is given by

$$W_M = \tfrac{1}{2} \int_{\text{all space}} \mathbf{B} \cdot \mathbf{H} \; dv \qquad \mathbf{J} \tag{6.1}$$

Recall that we define the vector magnetic potential $\mathbf{A}$ in terms of the magnetic flux density vector $\mathbf{B}$ as $\mathbf{B} = \nabla \times \mathbf{A}$. Substituting into 6.1 gives

$$W_M = \tfrac{1}{2} \int_{\text{all space}} (\nabla \times \mathbf{A}) \cdot \mathbf{H} \; dv \tag{6.2}$$

Substituting the vector identity [3]

$$(\nabla \times \mathbf{A}) \cdot \mathbf{H} = \nabla \cdot (\mathbf{A} \times \mathbf{H}) + \mathbf{A} \cdot (\nabla \times \mathbf{H}) \tag{6.3}$$

into (6.2) and using Ampère's law for dc currents, $\nabla \times \mathbf{H} = \mathbf{J}$, gives

$$W_M = \tfrac{1}{2} \int_{\text{all space}} \nabla \cdot (\mathbf{A} \times \mathbf{H}) \; dv + \tfrac{1}{2} \int_{\text{all space}} \mathbf{A} \cdot \mathbf{J} \; dv \tag{6.4}$$

Applying the divergence theorem (see the Appendix) to the first integral gives

$$W_M = \tfrac{1}{2} \oint_{s_\infty} (\mathbf{A} \times \mathbf{H}) \cdot ds + \tfrac{1}{2} \int_{\text{all space}} \mathbf{A} \cdot \mathbf{J} \; dv \tag{6.5}$$

where s_∞ is the closed surface at infinity. Now recall the Biot–Savart law in (2.11) and the equation for computing the vector magnetic potential in (2.47a). These show that at points far from a finite current distribution, the magnitude of $\mathbf{A}$ will decrease at a rate greater than or equal to the inverse distance $(1/r)$, and the magnitude of $\mathbf{B}$ will decrease at a rate greater than or equal to the inverse distance squared $(1/r^2)$. The $\mathbf{A}$ and $\mathbf{B}$ fields of some current distributions, such as the current loop in Fig. 2.25 whose fields were derived in Chapter 2, decrease at large distances from the loop at a greater rate: $1/r^2$ and $1/r^3$, respectively. In spherical coordinates, the differential surface is $ds = r^2 \sin \theta \, d\theta \, d\phi$. Since the product of the magnitudes of $\mathbf{A}$ and $\mathbf{H}$ decrease at a rate no less than $1/r^3$, the first integral in (6.5) over the surface at infinity, s_∞, will go to zero. The remaining integral is over all space, but the current density $\mathbf{J}$ is zero *except* where the current is located. Hence, the result for the stored energy is an integral only over the volume containing the current:

$$\boxed{W_M = \tfrac{1}{2} \int_{\substack{\text{throughout} \\ \text{the volume} \\ \text{containing the} \\ \text{current}}} \mathbf{A} \cdot \mathbf{J} \, dv} \tag{6.6}$$

Now consider applying the result in (6.6) to the case of two lands. Each land carries a total current I_1 and I_2, respectively. The total vector magnetic potential in the space around the two lands is $\mathbf{A} = \mathbf{A}_1 + \mathbf{A}_2$, where $\mathbf{A}_1$ is due to current I_1 that is carried by land 1, and $\mathbf{A}_2$ is due to current I_2 that is carried by land 2. The total magnetic energy in the field surrounding the lands is

$$W_M = \frac{1}{2} \int_{v_1} \mathbf{A} \cdot \mathbf{J}_1 \, dv_1 + \frac{1}{2} \int_{v_2} \mathbf{A} \cdot \mathbf{J}_2 \, dv_2 \tag{6.7}$$

where volumes v_1 and v_2 are the volumes of the respective lands that enclose the respective current densities, $\mathbf{J}_1$ and $\mathbf{J}_2$. Substituting the total vector magnetic potential in the space surrounding the two lands, $\mathbf{A} = \mathbf{A}_1 + \mathbf{A}_2$, gives

$$W_M = \frac{1}{2} \int_{v_1} \mathbf{A}_1 \cdot \mathbf{J}_1 \, dv_1 + \frac{1}{2} \int_{v_2} \mathbf{A}_1 \cdot \mathbf{J}_2 \, dv_2 + \frac{1}{2} \int_{v_2} \mathbf{A}_2 \cdot \mathbf{J}_2 \, dv_2$$

$$+ \frac{1}{2} \int_{v_1} \mathbf{A}_2 \cdot \mathbf{J}_1 \, dv_1 \tag{6.8a}$$

Our intent is to derive self and mutual partial inductances that can be used to model these two lands, as illustrated in Fig. 5.6. The total magnetic energy in the field is represented in terms of self and mutual partial inductances of the two lands as

$$W_M = \frac{1}{2} L_{p1} I_1^2 + \frac{1}{2} M_{p12} I_1 I_2 + \frac{1}{2} L_{p2} I_2^2 + \frac{1}{2} M_{p21} I_2 I_1 \tag{6.8b}$$

and of course the mutual partial inductances are reciprocal (i.e., $M_{p12} = M_{p21}$). Comparing (6.8a) and (6.8b), the various self and mutual partial inductances can be found from

$$L_{p1} = \frac{1}{I_1^2} \int_{v_1} \mathbf{A}_1 \cdot \mathbf{J}_1 \, dv_1 \tag{6.9a}$$

$$L_{p2} = \frac{1}{I_2^2} \int_{v_2} \mathbf{A}_2 \cdot \mathbf{J}_2 \, dv_2 \tag{6.9b}$$

$$M_{p12} = \frac{1}{I_1 I_2} \int_{v_2} \mathbf{A}_1 \cdot \mathbf{J}_2 \, dv_2 \tag{6.9c}$$

$$M_{p21} = \frac{1}{I_1 I_2} \int_{v_1} \mathbf{A}_2 \cdot \mathbf{J}_1 \, dv_1 \tag{6.9d}$$

Next we substitute the equation for the vector magnetic potential given in (2.47a),

$$\mathbf{A} = \frac{\mu_0}{4\pi} \int_v \frac{\mathbf{J} \, dv}{R} \tag{2.47a}$$

giving

$$L_{p1} = \frac{1}{I_1^2} \int_{v_1} \int_{v_1} \frac{\mu_0}{4\pi} \frac{\mathbf{J}_1 \cdot \mathbf{J}'_1}{R} \, dv'_1 \, dv_1 \qquad (6.10a)$$

$$L_{p2} = \frac{1}{I_2^2} \int_{v_2} \int_{v_2} \frac{\mu_0}{4\pi} \frac{\mathbf{J}_2 \cdot \mathbf{J}'_2}{R} \, dv'_2 \, dv_2 \qquad (6.10b)$$

$$M_{p12} = \frac{1}{I_1 I_2} \int_{v_1} \int_{v_2} \frac{\mu_0}{4\pi} \frac{\mathbf{J}'_1 \cdot \mathbf{J}_2}{R} \, dv_2 \, dv'_1 \qquad (6.10c)$$

$$M_{p21} = \frac{1}{I_1 I_2} \int_{v_2} \int_{v_1} \frac{\mu_0}{4\pi} \frac{\mathbf{J}'_2 \cdot \mathbf{J}_1}{R} \, dv_1 \, dv'_2 \qquad (6.10d)$$

where the term R is the distance between two differential chunks of current $\mathbf{J}_i \, dv_i$ and $\mathbf{J}'_j \, dv'_j$. We have denoted one current density with a prime to denote that chunk of current $\mathbf{J}' \, dv'$ as being the cause of $\mathbf{A}$ and it lies in the differential volume dv', whereas the differential chunk of current $\mathbf{J} \, dv$ lies in the differential volume dv.

It is important to point out the following observation about the internal inductances of lands:

Since the formulations for determining the self and mutual partial inductances in (6.10) result from the integration of the magnetic energy density over all space (including that internal to the lands), the results for the self partial inductances in (6.10a) and (6.10b) include the internal self inductances of the lands due to the magnetic fields internal to them.

As the frequency of the current increases from dc, the current distribution over the land cross sections migrate to the outer edges of the lands, and hence the internal inductances go to zero, so that *(6.10a) and (6.10b) for higher frequencies represent the external self partial inductances*. This will be shown through numerical computations later. However, it would not be simple to determine the external self partial inductances by integrating (6.10) only over a thin volume near the land surfaces that contains the total land current, since the high-frequency current distribution peaks at the corners of the lands, and this would generate a very difficult computation. It turns out that Holloway and Kuester have recently derived the result for the internal self inductance of a PCB land [18]. Their result for a square land was 48.3 nH/m, which had been confirmed with numerical computation [19]. This is on the order of the internal inductance of a circular wire (50 nH/m), which is independent of the wire radius.

Now we make two crucial assumptions in order to make the results in (6.10) useful for computing the self and mutual partial inductances of PCB lands.

1. Assume that the currents I with density $\mathbf{J}$ in volume v that are carried by the lands are *uniformly distributed over the cross sections of those lands*. Hence, we can simply write the magnitudes of the current distributions as $J = I/A$, where A denotes the cross-sectional area of the respective lands.

2. Assume that the currents $\mathbf{J}$ in v that are carried by the lands are also *uniformly distributed along the lengths of the lands*. This agrees with the fundamental limitation that the conductors must be electrically short for currents of nonzero frequency in order to represent those conductors with lumped-circuit elements such as an inductance. Hence, we can write

$$J \, dv = \frac{I}{A} \, dA \, dl$$

With these two assumptions the magnitudes of the current densities, J, are constants (independent of the longitudinal as well as the cross-sectional variables of the lands) and can therefore be removed from the integrals in (6.10). Hence, the results in (6.10) simplify considerably. The self partial inductances are then computed from

$$\boxed{L_p = \frac{1}{A \, A'} \int_A \int_{A'} M_f \, dA' \, dA} \qquad (6.11a)$$

where A and A' are over the same land, and M_f is the Neumann integral denoting the mutual partial inductance between two filamentary currents (within the same land) that are separated by distance R:

$$\boxed{M_f = \frac{\mu_0}{4\pi} \int_l \int_l \frac{d\mathbf{l}' \cdot d\mathbf{l}}{R}} \qquad (6.12)$$

In the case of two different lands, the mutual partial inductance between them is obtained from

$$\boxed{M_p = \frac{1}{A \, A'} \int_A \int_{A'} M_f \, dA' \, dA} \qquad (6.11b)$$

where A and A' are over different lands, and M_f in (6.12) denotes the mutual partial inductance between two filamentary currents (in different lands) that are separated by distance R.

So the computation of the self and mutual partial inductances essentially involves representing the currents of each land as being composed of filaments. We obtain the total self and mutual partial inductances of the lands as the summation, over the cross-sectional areas of the lands, of these mutual partial inductances between the filaments. Note that (6.11a) and (6.11b) for the self and mutual partial inductances seem to indicate an "averaging" over the cross-sectional areas of the lands. This averaging is not in the derivation of the result from the outset (i.e., it is not an approximation, but just comes out of the formal derivation). The computations in (6.11a) and (6.11b) involve sixfold integrals: two over the filament lengths and four over the cross-sectional areas of the lands. But the mutual partial inductance between filaments was derived in Chapter 5 and given in Section 5.5. For parallel but offset filaments of length l and m separated by distance d and whose endpoints are offset by distance s with reference to Fig. 5.11, the mutual partial inductance between the two parallel but offset filaments is given as

$$
\begin{aligned}
M_f = \frac{\mu_0}{4\pi} \Bigg[&(l+s+m)\sinh^{-1}\frac{l+s+m}{d} - (s+m)\sinh^{-1}\frac{s+m}{d} \\
&- (l+s)\sinh^{-1}\frac{l+s}{d} + s\sinh^{-1}\frac{s}{d} - \sqrt{(l+s+m)^2 + d^2} \\
&+ \sqrt{(s+m)^2 + d^2} + \sqrt{(l+s)^2 + d^2} - \sqrt{s^2 + d^2} \Bigg]
\end{aligned}
\tag{5.28}
$$

This can be put into an alternative form by using the identity for the inverse hyperbolic sine:

$$
\sinh^{-1}\frac{z}{d} = \ln\left[\frac{z}{d} + \sqrt{\left(\frac{z}{d}\right)^2 + 1} \right]
$$

$$
= \ln\left(z + \sqrt{z^2 + d^2} \right) - \ln d
\tag{D700.1}
$$

as

$$
\boxed{
\begin{aligned}
M_f &= \frac{\mu_0}{4\pi} \left[f(z) \right] \Bigg|_{(s+m),(l+s)}^{(l+s+m),s} \\
&= \frac{\mu_0}{4\pi} \left[f(l+s+m) - f(s+m) + f(s) - f(l+s) \right]
\end{aligned}
}
\tag{6.13a}
$$

where

$$f(z) = z \ln \left(z + \sqrt{z^2 + d^2} \right) - \sqrt{z^2 + d^2} \qquad (6.13b)$$

Note that the $- \ln d$ term in the identity in (D700.1) cancels out because we add and subtract this term twice in (5.28).

6.2 SELF PARTIAL INDUCTANCE OF PCB LANDS

For self partial inductance calculations, the mutual partial inductance between two parallel filaments in (6.13) simplifies since the two filaments are of *identical length* l and their endpoints coincide, giving $m = l$ and $s = -l$. Hence, (6.13) reduces to

$$
\begin{aligned}
M_f &= \frac{\mu_0}{4\pi} \left[f(z) \right] \Big|_{0,0}^{l,-l} (z) \\
&= \frac{\mu_0}{4\pi} \left[f(l) - f(0) + f(-l) - f(0) \right] \\
&= \frac{\mu_0}{4\pi} \left[l \ln \left(l + \sqrt{l^2 + d^2} \right) - l \ln \left(-l + \sqrt{l^2 + d^2} \right) \right. \\
&\qquad \left. - 2\sqrt{l^2 + d^2} + 2d \right] \\
&= \frac{\mu_0}{2\pi} \left(l \sinh^{-1} \frac{l}{d} - \sqrt{l^2 + d^2} + d \right) \qquad m = l, s = -l
\end{aligned}
$$

$$(6.14)$$

which agrees with (5.21b), which was derived directly in Chapter 5.

We first compute the self partial inductance for a land of zero thickness, $t = 0$. Hence, we integrate the mutual partial inductances of filaments in (6.14) that are separated by distance $d^2 = (x_2 - x_1)^2$ first from $x_1 = 0$ to $x_1 = w$ and then integrate that result from $x_2 = 0$ to $x_2 = w$. Hence, we integrate (6.11a) over the same cross-section land to give

$$L_{p(t=0)} = \frac{\mu_0}{4\pi} \frac{1}{w^2} \int_{x_2=0}^{w} \int_{x_1=0}^{w} M_f \, dx_1 \, dx_2 \qquad t = 0 \qquad (6.15)$$

as illustrated in Fig. 6.2. The inner integral is evaluated first:

$$(\text{I}) = \int_{x_1=0}^{w} \left[z \ln \left(z + \sqrt{z^2 + d^2} \right) - \sqrt{z^2 + d^2} \right] \, dx_1$$

and $d^2 = (x_1 - x_2)^2$. Since z here is treated as a constant, we evaluate $f(z)$ in the integrand at the four limits of $z = l, -l, 0, 0$ as in (6.14) after we

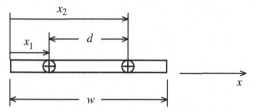

FIGURE 6.2. Computation of the self partial inductance for a land of zero thickness ($t = 0$).

have finished the integrations. Making a change of variables as $\lambda = x_1 - x_2$, $d\lambda = dx_1$ gives

$$\text{(I)} = \int_{\lambda = -x_2}^{W - x_2} \left[z \ln \left(z + \sqrt{z^2 + \lambda^2} \right) - \sqrt{z^2 + \lambda^2} \right] d\lambda$$

which is equivalent to

$$\text{(I)} = \int_{\lambda = x_2 - W}^{x_2} \left[z \ln \left(z + \sqrt{z^2 + \lambda^2} \right) - \sqrt{z^2 + \lambda^2} \right] d\lambda$$

This can be integrated using integrals from Dwight [7]:

$$\int \ln \left(z + \sqrt{z^2 + \lambda^2} \right) d\lambda = \lambda \ln \left(z + \sqrt{z^2 + \lambda^2} \right)$$

$$- \lambda + z \ln \left(\lambda + \sqrt{z^2 + \lambda^2} \right) \qquad \text{(D740)}$$

$$\int \sqrt{z^2 + \lambda^2} \, d\lambda = \frac{z^2}{2} \ln \left(\lambda + \sqrt{z^2 + \lambda^2} \right)$$

$$+ \frac{\lambda}{2} \sqrt{z^2 + \lambda^2} \qquad \text{(D230.01)}$$

Hence, the inner integral evaluates to

$$\text{(I)} = \left[\lambda z \ln \left(z + \sqrt{z^2 + \lambda^2} \right) - \lambda z + \frac{z^2}{2} \ln \left(\lambda + \sqrt{z^2 + \lambda^2} \right) \right.$$

$$\left. - \frac{\lambda}{2} \sqrt{z^2 + \lambda^2} \right]_{\lambda = x_2 - W}^{x_2}$$

Note that this result contains a term $-\lambda z$. But when this is evaluated for the four values of z, $z = l$, $-l$, 0, 0, it will cancel out:

$$\left[f(z) \right] \mathop{(z)}_{0,0}^{l,l} = \left[f(l) - f(0) + f(-l) - f(0) \right]$$

$$= -\lambda l + \lambda l$$

$$= 0$$

Hence, we ignore this term. This gives

$$(\text{I}) = \left[\lambda\, z\, \ln\left(z + \sqrt{z^2 + \lambda^2} \right) + \frac{z^2}{2} \ln\left(\lambda + \sqrt{z^2 + \lambda^2} \right) \right.$$

$$\left. - \frac{\lambda}{2}\sqrt{z^2 + \lambda^2} \right]_{\lambda = x_2 - w}^{x_2}$$

$$= \left[x_2\, z\, \ln\left(z + \sqrt{z^2 + x_2^2} \right) + \frac{z^2}{2} \ln\left(x_2 + \sqrt{z^2 + x_2^2} \right) - \frac{x_2}{2}\sqrt{z^2 + x_2^2} \right]$$

$$- \left[(x_2 - w)\, z\, \ln\left(z + \sqrt{z^2 + (x_2 - w)^2} \right) \right.$$

$$+ \frac{z^2}{2} \ln\left((x_2 - w) + \sqrt{z^2 + (x_2 - w)^2} \right)$$

$$\left. - \frac{(x_2 - w)}{2}\sqrt{z^2 + (x_2 - w)^2} \right]$$

Next, we integrate (I) from $x_2 = 0$ to $x_2 = w$:

$$(\text{II}) = \int_{x_2 = 0}^{w} (\text{I})\ dx_2$$

giving

$$(\text{II}) = \int_{x_2 = 0}^{w} \left[x_2\, z\, \ln\left(z + \sqrt{z^2 + x_2^2} \right) + \frac{z^2}{2} \ln\left(x_2 + \sqrt{z^2 + x_2^2} \right) \right.$$

$$\left. - \frac{x_2}{2}\sqrt{z^2 + x_2^2} \right] dx_2$$

$$+ \int_{\lambda = 0}^{-w} \left[\lambda\, z\, \ln\left(z + \sqrt{z^2 + \lambda^2} \right) \right.$$

$$\left. + \frac{z^2}{2} \ln\left(\lambda + \sqrt{z^2 + \lambda^2} \right) - \frac{\lambda}{2}\sqrt{z^2 + \lambda^2} \right] d\lambda$$

and we have used a change of variables $\lambda = x_2 - w$, $d\lambda = dx_2$ in the second integral. These integrals can be evaluated using Dwight [7]:

$$\int \lambda \ln\left(z + \sqrt{z^2 + \lambda^2} \right) d\lambda = -\frac{\lambda^2}{4} + \frac{z}{2}\sqrt{z^2 + \lambda^2}$$

$$+ \frac{\lambda^2}{2} \ln\left(z + \sqrt{z^2 + \lambda^2} \right) \tag{D602.5}$$

$$\int \ln\left(\lambda + \sqrt{z^2 + \lambda^2} \right) d\lambda = \lambda \ln\left(\lambda + \sqrt{z^2 + \lambda^2} \right) - \sqrt{z^2 + \lambda^2} \tag{D625}$$

$$\int \lambda \sqrt{z^2 + \lambda^2} \, d\lambda = \tfrac{1}{3} \left(z^2 + \lambda^2 \right)^{3/2} \qquad \text{(D231.01)}$$

to give

$$(\text{II}) = \left[-\frac{\lambda^2 z}{4} + \frac{z\lambda^2}{2} \ln \left(z + \sqrt{z^2 + \lambda^2} \right) + \frac{z^2 \lambda}{2} \ln \left(\lambda + \sqrt{z^2 + \lambda^2} \right) \right.$$
$$\left. - \frac{1}{6} \left(z^2 + \lambda^2 \right)^{3/2} \right]_{\lambda = 0, 0}^{w, -w}$$

Once again this contains a term $-\lambda^2 z/4$, which will be canceled out when this is evaluated at the four limits of z, $z = l$, $-l$, 0, 0, so it will also be ignored.

Hence, the result gives the self partial inductance of a land of zero thickness $(t = 0)$ as

$$L_{p(t=0)} = \frac{\mu_0}{4\pi} \frac{1}{w^2} \left[f(\lambda, z) \right] \Bigg|_{\substack{(\lambda) \\ 0,0}}^{\substack{w, -w}} \Bigg|_{\substack{(z) \\ 0,0}}^{\substack{l, -l}} \qquad t = 0 \qquad \text{(6.16a)}$$

where

$$f(\lambda, z) = \frac{z\lambda^2}{2} \ln \left(z + \sqrt{z^2 + \lambda^2} \right) + \frac{z^2 \lambda}{2} \ln \left(\lambda + \sqrt{z^2 + \lambda^2} \right)$$
$$- \frac{1}{6} \left(z^2 + \lambda^2 \right)^{3/2} \qquad \text{(6.16b)}$$

Evaluating this gives

$$L_{p(t=0)} = \frac{\mu_0}{4\pi} \frac{1}{w^2} \left[f(\lambda, z) \right] \Bigg|_{\substack{(\lambda) \\ 0,0}}^{\substack{w, -w}} \Bigg|_{\substack{(z) \\ 0,0}}^{\substack{l, -l}} \qquad t = 0$$

$$= \frac{\mu_0}{4\pi} \frac{1}{w^2} \begin{bmatrix} +f(w, l) - f(w, 0) + f(w, -l) - f(w, 0) \\ -f(0, l) + f(0, 0) - f(0, -l) + f(0, 0) \\ +f(-w, l) - f(-w, 0) + f(-w, -l) - f(-w, 0) \\ -f(0, l) + f(0, 0) - f(0, -l) + f(0, 0) \end{bmatrix}$$

$$\text{(6.17)}$$

Note that $f(0, 0) = 0$. In the sum in (6.17), the following identity may be used:

$$\ln \left(a + \sqrt{a^2 + b^2} \right) - \ln \left(-a + \sqrt{a^2 + b^2} \right) = 2 \ln \left[\frac{a}{b} + \sqrt{\left(\frac{a}{b} \right)^2 + 1} \right]$$

This identity can be confirmed by writing the left side as $\ln A - \ln B = \ln(A/B)$ and comparing the arguments of the natural logarithms on both sides. Caution should be observed in evaluating the term $\left(z^2 + \lambda^2 \right)^{3/2}$. When either $z = 0$ or $\lambda = 0$, the absolute values of z or λ should be used:

$$\left[\left(z^2 + \lambda^2\right)^{3/2}\right]_{z=0 \ \text{or} \ \lambda=0} = |\lambda|^3 \ \text{or} \ |z|^3$$

Finally, in evaluating the terms in (6.17) we should note that

$$\lim_{x \to 0} x \ln [ax] = 0$$

as can be proven using l'Hôpital's rule. This gives the self partial inductance of a PCB land of zero thickness, $t = 0$, as

$$
\begin{aligned}
L_{p(t=0)} = \frac{\mu_0}{2\pi} \frac{1}{w^2} \Bigg[& lw^2 \ \ln\left(\frac{l}{w} + \sqrt{\left(\frac{l}{w}\right)^2 + 1}\right) \\
& + l^2 w \ \ln\left(\frac{w}{l} + \sqrt{\left(\frac{w}{l}\right)^2 + 1}\right) \\
& + \frac{1}{3}\left(l^3 + w^3\right) - \frac{1}{3}\left(l^2 + w^2\right)^{3/2} \Bigg] \qquad t = 0
\end{aligned}
\tag{6.18}
$$

which agrees with the formula given by Hoer and Love [16].

The result for infinitesimally thin lands in (6.18) can be written compactly in terms of the "aspect ratio" of the land as the ratio of the land length to land width, $u = l/w$:

$$
\begin{aligned}
\frac{L_{p(t=0)}}{l} = \frac{\mu_0}{2\pi} \Bigg[& \ln\left(u + \sqrt{u^2 + 1}\right) + u \ \ln\left(\frac{1}{u} + \sqrt{\left(\frac{1}{u}\right)^2 + 1}\right) \\
& + \frac{1}{3}\left(u^2 + \frac{1}{u} - \frac{(u^2 + 1)^{3/2}}{u}\right) \Bigg] \qquad u = \frac{l}{w}, \quad t = 0
\end{aligned}
$$

$$\tag{6.19}$$

This can be simplified for extreme values of the "aspect ratio" of the lands: $u \gg 1$ (very "long" lands, $l \gg w$) or $u \ll 1$ (very "wide" lands, $l \ll w$). We have the following identities [7]:

$$
\ln(u + \sqrt{u^2 + 1}) =
\begin{cases}
\ln 2u + \dfrac{1}{4\,u^2} - \dfrac{3}{32\,u^4} + \cdots & u > 1 \\[2ex]
u - \dfrac{1}{6}u^3 + \dfrac{3}{40}u^5 + \cdots & u < 1
\end{cases}
\tag{D602.1}
$$

$$\ln\left[\frac{1}{u} + \sqrt{\left(\frac{1}{u}\right)^2 + 1}\right] = \begin{cases} \dfrac{1}{u} - \dfrac{1}{6u^3} + \dfrac{3}{40u^5} + \cdots & u > 1 \\[3mm] \ln\dfrac{2}{u} + \dfrac{u^2}{4} - \dfrac{3u^4}{32} + \cdots & u < 1 \end{cases} \tag{D602.5}$$

The result in (D602.5) follows from (D602.1) by substituting $u = 1/x$ into (D602.1). In addition, we have the series expansion

$$(1 + x)^n = 1 + nx + \frac{n(n-1)}{2!}x^2 + \frac{n(n-1)(n-2)}{3!} + \cdots \qquad x^2 \le 1 \tag{D1}$$

Using (D1) we obtain approximations for the term $(u^2 + 1)^{3/2}$ for $u \le 1$:

$$(u^2 + 1)^{3/2} = 1 + \frac{3}{2}u^2 + \frac{3}{8}u^4 + \cdots \qquad u \le 1$$

$$\cong 1 + \frac{3}{2}u^2 \qquad u \ll 1$$

and for $u \ge 1$:

$$(u^2 + 1)^{3/2} = u^3 \left(\frac{1}{u^2} + 1\right)^{3/2}$$

$$= u^3 \left(1 + \frac{3}{2}\frac{1}{u^2} + \frac{3}{8}\frac{1}{u^4} + \cdots + \right) \qquad u \ge 1$$

$$\cong u^3 + \frac{3}{2}u \qquad u \gg 1$$

Hence the last term in (6.19) becomes

$$\frac{1}{3}\left(u^2 + \frac{1}{u} - \frac{(u^2+1)^{3/2}}{u}\right) \cong \begin{cases} -\dfrac{1}{2} + \dfrac{1}{3u} & u \gg 1 \\[3mm] -\dfrac{1}{2}u + \dfrac{1}{3}u^2 & u \ll 1 \end{cases}$$

Hence, (6.19) should approach

$$\frac{L_{p(t=0)}}{l} \cong \frac{\mu_0}{2\pi} \begin{cases} \ln 2u + \dfrac{1}{2} + \dfrac{1}{3u} & u \gg 1, \; l \gg w \\[3mm] u\left(\ln\dfrac{2}{u} + \dfrac{1}{2} + \dfrac{u}{3}\right) & u \ll 1, \; l \ll w \end{cases} \tag{6.20}$$

The following table summarizes some computed data comparing (6.19) and (6.20).

u	(6.19)	(6.20)
10	17.925 nH/in.	17.928 nH/in.
100	29.472 nH/in.	29.473 nH/in.
$\frac{1}{10}$	1.7927 nH/in.	1.7928 nH/in.
$\frac{1}{100}$	0.2946 nH/in.	0.2947 nH/in.

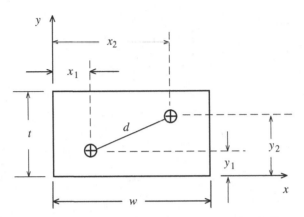

FIGURE 6.3. Computing the self partial inductance of a PCB land of nonzero thickness.

For very large aspect ratios, $l \gg w$, the self partial inductance approaches a variation of $5.08 \left[\ln \left(2l/w \right) + \frac{1}{2} \right]$ nH/in. For this to give the same result as in (5.18c) for a "long" wire would require the land to be replaced by a wire whose diameter is $0.446w$. This tends to support a commonly held design rule that rectangular straps have less "inductance" than do comparable sized wires.

Now we turn our attention to PCB lands with nonzero thicknesses, as illustrated in Fig. 6.3. In this case we must integrate (6.11b) over the entire cross section:

$$M_p = \frac{1}{A A'} \int_A \int_{A'} M_f \, dA' \, dA \qquad (6.11b)$$

The mutual inductance between the filaments, M_f, is the same as in (6.14) but with

$$d^2 = (x_2 - x_1)^2 + (y_2 - y_1)^2 \qquad (6.21)$$

Hence, the integral we must evaluate is

$$L_p = \frac{1}{t^2 w^2} \int_{y_2=0}^{t} \int_{y_1=0}^{t} \int_{x_2=0}^{w} \int_{x_1=0}^{w} M_f \, dx_1 \, dx_2 \, dy_1 \, dy_2 \qquad (6.22)$$

and, since z in M_f is treated here essentially as a constant, we evaluate the result at the limits of $z = l, -l, 0, 0$ according to (6.14) after we finish the integration. The interior integrals with respect to x represent the self partial inductances of two identical lands of zero thickness that we integrated before, but we must repeat that because d^2 in M_f is no longer just $(x_2 - x_1)^2$ but is given now by (6.21). The first two integrals with respect to x_1 and x_2 are fairly simple to integrate, and the process is very similar to what was done to obtain the result for a land of zero thickness in (6.15). Once these are performed, we are left with the integrals with respect to y_1 and y_2. The entire process, although straightforward, is exceedingly tedious. Hoer and Love give that result [16] as

$$L_p = \frac{\mu_0}{4\pi} \frac{1}{t^2 w^2} \left[\left[[f(x, y, z)] \begin{matrix} w \\ (x) \\ 0 \end{matrix} \right] \begin{matrix} t \\ (y) \\ 0 \end{matrix} \right] \begin{matrix} l \\ (z) \\ 0 \end{matrix} \tag{6.23a}$$

where

$$\left[\left[[f(x, y, z)] \begin{matrix} q_1 \\ (x) \\ q_2 \end{matrix} \right] \begin{matrix} r_1 \\ (y) \\ r_2 \end{matrix} \right] \begin{matrix} s_1 \\ (z) \\ s_2 \end{matrix} = \sum_{i=1}^{2} \sum_{j=1}^{2} \sum_{k=1}^{2} (-1)^{i+j+k+1} \; f(q_i, r_j, s_k) \tag{6.23b}$$

and $f(x, y, z)$ is given by

$$
\begin{aligned}
f(x, y, z) = & \left(\frac{y^2 z^2}{4} - \frac{y^4}{24} - \frac{z^4}{24} \right) x \ln \frac{x + \sqrt{x^2 + y^2 + z^2}}{\sqrt{y^2 + z^2}} \\
& + \left(\frac{x^2 z^2}{4} - \frac{x^4}{24} - \frac{z^4}{24} \right) y \ln \frac{y + \sqrt{x^2 + y^2 + z^2}}{\sqrt{x^2 + z^2}} \\
& + \left(\frac{x^2 y^2}{4} - \frac{x^4}{24} - \frac{y^4}{24} \right) z \ln \frac{z + \sqrt{x^2 + y^2 + z^2}}{\sqrt{x^2 + y^2}} \\
& + \frac{1}{60} \left(x^4 + y^4 + z^4 - 3x^2 y^2 - 3y^2 z^2 - 3x^2 z^2 \right) \sqrt{x^2 + y^2 + z^2} \\
& - \frac{xyz^3}{6} \tan^{-1} \frac{xy}{z\sqrt{x^2 + y^2 + z^2}} \\
& - \frac{xy^3 z}{6} \tan^{-1} \frac{xz}{y\sqrt{x^2 + y^2 + z^2}} \\
& - \frac{x^3 yz}{6} \tan^{-1} \frac{yz}{x\sqrt{x^2 + y^2 + z^2}}
\end{aligned}
$$

$$\tag{6.23c}$$

Ruehli shows [15] that the general result in (6.23) can be written solely in terms of $u = l/w$ and $v = t/w$. One would expect that the self partial inductance should vary in a smooth fashion as a function of $u = l/w$ and $v = t/w$. Hence, it should be possible to obtain a general formula that is much simpler than (6.23) by curve fitting to computed data from (6.23) or Ruehli's version of it. Ruehli in [15] also points out that for extreme values of the aspect ratios of $u = l/w$ and $v = t/w$, the general result in (6.23) may involve subtraction of terms of similar magnitude, thereby giving numerical errors. He gives a more stable form of this result in [15]. His computations also show that the result for a land of zero thickness in (6.19) gives reasonable accuracy for $v < 0.1$. In addition, his results show that the zero-thickness land result in (6.19) gives approximately the same result as (6.23) for all values of v and $u > 100$ (i.e., for "very long" lands). Hoer and Love [16] give an approximation for the general result in (6.23) for $v = t/w \leq 0.1$ using the zero-thickness land result in (6.19) as

$$L_p \cong L_{p(t=0)} - 2 \times 10^{-7} \frac{t}{w} l \qquad w \geq 10t \qquad (6.24)$$

6.3   MUTUAL PARTIAL INDUCTANCE BETWEEN PCB LANDS

Calculating the mutual partial inductance between lands follows a pattern similar to that for self partial inductance. Treat each land as a set of filaments and use the basic result

$$M_p = \frac{1}{A A'} \int_A \int_{A'} M_f \, dA' \, dA \qquad (6.11b)$$

In this case the lands may be offset from each other and hence we use the general relation for the mutual partial inductances between two filaments of lengths l and m which are offset from each other by a distance s as shown in Fig. 5.11. Hence, the mutual partial inductance between the two filaments is given by (5.28), which may be written as

$$M_f = \frac{\mu_0}{4\pi} \left[f(z) \right] \Big|_{(s+m),(l+s)}^{(l+s+m),s}$$

$$= \frac{\mu_0}{4\pi} \left[f(l+s+m) - f(s+m) + f(s) - f(l+s) \right] \quad (6.13a)$$

where

$$f(z) = z \ln \left(z + \sqrt{z^2 + d^2} \right) - \sqrt{z^2 + d^2} \qquad (6.13b)$$

If the lands (and their associated filaments) overlap, s will be negative by the amount of overlap.

We first obtain the mutual partial inductance between two lands of zero thickness whose lengths in the z direction are l and m and whose surfaces are parallel to each other (but perhaps offset by distance s) as shown in Fig. 6.4(a). For this case, d in (6.13b) is the distance in the xy plane between the filaments composing each land and is given by

$$d^2 = (x_2 - x_1)^2 + b^2 \qquad (6.25)$$

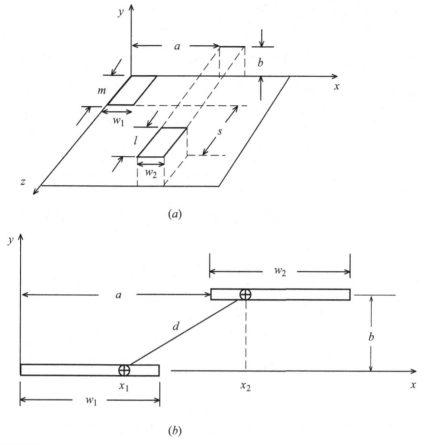

FIGURE 6.4. Computing the mutual partial inductance between two PCB lands of zero thickness.

where b is the vertical separation in the xy plane between the lands as shown in Fig. 6.4. Hence, the integral we must evaluate is

$$M_p = \frac{1}{w_1 w_2} \int_{x_2=a}^{a+w_2} \int_{x_1=0}^{w_1} M_f \, dx_1 \, dx_2 \qquad (6.26)$$

as illustrated in Fig. 6.4(b), and since z in M_f is essentially treated as a constant here, we evaluate the result at the limits of $z = (l + s + m), (s + m), s, (l + s)$ according to (6.13a) after we finish the integration. Hence, we first integrate

$$(I) = \int_{x_1=0}^{w_1} \left[z \ln \left(z + \sqrt{z^2 + b^2 + (x_2 - x_1)^2} \right) \right.$$
$$\left. - \sqrt{z^2 + b^2 + (x_2 - x_1)^2} \right] dx_1$$

Making a change of variables as $\lambda = (x_2 - x_1)$, $d\lambda = -dx_1$ gives

$$(\text{I}) = \int_{\lambda = x_2 - W_1}^{x_2} \left[z \ln \left(z + \sqrt{z^2 + b^2 + \lambda^2} \right) - \sqrt{z^2 + b^2 + \lambda^2} \right] d\lambda$$

This can be integrated as [16]

$$(\text{I}) = \left[zx_2 \ln \left(z + \sqrt{z^2 + b^2 + x_2^2} \right) - zb \tan^{-1} \frac{zx_2}{b\sqrt{z^2 + b^2 + x_2^2}} \right.$$

$$+ \frac{z^2 - b^2}{2} \ln \left(x_2 + \sqrt{z^2 + b^2 + x_2^2} \right) - \left. \frac{x_2}{2} \sqrt{z^2 + b^2 + x_2^2} \right]$$

$$- \left[z(x_2 - w_1) \ln \left(z + \sqrt{z^2 + b^2 + (x_2 - w_1)^2} \right) \right.$$

$$- zb \tan^{-1} \frac{z(x_2 - w_1)}{b\sqrt{z^2 + b^2 + (x_2 - w_1)^2}}$$

$$- \frac{z^2 - b^2}{2} \ln \left((x_2 - w_1) + \sqrt{z^2 + b^2 + (x_2 - w_1)^2} \right)$$

$$- \left. \frac{x_2 - w_1}{2} \sqrt{z^2 + b^2 + (x_2 - w_1)^2} \right]$$

Hence, it can be written symbolically as

$$(\text{I}) = \left[z\lambda \ln \left(z + \sqrt{z^2 + b^2 + \lambda^2} \right) - zb \tan^{-1} \frac{z\lambda}{b\sqrt{z^2 + b^2 + \lambda^2}} \right.$$

$$+ \frac{z^2 - b^2}{2} \ln \left(\lambda + \sqrt{z^2 + b^2 + \lambda^2} \right) - \left. \frac{\lambda}{2} \sqrt{z^2 + b^2 + \lambda^2} \right]_{\lambda = x_2 - W_1}^{x_2}$$

which, when multiplied by $(\mu_0/4\pi)(1/w_1)$ and evaluated at the four limits of z according to (6.13a), represents the mutual partial inductance between the first land and a filament in the second land at $(y, x) = (b, x_2)$ [16].

The second integral in (6.26) becomes

$$(\text{II}) = \int_{x_2 = a}^{a + W_2} (\text{I}) \, dx_2$$

Making a change of variables in the second half of the result for the first integral of $\lambda = (x_2 - w_1)$, $d\lambda = dx_2$ gives

$$
\text{(II)} = \int_{x_2=a}^{a+W_2} \left[zx_2 \ln \left(z + \sqrt{z^2 + b^2 + x_2^2} \right) - zb \tan^{-1} \frac{zx_2}{b\sqrt{z^2 + b^2 + x_2^2}} \right.
$$

$$
\left. + \frac{z^2 - b^2}{2} \ln \left(x_2 + \sqrt{z^2 + b^2 + x_2^2} \right) - \frac{x_2}{2} \sqrt{z^2 + b^2 + x_2^2} \right] dx_2
$$

$$
+ \int_{\lambda=a+W_2-W_1}^{a-W_1} \left[z\lambda \ln \left(z + \sqrt{z^2 + b^2 + \lambda^2} \right) - zb \tan^{-1} \frac{z\lambda}{b\sqrt{z^2 + b^2 + \lambda^2}} \right.
$$

$$
\left. + \frac{z^2 - b^2}{2} \ln \left(\lambda + \sqrt{z^2 + b^2 + \lambda^2} \right) - \frac{\lambda}{2} \sqrt{z^2 + b^2 + \lambda^2} \right] dx_2
$$

These integrals can be evaluated, giving the mutual partial inductance between two lands of zero thickness as [16]

$$
M_p = \frac{\mu_0}{4\pi} \frac{1}{w_1 w_2} \left[f(x, z) \right] \begin{vmatrix} a+W_2, a-W_1 \\ (x) \\ a, a+W_2-W_1 \end{vmatrix} \begin{vmatrix} (l+s+m), s \\ (z) \\ (s+m), (l+s) \end{vmatrix} \tag{6.27a}
$$

where

$$
\left[f(x, z) \right] \begin{vmatrix} q_1, q_3 \\ (x) \\ q_2, q_4 \end{vmatrix} \begin{vmatrix} s_1, s_3 \\ (z) \\ s_2, s_4 \end{vmatrix} = \sum_{i=1}^{4} \sum_{j=1}^{4} (-1)^{i+j} f(q_i, s_j) \tag{6.27b}
$$

and

$$
f(x, z) = \frac{x^2 - b^2}{2} z \ln \left(z + \sqrt{z^2 + b^2 + x^2} \right)
$$

$$
- \frac{1}{6} \left(z^2 - 2b^2 + x^2 \right) \sqrt{z^2 + b^2 + x^2}
$$

$$
+ \frac{z^2 - b^2}{2} x \ln \left(x + \sqrt{z^2 + b^2 + x^2} \right)
$$

$$
- zbx \tan^{-1} \frac{zx}{b\sqrt{z^2 + b^2 + x^2}} \tag{6.27c}
$$

The mutual partial inductance between two parallel lands of widths w_1, w_2 and corresponding thicknesses of t_1, t_2 is extraordinarily complicated and becomes [16]

$$M_p = \frac{\mu_0}{4\pi} \frac{1}{w_1 t_1 w_2 t_2} \left[\left[[f(x, y, z)]^{a-w_1,a+w_2}_{a+w_2-w_1,a}(x) \right]^{b-t_1,b+t_2}_{b+t_2-t_1,b}(y) \right]^{(l+s+m),s}_{(s+m),(l+s)}(z)$$

$$(6.28a)$$

where

$$\left[\left[[f(x, y, z)]^{q_1,q_3}_{q_2,q_4}(x) \right]^{r_1,r_3}_{r_2,r_4}(y) \right]^{s_1,s_3}_{s_2,s_4}(z) = \sum_{i=1}^{4}\sum_{j=1}^{4}\sum_{k=1}^{4}(-1)^{i+j+k+1}\ f(q_i, r_j, s_k)$$

$$(6.28b)$$

and $f(x, y, z)$ is given in (6.23c).

6.4   CONCEPT OF GEOMETRIC MEAN DISTANCE

The concept of the geometric mean distance (GMD) between two objects gives a method for obtaining simplified (but approximate) calculations of the mutual partial inductance between those objects [14,20–22]. The origin of the name is illustrated in Fig. 6.5, where we have shown a point P and a line. Distances d_i are drawn from the point to the line. The *geometric mean distance* D between the point and the line is the nth root of the product of the distances

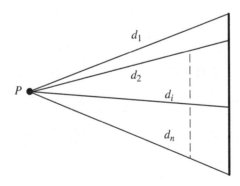

FIGURE 6.5. Concept of geometrical mean distance.

(their *geometric mean*) as the number of distances n increases without bound:

$$D = \lim_{n \to \infty} (d_1 \, d_2 \cdots d_i \cdots d_n)^{1/n} \qquad (6.29)$$

Taking the natural logarithm of (6.29) gives

$$\ln D = \lim_{n \to \infty} \frac{\ln d_1 + \ln d_2 + \cdots + \ln d_i + \cdots + \ln d_n}{n} \qquad (6.30)$$

Hence, the natural logarithm of the *geometric mean distance* between the point and the line is the *arithmetic mean* of the natural logarithms of the distances from the point to the line.

The concept of geometric mean distance has a particularly beneficial application in computing the mutual partial inductance between a pair of two-dimensional shapes representing the cross sections of two conductors when *the lengths of the two conductors are much greater than their separation*. The concept of geometric mean distance was originally developed by Maxwell in the late nineteenth century [23]. It is used routinely in the electric power distribution area to compute the self inductance of a bundle of wires, as well as the mutual inductances between sets of multiphase, high-voltage power transmission lines [24]. A very large number of formulas for the GMD of various shapes was published by Rosa and his colleagues in the *Bulletin of the National Bureau of Standards* in the period 1900–1910 [25–28]. A magnificent book by Andrew Gray gives a very thorough discussion of GMD, and it was written in 1893 [29]!

We computed the mutual partial inductance by considering the conductors to be composed of parallel current filaments via (6.11b):

$$M_p = \frac{1}{AA'} \int_A \int_{A'} M_f \, dA' \, dA \qquad (6.11b)$$

where M_f is the mutual partial inductance between two filaments of current within the cross-sectional areas A and A' given in (5.21a). The basic idea here is to treat the currents (assumed to be uniformly distributed over their cross sections) as being composed of filaments of current, sweep the filaments over the two cross sections, and then average the result over the cross sections as illustrated in Fig. 6.6.

The basic result for the mutual partial inductance between two filaments of current of length l (into the page) separated by distance d is given in (5.21a). If the lengths of the filaments are much greater that their separation, (5.21a)

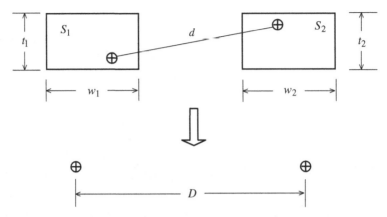

FIGURE 6.6. Computing the mutual partial inductance using the GMD between two shapes.

approximates to

$$M_f = \frac{\mu_0}{2\pi} l \left[\ln \frac{2l}{d} - 1 + \frac{d}{l} - \frac{1}{4} \left(\frac{d}{l} \right)^2 + \cdots \right]$$

$$\cong \frac{\mu_0}{2\pi} l \left(\ln \frac{2l}{d} - 1 \right) \qquad l \gg d \qquad (5.21c)$$

Note that only d will vary as we sweep the filaments over the surfaces as in (6.11b). Substituting (5.21c) into (6.11b) gives

$$M_p \cong \frac{1}{A\,A'} \int_A \int_{A'} \frac{\mu_0}{2\pi} l \left(\ln \frac{2l}{d} - 1 \right) dA'\, dA$$

$$= \frac{\mu_0}{2\pi} l \frac{1}{A\,A'} \int_A \int_{A'} (\ln 2l - \ln d - 1)\, dA'\, dA$$

$$= \frac{\mu_0}{2\pi} l \,(\ln 2l - 1) - \frac{\mu_0}{2\pi} l \frac{1}{A\,A'} \int_A \int_{A'} (\ln d)\, dA'\, dA \qquad (6.31)$$

The goal here is to *determine an equivalent distance between two filaments, D, which will have the same mutual partial inductance between them as between the two surfaces*, as illustrated in Fig. 6.6. Note that for very long filaments $l \gg d$, the only parameter in M_f in (5.21c) that varies is d. Hence, we may instead determine a D (the GMD between the two surfaces) such that

$$\boxed{\ln D = \frac{1}{A\,A'} \int_A \int_{A'} (\ln d)\, dA'\, dA} \qquad (6.32)$$

If this computation is carried out, two (very long) filaments of length l spaced a distance equal to the *geometric mean distance* between the two cross-sectional shapes of the two conductors, D, will have a mutual partial inductance between those conductors of

$$M_p \cong \frac{\mu_0}{2\pi} l \left(\ln \frac{2l}{D} - 1 \right) \qquad l \gg D \qquad (6.33)$$

which will give the same mutual partial inductance as between the (very long) conductors originally desired and obtained with (6.11b). If we include the third term of (5.21c), d/l, the d should properly be the *arithmetic mean distance* between the filaments used to represent the conductors. But usually this term is inconsequential and will be neglected. So (6.33) is an approximate solution to (6.11b) which is reasonably valid only for "very long" conductors (i.e., $D \ll l$). Of course, the work in computing the GMD in (6.32) can be as tedious as that in directly computing the original integral in (6.11b), but the GMD for various shapes has been tabulated over the years in various publications [14,20,21].

The discussion above regarding the use of the GMD in computing the mutual partial inductance applies to computation of the self partial inductance in (6.11a). The GMD here is said to be between the shape and itself.

Another interpretation of the utility of the GMD is computation of the vector magnetic potential due to a very long conductor of rectangular cross section whose current I is *distributed uniformly over the cross section of the conductor* as illustrated in Fig. 6.7 [22]. The vector magnetic potential of a current filament of infinite length (pointing in the z direction, into the page) is given as

$$A_z = -\frac{\mu_0 I}{2\pi} \ln d \qquad (2.53)$$

where d is the distance between that filament of current and the point $P(X, Y)$ at which we desire to determine the vector magnetic potential. This is unique within a constant. Assuming that *the current I of the rectangular bar is distributed uniformly over the bar cross section*, the current density over the bar is $J = I/A$, where A is the cross-sectional area of the bar. Hence, the current over the bar can be concentrated into filaments, giving differential contributions to $A_z(X, Y)$ at point P located at $x = X$ and $y = Y$ as

$$dA_z = -\frac{\mu_0 J}{2\pi} \ln d \, dA \qquad (6.34)$$

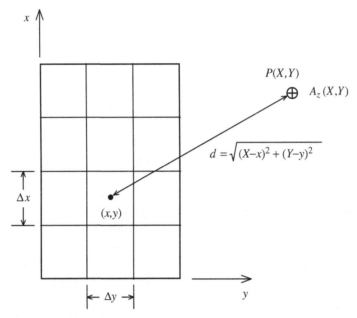

FIGURE 6.7. Another interpretation of geometric mean distance between a point and a surface.

and dA is a differential area of the cross section. Hence, the magnetic vector potential at point P is obtained as

$$A_z(X, Y) = -\frac{\mu_0 J}{2\pi} \int_y \int_x \ln\left[\sqrt{(X - x)^2 + (Y - y)^2}\right] dx\, dy$$

$$= -\frac{\mu_0 I}{2\pi} \frac{1}{A} \int_y \int_x \ln\left[\sqrt{(X - x)^2 + (Y - y)^2}\right] dx\, dy$$

$$(6.35a)$$

where the distance between the filament and the point is

$$d = \sqrt{(X - x)^2 + (Y - y)^2} \qquad (6.35b)$$

and the integral is to be taken over the coordinates of the conductor cross section.

We can evaluate the integral in (6.35) in an approximate fashion via numerical means simply by dividing the cross section into N subrectangles of area $\Delta x\, \Delta y$ as illustrated in Fig. 6.7, performing (6.35) over the individual subrectangles, and summing the results:

$$A_z(X, Y) \cong -\frac{\mu_0 I}{2\pi} \frac{1}{A} \sum_{i=1}^{N} \int_{y_i} \int_{x_i} \ln\left[\sqrt{(X - x_i)^2 + (Y - y_i)^2}\right] dx_i \, dy_i$$

(6.36)

where x_i and y_i are the coordinates of the subrectangles. If this division is such that the dimensions of the subrectangles are sufficiently small, the integral in (6.36) over each subrectangle can be further approximated as

$$\int_{y_i} \int_{x_i} \ln\left[\sqrt{(X - x_i)^2 + (Y - y_i)^2}\right] dx_i \, dy_i$$

$$\cong \Delta x \, \Delta y \ln\left[\sqrt{(X - x_n)^2 + (Y - y_n)^2}\right]$$

(6.37)

This amounts to replacing the smooth and uniform current distribution by filaments at the centers of the subrectangles located at (x_n, y_n). Hence, (6.35) is approximated as (the area of the conductor cross section is $A = N\Delta x \, \Delta y$)

$$A_z(X, Y) \cong -\frac{\mu_0 I}{2\pi} \frac{1}{N} \sum_{n=1}^{N} \ln\left[\sqrt{(X - x_n)^2 + (Y - y_n)^2}\right]$$

(6.38)

By finely dividing the cross section, we can obtain a reasonably accurate approximation to (6.35).

The computation in (6.38) requires an excessive number of operations to (1) take the square root, (2) take the natural logarithm of that result, and (3) sum all of these N resulting contributions. Alternatively, we can write this result in a more computationally efficient form as

$$A_z(X, Y) \cong -\frac{\mu_0 I}{2\pi} \frac{1}{N} \sum_{n=1}^{N} \ln\left[\sqrt{(X - x_n)^2 + (Y - y_n)^2}\right]$$

$$= -\frac{\mu_0 I}{4\pi} \frac{1}{N} \ln\left\{\prod_{n=1}^{N} \left[(X - x_n)^2 + (Y - y_n)^2\right]\right\}$$

(6.39)

and we have used the result that $\ln x_1 + \ln x_2 + \cdots + \ln x_N = \ln(x_1 x_2 \cdots x_N)$ along with $\ln \sqrt{a} = \frac{1}{2} \ln a$. Hence, the result in (6.35) can be approximated as

$$A_z(X, Y) \cong -\frac{\mu_0 I}{2\pi} \ln D$$

(6.40a)

where the *geometric mean distance D between the point P and the rectangle* is

$$\ln D = \frac{1}{N} \sum_{n=1}^{N} \ln \left[\sqrt{(X - x_n)^2 + (Y - y_n)^2} \right]$$

$$= \frac{1}{2N} \ln \left\{ \prod_{n=1}^{N} \left[(X - x_n)^2 + (Y - y_n)^2 \right] \right\} \qquad (6.40b)$$

or

$$D = 2N \sqrt{\prod_{n=1}^{N} \left[(X - x_n)^2 + (Y - y_n)^2 \right]} \qquad (6.40c)$$

So the vector magnetic potential at a point P from a conductor of rectangular cross section and very long length compared to D, $l \gg D$, can be computed alternatively as being the same as that due to a filament containing the total current I that is separated from the point P by the distance D, which is the *geometric mean distance between the rectangle and the point.*

This works well for points P outside the conductor. In using it to determine $A_z(X, Y)$ at points within the conductor, we run into an obvious problem when the desired point is at the center of a subrectangle, $X = x_n$, $Y = y_n$. For this particular "self term" in (6.38), we integrate over the subrectangle and average to give

$$\frac{1}{\Delta^2 x \, \Delta^2 y} \int_y \int_x \int_\eta \int_\xi \ln \left[\sqrt{(x - \xi)^2 + (y - \eta)^2} \right] d\xi \, d\eta \, dx \, dy \quad (6.41)$$

The GMD between the subrectangle and itself is proportional to the perimeter of the subrectangle and evaluates to $\ln D_n = \ln [0.223525 (W + H)]$ [14,20]. This represents a special case of the *geometric mean distance between a shape and itself.*

The numerical solution above was used by Antonini et al. [19] to compute the internal self partial inductance of a conductor of rectangular cross section. We discuss those results in Section 6.5.

The GMD of the combination of two or more surfaces $S_1, S_2, \ldots$ with another surface S can be found from

$$\boxed{\ln D_S = \frac{A_1 \ln D_{S_1} + A_2 \ln D_{S_2} + \cdots}{A_1 + A_2 + \cdots}} \qquad (6.42)$$

which follows from the definition of the GMD. The notation D_{S_i} denotes the GMD from surface S_i to the desired surface S, and A_i denotes the cross-sectional area of surface S_i.

6.4.1   Geometrical Mean Distance Between a Shape and Itself and the Self Partial Inductance of a Shape

The GMD between a shape and itself is given by (6.32):

$$\ln(D) = \frac{1}{A\,A'} \int_A \int_{A'} (\ln d)\ dA'\,dA \tag{6.32}$$

where A and A' are the *same* surface area. When the conductor length is much greater than the GMD given by (6.32), the self partial inductance is approximately

$$L_p \cong \frac{\mu_0}{2\pi} l \left(\ln \frac{2l}{D} - 1 \right) \qquad l \gg D \tag{6.33}$$

Since the integration for the GMD in (6.32) is over the entire cross section of the conductor, the self partial inductance in (6.33) includes the *internal* self partial inductance of the conductor due to magnetic flux internal to it. Although there are innumerable shapes having a GMD with itself, we consider only the GMDs of the common and useful shapes shown in Fig. 6.8. GMDs of other shapes may be found in Grover [14], Rosa and Grover [20], Gray [29], and Higgins [30].

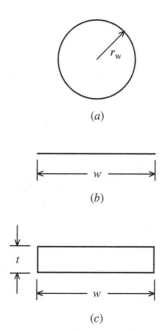

(a)

(b)

(c)

FIGURE 6.8. GMDs of various shapes from which the self partial inductance can be obtained.

GMD of a Circular Shape from Itself The first shape is the circle shown in Fig. 6.8(a), representing the cross section of a wire. The self partial inductance due to magnetic flux *external to the wire* was computed directly in Chapter 4 by replacing the wire with a filament on its axis carrying the total wire current *I*. This was permissible by the fundamental assumption that *the current of the wire is distributed uniformly over its cross section.* Hence, the *external* self partial inductance of a wire of length *l* and radius r_w was obtained in Chapter 5 and is given by

$$L_{p,\text{external}} = \frac{\psi_\infty}{I}$$

$$= \frac{\mu_0}{2\pi} l \left[\ln\left(\frac{l}{r_w} + \sqrt{\left(\frac{l}{r_w}\right)^2 + 1} \right) - \sqrt{1 + \left(\frac{r_w}{l}\right)^2} + \frac{r_w}{l} \right]$$

$$(5.18a)$$

For a long wire such that $l \gg r_w$ (a very reasonable assumption) this approximates to

$$L_{p,\text{external}} = \frac{\mu_0}{2\pi} l \left[\ln\frac{2l}{r_w} - 1 + \frac{r_w}{l} - \frac{1}{4}\left(\frac{r_w}{l}\right)^2 + \cdots \right]$$

$$\cong \frac{\mu_0}{2\pi} l \left(\ln\frac{2l}{r_w} - 1 \right) \qquad l \gg r_w \qquad (5.18c)$$

But this only gives the self partial inductance due to the magnetic flux external to the wire. The contribution to the self partial inductance due to the magnetic flux internal to the wire was obtained as

$$L_{p,\text{internal}} = \frac{\mu_0}{8\pi} l \qquad (4.70)$$

Adding the external self partial inductance in (5.18c) to the internal self partial inductance in (4.70) gives the total self partial inductance of a length of wire *that is due to magnetic flux both external to the wire and internal to the wire* as

$$L_p = L_{p,\text{external}} + L_{p,\text{internal}}$$

$$= \frac{\mu_0}{2\pi} l \left(\ln\frac{2l}{r_w} - 1 + \frac{1}{4} \right)$$

$$= \frac{\mu_0}{2\pi} l \left(\ln\frac{2l}{r_w} - \frac{3}{4} \right) \qquad (6.43a)$$

But this can be written as

$$L_p = L_{p,\text{external}} + L_{p,\text{internal}}$$

$$= \frac{\mu_0}{2\pi} l \left(\ln \frac{2l}{r_w} - 1 + \frac{1}{4} \right)$$

$$= \frac{\mu_0}{2\pi} l \left(\ln \frac{2l}{r_w \, e^{-1/4}} - 1 \right) \qquad (6.43b)$$

Hence, the GMD of a circular area of radius r_w (a wire cross section) from itself is

$$\boxed{\begin{aligned} D &= r_w e^{-1/4} \\ &= 0.7788 \, r_w \end{aligned}} \qquad (6.44a)$$

or

$$\boxed{\ln D = \ln r_w - \tfrac{1}{4}} \qquad (6.44b)$$

and the self partial inductance of a cylinder (a wire) of radius r_w and length l is

$$\boxed{\begin{aligned} L_p &\cong \frac{\mu_0}{2\pi} l \left(\ln \frac{2l}{D} - 1 \right) \\ &= \frac{\mu_0}{2\pi} l \left(\ln \frac{2l}{r_w} - \frac{3}{4} \right) \qquad l \gg D \end{aligned}} \qquad (6.44c)$$

Note that the self partial inductance here *includes the internal inductance of the wire due to magnetic flux internal to the wire.*

To compute the GMD of a circular area with itself directly, we first compute the GMD between a point P and a circular area of radius r_w as shown in Fig. 6.9. The GMD becomes, from (6.32),

$$\ln D = \frac{1}{\pi r_w^2} \int_{r=0}^{r_w} \int_{\theta=0}^{2\pi} \ln \left(\sqrt{R^2 + r^2 - 2rR\cos\theta} \right) r \, d\theta \, dr$$

$$= \frac{1}{2} \frac{1}{\pi r_w^2} \int_{r=0}^{r_w} \int_{\theta=0}^{2\pi} \ln \left(R^2 + r^2 - 2rR\cos\theta \right) r \, d\theta \, dr \qquad (6.45a)$$

and we have used the law of consines to write

$$d^2 = R^2 + r^2 - 2rR\cos\theta \qquad (6.45b)$$

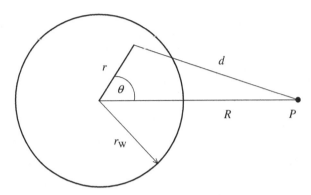

FIGURE 6.9. Computing the GMD between a point and a circular area.

The integral with respect to θ can be written as

$$(I) = \int_{\theta=0}^{2\pi} \ln \left(R^2 + r^2 - 2rR \cos \theta \right) d\theta$$

$$= \int_{\theta=0}^{2\pi} \ln \left[1 + \left(\frac{r}{R} \right)^2 - 2 \left(\frac{r}{R} \right) \cos \theta \right] d\theta + 2 \ln R \int_{\theta=0}^{2\pi} d\theta$$

$$= \int_{\theta=0}^{2\pi} \ln \left[1 + \left(\frac{r}{R} \right)^2 - 2 \left(\frac{r}{R} \right) \cos \theta \right] d\theta + 4\pi \ln R \qquad (6.46)$$

This integral can be evaluated using an integral from Dwight [7]:

$$\int_{\theta=0}^{2\pi} \ln \left(1 + a^2 - 2a \cos x \right) dx = \begin{cases} 4\pi \ln a & a > 1 \\ 0 & a < 1 \end{cases} \qquad \text{(D865.73c)}$$

giving $(I) = 4\pi \ln(r/R) + 4\pi \ln R = 4\pi \ln r$ for the point inside the circle, $R < r$, and $(I) = 4\pi \ln R$ for the point outside the circle, $R > r$. Evaluating the second integral with respect to r yields

$$\ln D = \frac{4\pi}{2\pi r_{\mathrm{w}}^2} \int_{r=0}^{r_{\mathrm{w}}} \ln (r) r \, dr \qquad R < r_{\mathrm{w}}$$

$$= \frac{4\pi}{2\pi r_{\mathrm{w}}^2} \left[\frac{r^2}{2} \ln r - \frac{r^2}{4} \right]_{r=0}^{r_{\mathrm{w}}}$$

$$= \left(\ln r_{\mathrm{w}} - \frac{1}{2} \right) \qquad R < r_{\mathrm{w}} \qquad (6.47a)$$

and

$$\ln D = \frac{4\pi}{2\pi r_w^2} \ln R \int_{r=0}^{r_w} r \, dr \qquad R > r_w$$

$$= \frac{4\pi}{2\pi r_w^2} \ln R \left[\frac{r^2}{2} \right]_{r=0}^{r_w}$$

$$= \ln R \qquad R > r_w \tag{6.47b}$$

and we have used an integral from Dwight [7]:

$$\int x \ln x \, dx = \frac{x^2}{2} \ln x - \frac{x^2}{4} \tag{D610.1}$$

in (6.47a). The result in (6.47a) shows that *the natural logarithm of the GMD from a circular area to any point inside it is the natural logarithm of the radius of the circular area, r_w, minus $\frac{1}{2}$.* The result in (6.47b) shows that *the GMD from a circular area to any point P outside it is simply the distance between the point and the center of the circular area.*

Now consider an annulus of differential thickness dr and radius r and a point P within the annulus as shown in Fig. 6.10. The total circular surface is divided into a part internal to the point and the annulus, and a part external to the point and the annulus. The GMD of the combination of two or more surfaces $S_1, S_2, \ldots$ with another surface S can be found from

$$\ln D_S = \frac{A_1 \ln D_{S_1} + A_2 \ln D_{S_2} + \cdots}{A_1 + A_2 + \cdots} \tag{6.42}$$

which follows from the definition of the GMD. The notation D_{S_i} denotes the GMD from surface S_i to the desired surface S, and A_i denotes the

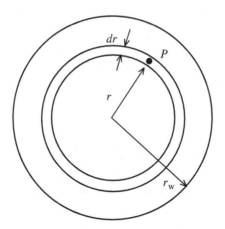

FIGURE 6.10. GMD of a circle from itself.

cross-sectional area of surface S_i. Hence, the GMD of the entire circle to point P is

$$\ln D_P = \frac{\pi r_w^2 \left(\ln r_w - \frac{1}{2} \right) - \pi r^2 \left(\ln r - \frac{1}{2} \right) + \pi r^2 \ln r}{\pi r_w^2}$$

$$= \left(\ln r_w - \frac{1}{2} \right) + \frac{r^2}{r_w^2} \frac{1}{2} \tag{6.48}$$

Hence, the GMD of the circle from itself is

$$\ln D = \frac{1}{\pi r_w^2} 2\pi \int_{r=0}^{r_w} \ln (D_P) r \, dr$$

$$= \frac{2}{r_w^2} \int_{r=0}^{r_w} \left[\left(\ln r_w - \frac{1}{2} \right) + \frac{r^2}{r_w^2} \frac{1}{2} \right] r \, dr$$

$$= \left(\ln r_w - \frac{1}{2} \right) + \frac{1}{4}$$

$$= \ln r_w - \frac{1}{4} \tag{6.49}$$

as before.

GMD of a Line from Itself   Next we determine the GMD of a line of width w and zero thickness from itself as illustrated in Fig. 6.11. The GMD of a line from itself is again obtained from the basic definition in (6.32) as

$$\ln D = \frac{1}{w^2} \int_{x_2=0}^{w} \int_{x_1=0}^{w} \ln |x_2 - x_1| \, dx_1 \, dx_2$$

$$= \frac{1}{2w^2} \int_{x_2=0}^{w} \int_{x_1=0}^{w} \ln (x_2 - x_1)^2 \, dx_1 \, dx_2 \tag{6.50}$$

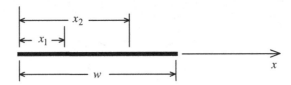

FIGURE 6.11. GMD of a line from itself.

The integral with respect to x_1 becomes

$$(\text{I}) = \int_{x_1=0}^{w} \ln(x_2 - x_1)^2 \, dx_1$$

$$= \int_{\lambda=x_2-w}^{x_2} \ln \lambda^2 \, d\lambda$$

$$= \left[\lambda \ln \lambda^2 - 2\lambda \right]_{\lambda=x_2-w}^{x_2}$$

$$= x_2 \ln x_2^2 - 2x_2 - (x_2 - w) \ln(x_2 - w)^2 + 2(x_2 - w) \quad (6.51)$$

and we have used the change of variables $\lambda = x_2 - x_1$, $d\lambda = -dx_1$ and an integral from Dwight [7]:

$$\int \ln\left(x^2 + a^2\right) dx = x \ln\left(x^2 + a^2\right) - 2x + 2a \tan^{-1} \frac{x}{a} \quad (\text{D623})$$

The integral with respect to x_2 becomes

$$(\text{II}) = \int_{x_2=0}^{w} (\text{I}) \, dx_2$$

$$= \left[\frac{\lambda^2}{2} \ln \lambda^2 - \frac{\lambda^2}{2} - \lambda^2 \right]_{\lambda=0}^{w} + \left[\frac{\lambda^2}{2} \ln \lambda^2 - \frac{\lambda^2}{2} - \lambda^2 \right]_{\lambda=0}^{-w}$$

$$= w^2 \ln w^2 - 3w^2 \quad (6.52)$$

and we have used a change of variables $\lambda = x_2 - w$, $d\lambda = dx_2$ in the second half of the integral and an integral from Dwight [7]:

$$\int x \ln\left(x^2 + a^2\right) dx = \tfrac{1}{2} \left(x^2 + a^2\right) \ln\left(x^2 + a^2\right) - \tfrac{1}{2}x^2 \quad (\text{D623.1})$$

Hence, dividing (6.52) by $2w^2$ according to (6.50) gives the GMD of a line with itself as

$$\boxed{\ln D = \ln w - \tfrac{3}{2}} \quad (6.53a)$$

or

$$\boxed{D = 0.22313w} \quad (6.53b)$$

and the self partial inductance of a line is

$$\boxed{\begin{aligned} L_p &\cong \frac{\mu_0}{2\pi} l \left(\ln \frac{2l}{D} - 1 \right) \\ &= \frac{\mu_0}{2\pi} l \left(\ln \frac{2l}{w} + \frac{1}{2} \right) \qquad l \gg D \end{aligned}} \quad (6.53c)$$

Compare this result to (6.20) for $u = l/w \gg 1$:

$$\frac{L_{p(t=0)}}{l} \cong \frac{\mu_0}{2\pi} \begin{cases} \ln 2u + \dfrac{1}{2} + \dfrac{1}{3u} & u \gg 1, \ l \gg w \\[2ex] u\left(\ln\dfrac{2}{u} + \dfrac{1}{2} + \dfrac{u}{3}\right) & u \ll 1, \ l \ll w \end{cases} \tag{6.20}$$

For $u = 1$ ($l = w$) the exact result in (6.19) gives $L_{p(t=0)} = 7.552\,\text{nH/in.}$, whereas (6.53) using the GMD gives $L_{p(t=0)} = 6.061\,\text{nH/in.}$, a difference of 24.6%. For $u = 10$ ($l = 10w$) the exact result in (6.19) gives $L_{p(t=0)} = 17.926\,\text{nH/in.}$, whereas (6.53) using the GMD gives $L_{p(t=0)} = 17.758\,\text{nH/in.}$, a difference of 0.9%. For a typical PCB land of width $w = 8$ mils, a ratio of $l/w = 10$ gives the length of the land as 0.08 in., which is not an unreasonably long length for a typical PCB land. For $u = 100$ ($l = 100w$) the exact result in (6.19) gives $L_{p(t=0)} = 29.472\,\text{nH/in.}$, whereas (6.53) using the GMD gives $L_{p(t=0)} = 29.455\,\text{nH/in.}$, a difference of 0.06%. For a typical PCB land of width $w = 8$ mils, a ratio of $l/w = 100$ gives the length of the land as 0.8 in., which is still not an unreasonably long length of a typical PCB land. This reinforces the restriction on the validity of the GMD concept to situations where the length of the land is much greater than its width. But the computation of the self partial inductance of a thin land via the GMD is considerably simpler than the exact formula given in (6.18) or (6.19).

GMD of a Rectangular Shape from Itself   The GMD of the rectangle shown in Fig. 6.8(c) can be derived in a fashion similar to that from the basic definition in (6.32) and is given in [19,30] as

$$\ln D = -\frac{25}{12} + \frac{1}{2t^2w^2} \sum_{i=1}^{4}\sum_{j=1}^{4}(-1)^{i+j} f(q_i, r_j) \tag{6.54a}$$

where

$$\begin{aligned} f(q,r) = &\left(\frac{q^2r^2}{4} - \frac{q^4}{24} - \frac{r^4}{24}\right)\ln(q^2+r^2) \\ &+ \frac{q^3r}{3}\tan^{-1}\frac{r}{q} + \frac{qr^3}{3}\tan^{-1}\frac{q}{r} \end{aligned} \tag{6.54b}$$

and

$$
\begin{aligned}
q_1 &= -w \\
q_2 &= q_4 = 0 \\
q_3 &= w \\
r_1 &= -t \\
r_2 &= r_4 = 0 \\
r_3 &= t
\end{aligned}
$$

(6.54c)

Writing this out and using the fact that $f(0,0) = 0$, $f(\pm w, \pm t) = f(w,t)$, $f(-w,0) = f(w,0)$, and $f(-t,0) = f(t,0)$ gives a clearer result as

$$
\begin{aligned}
\ln D =\ &-\frac{25}{12} + \frac{1}{2}\ln\left(w^2 + t^2\right) - \frac{1}{12}\frac{w^2}{t^2}\ln\left(1 + \frac{t^2}{w^2}\right) \\
&- \frac{1}{12}\frac{t^2}{w^2}\ln\left(1 + \frac{w^2}{t^2}\right) \\
&+ \frac{2}{3}\frac{w}{t}\tan^{-1}\frac{t}{w} + \frac{2}{3}\frac{t}{w}\tan^{-1}\frac{w}{t}
\end{aligned}
$$

(6.55a)

which is equivalent to the result given by Gray [29], p. 302, eq. (114). If the rectangle is square, $w = t$, its GMD is

$$
\begin{aligned}
\ln D &= -\frac{25}{12} + \frac{1}{2}\ln\left(2w^2\right) - \frac{1}{12}\ln(2) - \frac{1}{12}\ln(2) + \frac{2}{3}\frac{\pi}{4} + \frac{2}{3}\frac{\pi}{4} \\
&= -\frac{25}{12} + \ln w + \frac{1}{3}\ln(2) + \frac{\pi}{3} \\
&= \ln w - 0.80509 \qquad w = t
\end{aligned}
$$

(6.55b)

It turns out that the GMD of a rectangle shape (a bar) can be represented approximately in terms of the dimensions of its perimeter, its width w and its thickness t, as [14,20,29]

$$
\log D \cong \ln(w + t) - \tfrac{3}{2}
$$

(6.56a)

and the GMD is

$$
D \cong (0.2235)\,(w + t)
$$

(6.56b)

If the bar is square, $w = t$, this reduces approximately to (6.55b). So the self partial inductance of a conductor of rectangular cross section is approximately

$$
L_p \cong \frac{\mu_0}{2\pi} l \left(\ln \frac{2l}{w+t} + \frac{1}{2} \right) \qquad l \gg D \tag{6.56c}
$$

which reduces to (6.53c) for $t = 0$. Note that the self partial inductance here *includes the internal inductance of the bar due to magnetic flux internal to the bar.*

The result for the GMD of a rectangle given in (6.54) can be obtained by direct integration using the basic result in (6.32) as

$$
\ln D = \frac{1}{(wt)^2} \int_{y_2=0}^{t} \int_{y_1=0}^{t} \int_{x_2=0}^{w} \int_{x_1=0}^{w}
$$

$$
\ln \left[\sqrt{(y_2 - y_1)^2 + (x_2 - x_1)^2} \right] dx_1 \, dx_2 \, dy_1 \, dy_2
$$

$$
= \frac{1}{(wt)^2} \frac{1}{2} \int_{y_2=0}^{t} \int_{y_1=0}^{t} \int_{x_2=0}^{w} \int_{x_1=0}^{w}
$$

$$
\ln \left[(y_2 - y_1)^2 + (x_2 - x_1)^2 \right] dx_1 \, dx_2 \, dy_1 \, dy_2 \tag{6.57}
$$

The first integral with respect to x_1 is evaluated as

$$
\text{(I)} = \int_{x_1=0}^{w} \ln \left[(y_2 - y_1)^2 + (x_2 - x_1)^2 \right] dx_1
$$

$$
= \int_{\lambda=x_2-w}^{x_2} \ln \left[(y_2 - y_1)^2 + \lambda^2 \right] d\lambda
$$

where we have used a change of variables $\lambda = x_2 - x_1$, $d\lambda = -dx_1$. This can be evaluated using Dwight [7]:

$$
\int \ln \left(a^2 + x^2 \right) dx = x \ln \left(a^2 + x^2 \right) - 2x + 2a \tan^{-1} \frac{x}{a} \tag{D623}
$$

to give

$$
\text{(I)} = \left[\lambda \ln \left((y_2 - y_1)^2 + \lambda^2 \right) - 2\lambda + 2 (y_2 - y_1) \tan^{-1} \frac{\lambda}{y_2 - y_1} \right]_{\lambda=x_2-w}^{x_2}
$$

$$
= \left[x_2 \ln \left((y_2 - y_1)^2 + x_2{}^2 \right) - 2x_2 + 2 (y_2 - y_1) \tan^{-1} \frac{x_2}{y_2 - y_1} \right]
$$

$$-\left[(x_2 - w)\ln\left((y_2 - y_1)^2 + (x_2 - w)^2\right) - 2\left(x_2 - w\right)\right.$$

$$\left. + 2(y_2 - y_1)\tan^{-1}\frac{x_2 - w}{y_2 - y_1}\right]$$

The next integral with respect to x_2 is evaluated as

$$(\text{II}) = \int_{x_2=0}^{w} (\text{I})\, dx_2$$

$$= \int_{x_2=0}^{w}\left[x_2\ln\left((y_2 - y_1)^2 + x_2^2\right) - 2x_2 + 2(y_2 - y_1)\tan^{-1}\frac{x_2}{y_2 - y_1}\right] dx_2$$

$$+ \int_{\lambda=0}^{-w}\left[\lambda\ln\left((y_2 - y_1)^2 + \lambda^2\right) - 2\lambda + 2(y_2 - y_1)\tan^{-1}\frac{\lambda}{y_2 - y_1}\right] d\lambda$$

where we have made a change of variables $\lambda = x_2 - w$, $d\lambda = dx_2$ in the second integral. These can be evaluated using integrals from Dwight [7]:

$$\int x\ln\left(a^2 + x^2\right) dx = \frac{1}{2}\left(a^2 + x^2\right)\ln\left(a^2 + x^2\right) - \frac{x^2}{2} \quad \text{(D623.1)}$$

and

$$\int \tan^{-1}\frac{x}{a}\, dx = x\tan^{-1}\frac{x}{a} - \frac{a}{2}\ln\left(a^2 + x^2\right) \quad \text{(D525)}$$

to give

$$(\text{II}) = \left[\frac{\lambda^2}{2}\ln\left((y_2 - y_1)^2 + \lambda^2\right) - \frac{3\lambda^2}{2}\right.$$

$$\left. + 2(y_2 - y_1)\,\lambda\tan^{-1}\frac{\lambda}{y_2 - y_1} - \frac{1}{2}(y_2 - y_1)^2\ln\left((y_2 - y_1)^2 + \lambda^2\right)\right]\Big|_{0,0}^{w,-w}(\lambda)$$

where the notation is the same as in Sections 6.2 and 6.3. The third integral with respect to y_1 can be similarly integrated as

$$(\text{III}) = \int_{y_1=0}^{t} (\text{II})\, dy_1$$

$$= \int_{y_1=0}^{t}\left[\frac{\lambda^2}{2}\ln\left((y_2 - y_1)^2 + \lambda^2\right) - \frac{3\lambda^2}{2}\right.$$

$$+ 2(y_2 - y_1)\,\lambda\tan^{-1}\frac{\lambda}{y_2 - y_1}$$

$$\left. - \frac{1}{2}(y_2 - y_1)^2\ln\left((y_2 - y_1)^2 + \lambda^2\right)\right]\Big|_{0,0}^{w,-w}(\lambda)\, dy_1$$

$$= \int_{\xi=y_2-t}^{y_2} \left[\frac{\lambda^2}{2} \ln \left(\xi^2 + \lambda^2 \right) - \frac{3\lambda^2}{2} + 2\xi\lambda \tan^{-1} \frac{\lambda}{\xi} \right. $$

$$\left. - \frac{1}{2} \xi^2 \ln \left(\xi^2 + \lambda^2 \right) \right]_{0,0}^{w,-w} (\lambda) \, d\xi$$

and we have used a change of variables $\xi = y_2 - y_1$, $d\xi = -dy_1$. These can be evaluated using integrals from Dwight [7]:

$$\int \ln \left(a^2 + x^2 \right) dx = x \ln \left(a^2 + x^2 \right) - 2x + 2a \tan^{-1} \frac{x}{a} \qquad \text{(D623)}$$

$$\int x^2 \ln \left(a^2 + x^2 \right) dx = \frac{x^3}{3} \ln \left(a^2 + x^2 \right) - \frac{2}{9} x^3 + \frac{2}{3} xa^2 - \frac{2}{3} a^3 \tan^{-1} \frac{x}{a}$$

$$\text{(D623.2)}$$

$$\int x \tan^{-1} \frac{a}{x} dx = \frac{ax}{2} + \frac{x^2 + a^2}{2} \tan^{-1} \frac{a}{x} \qquad \text{(D528.1)}$$

giving

$$\text{(III)} = \left[\left[\frac{\lambda^2 \xi}{2} \ln \left(\xi^2 + \lambda^2 \right) - \frac{\xi^3}{6} \ln \left(\xi^2 + \lambda^2 \right) - \frac{\lambda^2 \xi}{3} - \frac{3}{2} \lambda^2 + \frac{\xi^2}{9} \right. \right.$$

$$\left. \left. + \frac{4}{3} \lambda^3 \tan^{-1} \left(\frac{\xi}{\lambda} \right) + \lambda \left(\xi^2 + \lambda^2 \right) \tan^{-1} \frac{\lambda}{\xi} \right]_{0,0}^{w,-w} (\lambda) \right]_{\xi=y_2-t}^{y_2}$$

$$= \left[\frac{\lambda^2 y_2}{2} \ln \left(y_2^2 + \lambda^2 \right) - \frac{y_2^3}{6} \ln \left(y_2^2 + \lambda^2 \right) - \frac{\lambda^2 y_2}{3} - \frac{3}{2} \lambda^2 + \frac{y_2^2}{9} \right.$$

$$\left. + \frac{4}{3} \lambda^3 \tan^{-1} \frac{y_2}{\lambda} + \lambda \left(y_2^2 + \lambda^2 \right) \tan^{-1} \frac{\lambda}{y_2} \right]_{0,0}^{w,-w} (\lambda)$$

$$- \left[\frac{\lambda^2 (y_2 - t)}{2} \ln \left((y_2 - t)^2 + \lambda^2 \right) - \frac{(y_2 - t)^3}{6} \ln \left((y_2 - t)^2 + \lambda^2 \right) \right.$$

$$- \frac{\lambda^2 (y_2 - t)}{3} - \frac{3}{2} \lambda^2$$

$$+ \frac{(y_2 - t)^2}{9} + \frac{4}{3} \lambda^3 \tan^{-1} \frac{y_2 - t}{\lambda} + \lambda \left((y_2 - t)^2 \right.$$

$$\left. \left. + \lambda^2 \right) \tan^{-1} \frac{\lambda}{y_2 - t} \right]_{0,0}^{w,-w} (\lambda)$$

or

$$(III) = \left[\left[\frac{\lambda^2 \eta}{2} \ln \left(\eta^2 + \lambda^2 \right) - \frac{\eta^3}{6} \ln \left(\eta^2 + \lambda^2 \right) - \frac{\lambda^2 \eta}{3} - \frac{3}{2} \lambda^2 + \frac{\eta^2}{9} \right. \right.$$

$$\left. \left. + \frac{4}{3} \lambda^3 \tan^{-1} \frac{\eta}{\lambda} + \lambda \left(\eta^2 + \lambda^2 \right) \tan^{-1} \frac{\lambda}{\eta} \right] (\lambda) \right]_{0,0}^{w,-w} \binom{y2}{(\eta)} \bigg|_{y_2-t}$$

In a similar fashion, the fourth integral can be evaluated as

$$(IV) = \int_{y_2=0}^{t} (III) \, dy_2$$

$$= \int_{\eta=0}^{t} \left[\frac{\lambda^2 \eta}{2} \ln \left(\eta^2 + \lambda^2 \right) - \frac{\eta^3}{6} \ln \left(\eta^2 + \lambda^2 \right) - \frac{\lambda^2 \eta}{3} - \frac{3}{2} \lambda^2 + \frac{\eta^2}{9} \right.$$

$$\left. + \frac{4}{3} \lambda^3 \tan^{-1} \frac{\eta}{\lambda} + \lambda \left(\eta^2 + \lambda^2 \right) \tan^{-1} \frac{\lambda}{\eta} \right]_{0,0}^{w,-w} (\lambda) \, d\eta$$

$$+ \int_{\eta=0}^{-t} \left[\frac{\lambda^2 \eta}{2} \ln \left(\eta^2 + \lambda^2 \right) - \frac{\eta^3}{6} \ln \left(\eta^2 + \lambda^2 \right) - \frac{\lambda^2 \eta}{3} - \frac{3}{2} \lambda^2 + \frac{\eta^2}{9} \right.$$

$$\left. + \frac{4}{3} \lambda^3 \tan^{-1} \frac{\eta}{\lambda} + \lambda \left(\eta^2 + \lambda^2 \right) \tan^{-1} \frac{\lambda}{\eta} \right]_{0,0}^{w,-w} (\lambda) \, d\eta$$

and using Dwight [7]:

$$\int x \ln \left(a^2 + x^2 \right) dx = \frac{1}{2} \left(a^2 + x^2 \right) \ln \left(a^2 + x^2 \right) - \frac{x^2}{2} \quad \text{(D623.1)}$$

$$\int x^3 \ln \left(x^2 + a^2 \right) dx = \frac{x^4 - a^4}{4} \ln \left(x^2 + a^2 \right) - \frac{x^4}{8} + \frac{x^2 a^2}{4} \quad \text{(D623.3)}$$

$$\int \tan^{-1} \frac{x}{a} \, dx = x \tan^{-1} \frac{x}{a} - \frac{a}{2} \ln \left(a^2 + x^2 \right) \quad \text{(D525)}$$

$$\int \tan^{-1} \frac{a}{x} \, dx = x \tan^{-1} \frac{a}{x} + \frac{a}{2} \ln \left(x^2 + a^2 \right) \quad \text{(D528)}$$

$$\int x^2 \tan^{-1} \frac{a}{x} \, dx = \frac{x^3}{3} \tan^{-1} \frac{a}{x} + \frac{a x^2}{6} - \frac{a^3}{6} \ln \left(x^2 + a^2 \right) \quad \text{(D528.2)}$$

giving the result in (6.54).

6.4.2   Geometrical Mean Distance and Mutual Partial Inductance Between Two Shapes

In this subsection we obtain the geometrical mean distances between the common shapes shown in Fig. 6.12. These represent (a) the cross sections of

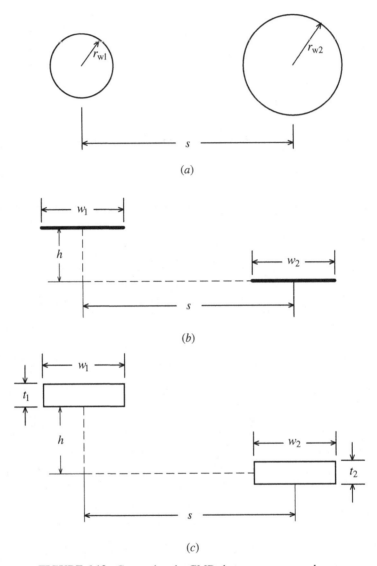

FIGURE 6.12. Computing the GMDs between common shapes.

two wires of radii r_{w1} and r_{w2} that are separated a distance s, (b) two strips of widths w_1 and w_2 and zero thickness, and (c) two rectangular bars of widths w_1 and w_2 and thicknesses t_1 and t_2 that represent the cross sections of PCB lands. Once again we stress that the GMDs for these shapes will be valid only as long as the lengths of the conductors whose cross sections these shapes represent are much longer than the separation between them.

GMD Between Two Circular Shapes The GMD between two circular shapes representing the cross sections of two wires shown in Fig. 6.12(a) is simply the distance between their centers, s:

$$\ln D = \ln s$$

This follows from the fact that the mutual partial inductance between between two parallel wires of equal length l having currents that are *uniformly distributed over their cross sections* can be obtained by replacing them with two filaments *located on the axes of the conductors*. This result is the basic result for the mutual partial inductance between two filaments separated a distance s and given as

$$M_p = \frac{\mu_0}{2\pi} l \left[\ln \frac{2l}{s} - 1 + \frac{s}{l} - \frac{1}{4}\left(\frac{s}{l}\right)^2 + \frac{1}{32}\left(\frac{s}{l}\right)^4 - \cdots \right]$$

$$\cong \frac{\mu_0}{2\pi} l \left(\ln \frac{2l}{s} - 1 \right) \qquad l \gg s \tag{5.21c}$$

GMD Between Two Lines The GMD between two parallel lines having widths w_1 and w_2 and zero thickness with vertical separation h and horizontal separation s between their centers as shown in Fig. 6.12(b) can again be obtained from the basic definition of the GMD in (6.32):

$$\ln D = \frac{1}{w_1 w_2} \int_{x_2=s+w_1/2-w_2/2}^{s+w_1/2+w_2/2} \int_{x_1=0}^{w_1} \ln\left[\sqrt{(x_2-x_1)^2+h^2}\right] dx_1\, dx_2$$

$$= \frac{1}{w_1 w_2} \frac{1}{2} \int_{x_2=s+w_1/2-w_2/2}^{s+w_1/2+w_2/2} \int_{x_1=0}^{w_1} \ln\left[(x_2-x_1)^2+h^2\right] dx_1\, dx_2 \tag{6.58}$$

The integral with respect to x_1 becomes

$$(\mathrm{I}) = \int_{x_1=0}^{w_1} \ln\left[(x_2-x_1)^2+h^2\right] dx_1$$

$$= \int_{\lambda=x_2-w_1}^{x_2} \ln\left(\lambda^2+h^2\right) d\lambda$$

$$= \left[\lambda \ln\left(\lambda^2+h^2\right) - 2\lambda + 2h \tan^{-1}\frac{\lambda}{h}\right]_{\lambda=x_2-w_1}^{x_2}$$

$$= x_2 \ln\left(x_2^2+h^2\right) - 2x_2 + 2h \tan^{-1}\frac{x_2}{h}$$

$$- (x_2-w_1)\ln\left((x_2-w_1)^2+h^2\right) + 2(x_2-w_1) - 2h\tan^{-1}\frac{x_2-w_1}{h}$$

$$\tag{6.59}$$

and we have used the change of variables $\lambda = x_2 - x_1$, $d\lambda = -dx_1$ and an integral from Dwight [7]:

$$\int \ln\left(x^2 + a^2\right) dx = x\ln\left(x^2 + a^2\right) - 2x + 2a\tan^{-1}\frac{x}{a} \quad \text{(D623c)}$$

The integral with respect to x_2 becomes

$$(\text{II}) = \int_{x_2=s+w_1/2-w_2/2}^{s+w_1/2+w_2/2} (\text{I})\, dx_2$$

$$= \left[\frac{x_2^2 - h^2}{2}\ln\left(x_2^2 + h^2\right) - \frac{3}{2}x_2^2 + 2hx_2\tan^{-1}\frac{x_2}{h}\right]_{x_2=s+w_1/2-w_2/2}^{s+w_1/2+w_2/2}$$

$$- \left[\frac{\lambda^2 - h^2}{2}\ln\left(\lambda^2 + h^2\right) - \frac{3}{2}\lambda^2 + 2h\lambda\tan^{-1}\frac{\lambda}{h}\right]_{\lambda=s-w_1/2-w_2/2}^{s-w_1/2+w_2/2}$$

$$\tag{6.60}$$

and we have used a change of variables $\lambda = x_2 - w_1$, $d\lambda = dx_2$ in the second half of the integral and integrals from Dwight [7]:

$$\int x\ln\left(x^2 + a^2\right) dx = \frac{x^2 + a^2}{2}\ln\left(x^2 + a^2\right) - \frac{x^2}{2} \quad \text{(D623.1)}$$

$$\int \tan^{-1}\frac{x}{a}\,dx = x\tan^{-1}\frac{x}{a} - \frac{a}{2}\ln\left(x^2 + a^2\right) \quad \text{(D525)}$$

Hence, dividing (6.60) by $2\,w_1 w_2$ according to (6.58) gives the GMD between the two lines. Although the integration is complete, this gives a complicated expression which we obtain in the next section.

A more interesting and simpler result is for the case where the line widths are identical, $w_1 = w_2 = w$, and the two lines are on line with each other, $h = 0$. The results above simplify to give the GMD as

$$\boxed{\begin{aligned} 2w^2\ln D &= (s+w)^2\ln(s+w) - \tfrac{3}{2}(s+w)^2 - 2s^2\ln s \\ &+ 3s^2 + (s-w)^2\ln(s-w) - \tfrac{3}{2}(s-w)^2 \qquad h = 0,\ w_1 = w_2 = w \end{aligned}}$$

$$\tag{6.61}$$

We can simplify this further. Suppose that the lines lie in the same plane, $h = 0$, and the separation between the lines is $s = nw$. The lines are touching when $n = 1$ and are separated edge to edge by one line width w when $n = 2$.

CONCEPT OF GEOMETRIC MEAN DISTANCE   **289**

Substituting $s = nw$ into (6.61) gives

$$
\begin{aligned}
\ln D &= \frac{(n+1)^2}{2}\ln\left[(n+1)\,w\right] - \frac{3}{4}(n+1)^2 - n^2\ln nw \\
&\quad + \frac{3}{2}n^2 + \frac{(n-1)^2}{2}\ln\left[(n-1)\,w\right] - \frac{3}{4}(n-1)^2 \\
&= \frac{(n+1)^2}{2}\ln\left[(n+1)\,w\right] - n^2\ln nw \\
&\quad + \frac{(n-1)^2}{2}\ln\left[(n-1)\,w\right] - \frac{3}{2} \qquad s = nw
\end{aligned}
$$

(6.62)

This matches the result given by Rosa [25], p. 164, eq. (11). Rosa also gives a convenient series expansion for this case of $h = 0$, $w_1 = w_2 = w$, and $s = nw$:

$$
\ln D = \ln nw - \left(\frac{1}{12n^2} + \frac{1}{60n^4} + \frac{1}{168n^6} + \frac{1}{360n^8} + \frac{1}{660n^{10}} + \cdots\right)
$$

(6.63)

which converges very rapidly.

In either case, the mutual partial inductance between the conductors is

$$
M_p \cong \frac{\mu_0}{2\pi}l\left(\ln\frac{2l}{D} - 1\right) \qquad l \gg D
$$

(6.33)

GMD Between Two Rectangles   Finally, we obtain the GMD for the general case shown in Fig. 6.12(c) of two parallel rectangles with widths w_1, w_2, thicknesses t_1, t_2, and vertical separation h and horizontal separation center to center of s. The result is given in [19,30] as

$$
\ln D = -\frac{25}{12} + \frac{1}{2\,(t_1 w_1)\,(t_2 w_2)}\sum_{i=1}^{4}\sum_{j=1}^{4}(-1)^{i+j}\,f\,(q_i, r_j)
$$

(6.64a)

where

$$
\begin{aligned}
f\,(q, r) &= \left(\frac{q^2 r^2}{4} - \frac{q^4}{24} - \frac{r^4}{24}\right)\ln\left(q^2 + r^2\right) \\
&\quad + \frac{q^3 r}{3}\tan^{-1}\frac{r}{q} + \frac{q r^3}{3}\tan^{-1}\frac{q}{r}
\end{aligned}
$$

(6.64b)

and

$$
\begin{aligned}
q_1 &= s - \frac{w_1}{2} - \frac{w_2}{2} \\
q_2 &= s + \frac{w_1}{2} - \frac{w_2}{2} \\
q_3 &= s + \frac{w_1}{2} + \frac{w_2}{2} \\
q_4 &= s - \frac{w_1}{2} + \frac{w_2}{2} \\
r_1 &= h - \frac{t_1}{2} - \frac{t_2}{2} \\
r_2 &= h + \frac{t_1}{2} - \frac{t_2}{2} \\
r_3 &= h + \frac{t_1}{2} + \frac{t_2}{2} \\
r_4 &= h - \frac{t_1}{2} + \frac{t_2}{2}
\end{aligned}
\tag{6.64c}
$$

Once again, for "long conductors," the mutual partial inductance between them is

$$
M_p \cong \frac{\mu_0}{2\pi} l \left(\ln \frac{2l}{D} - 1 \right) \qquad l \gg D
\tag{6.33}
$$

Grover [14] gives several tables for determining the GMD of two rectangles for the cases of Fig. 6.12(c) with various ratios of width to thickness and length to width for $h = 0$. Ruehli [15] has computed results for the mutual partial inductance between two parallel lands of equal width and equal thickness for various values of $u = l/w$ and $v = t/w$. He shows that there is little error between the exact result and the mutual partial inductance obtained by replacing the lands with one filament at the center of each land (i.e., $M_p \cong M_f$) as long as the lands are not very close to each other. This correlates with the exact solution for round wires discussed in Section 4.6, in that proximity effect and the associated redistribution of the current over the cross section to the facing surfaces of the wires is not significantly pronounced unless the wires are close enough that they are separated by a distance such that one wire will exactly fit between them. This is somewhat remarkable since as the lands are brought close together, their currents will no longer be *distributed uniformly over their cross sections but will migrate toward the facing sides,* as we show with numerical computations in the next section. It should be reiterated that *all our previous formulas were derived assuming that the current remains uniformly distributed over the conductor cross section.* Considering the nonuniform distribution of the current over the cross section of a PCB land does not seem to be feasible except with approximate numerical solutions which we address in the next section.

6.5 COMPUTING THE HIGH-FREQUENCY PARTIAL INDUCTANCES OF LANDS AND NUMERICAL METHODS

We have seen that computation of the self and mutual partial inductances of conductors having rectangular cross sections (PCB lands) is very complicated and the results are quite tedious. This is why the earlier textbooks and publications contained extensive tables for the calculation of these partial inductances; digital computers had not been invented. Today we enjoy the enormous computing power of computers, and involved formulas are no longer the obstacle they used to be. Still there remain many problems in partial inductance for which there are no feasible closed-form solutions. An example is the nonuniform current distribution over the conductor cross section caused either by (1) proximity effect, or (2) frequencies of excitation other than dc. In this section we develop methods for numerically computing the self and partial inductances of conductors of rectangular cross section that can be used to solve those difficult problems.

The primary restriction on all the previous results of this book is that *the current is assumed to be uniformly distributed over the conductor cross sections.* This condition is not satisfied for either proximity effect or high-frequency excitation. However, we can approximate a nonuniform current distribution in a discrete fashion, that is, by approximating the current distribution over the conductor cross section in a piecewise-constant or piecewise-linear manner. The actual parameters in that distribution are unknown but will be determined by enforcing the constraints that the currents must satisfy, resulting in the simultaneous solution of a large set of simultaneous equations which computers can readily handle.

For example, consider breaking the cross section of a rectangular conductor into individual "subbars" of rectangular cross section, as illustrated in Fig. 6.13(a). The number of divisions along the width is NW and the number of divisions along the thickness is NT. Each subbar dimension is $\Delta t = t/\text{NT}$ and $\Delta w = w/\text{NW}$. The total number of subbars is therefore $N = \text{NW} \times \text{NT}$. The bar can be modeled as an equivalent circuit shown in Fig. 6.13(b). Represent each subbar with its dc resistance $R_i = l/\sigma \, \Delta w \, \Delta t$, where σ is the conductivity of the metal, which we will assume is copper (having $\sigma_{\text{Cu}} = 5.8 \times 10^7$). The L_{pi} are the exact self partial inductances of the subbar, given by Hoer and

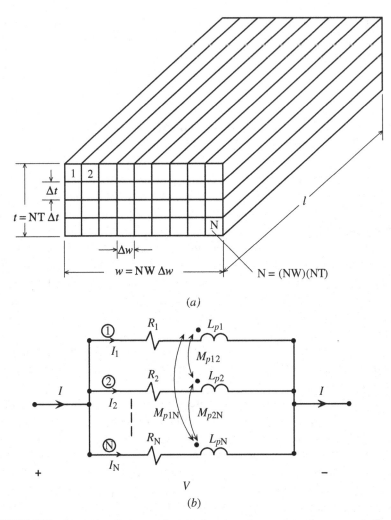

FIGURE 6.13. Approximating a conductor of rectangular cross section as a set of parallel "subbars" of rectangular cross section.

Love in (6.23), which Ruehli ([15], eq. (15)) put into a more stable numerical form. The mutual partial inductances between the subbars is approximated as being between filaments at the centers of the subbars and is given by

$$M_{p_{ij}} = \frac{\mu_0}{2\pi} l \left[\ln \left(\frac{l}{d_{ij}} + \sqrt{\left(\frac{l}{d_{ij}} \right)^2 + 1} \right) - \sqrt{1 + \left(\frac{d_{ij}}{l} \right)^2} + \frac{d_{ij}}{l} \right] \quad (5.21a)$$

where d_{ij} is the distance between the centers of subbars i and j.

All the subbars are connected in parallel so that the voltage across each subbar is the same and is the voltage between the ends of the entire conductor and denoted as V. The total current drawn by the conductor is denoted as I, and we determine and plot over the conductor cross section the individual currents of the subbars, I_i, for $i = 1, 2, \ldots, N$ where $N = NW \times NT$. This demonstrates two important aspects of this problem. For increasing frequencies of excitation (1) the currents crowd to the edges of the conductor, and (2) the currents peak at the corners of the conductor, giving the so-called "bedpost" distribution over the conductor cross section. The equivalent circuit of the conductor and its subbars is shown in Fig. 6.13(b). The voltage across each subbar, V_i, is related to the current through it, I_i, as

$$V_i = (R_i + j\omega L_{pi}) I_i + j\omega \sum_{\substack{j=1 \\ j \neq i}}^{N} M_{pij} I_j \qquad (6.65a)$$

In matrix notation this becomes

$$\mathbf{V} = \mathbf{Z}\mathbf{I} \qquad (6.65b)$$

where $\mathbf{V}$ and $\mathbf{I}$ are vectors of N rows and one column containing the voltages and currents of the N individual subbars as

$$\mathbf{V} = \begin{bmatrix} V_1 \\ V_2 \\ \vdots \\ V_N \end{bmatrix}, \quad \mathbf{I} = \begin{bmatrix} I_1 \\ I_2 \\ \vdots \\ I_N \end{bmatrix} \qquad (6.65c)$$

The "impedance matrix" $\mathbf{Z}$ is square with $N = NW \times NT$ rows and $N = NW \times NT$ columns and contains the self impedances of the individual subbars and the mutual impedances between the subbars. This matrix of subbar impedances can be separated into two parts:

$$\mathbf{Z} = \mathbf{Z}_s + \mathbf{Z}_m \qquad (6.65d)$$

The "self impedance" matrix is a diagonal matrix as

$$\mathbf{Z}_s = \begin{bmatrix} Z_{s1} & 0 & 0 & \cdots & 0 \\ 0 & Z_{s2} & 0 & \cdots & 0 \\ 0 & 0 & \ddots & \cdots & \vdots \\ \vdots & \vdots & \vdots & \ddots & 0 \\ 0 & 0 & \cdots & 0 & Z_{sN} \end{bmatrix} \qquad (6.65e)$$

where

$$Z_{si} = R_i + j\omega L_{pi} \qquad (6.65f)$$

The "mutual impedance" matrix is

$$\mathbf{Z}_m = j\omega \begin{bmatrix} 0 & M_{p12} & M_{p13} & \cdots & M_{p1N} \\ M_{p12} & 0 & M_{p23} & \cdots & M_{p2N} \\ M_{p13} & M_{p23} & \ddots & \cdots & \vdots \\ \vdots & \vdots & \vdots & \ddots & \vdots \\ M_{p1N} & M_{p2N} & \cdots & \cdots & 0 \end{bmatrix} \qquad (6.65g)$$

We solve these equations by inverting (6.65b) to give, at each frequency of excitation,

$$\mathbf{I} = \mathbf{Z}^{-1}\mathbf{V} \qquad (6.66)$$

But the subbars are all connected in parallel, so that $V = V_1 = V_2 = \cdots = V_N$. Hence, (6.66) becomes

$$\begin{bmatrix} I_1 \\ I_2 \\ \vdots \\ I_N \end{bmatrix} = \mathbf{Z}^{-1} \begin{bmatrix} 1 \\ 1 \\ \vdots \\ 1 \end{bmatrix} V \qquad (6.67)$$

Hence, we sum the *columns* of $\mathbf{Z}^{-1}$ to obtain the individual subbar currents:

$$I_i = \sum_{j=1}^{N} \left[\mathbf{Z}^{-1}\right]_{ij} V \qquad (6.68)$$

Choosing $V = 1\,\mathrm{V}$ gives the explicit currents of the subbars. We can also obtain the overall impedance (resistance and self partial inductance) of the entire conductor by noting that the total conductor current is the sum of the subbar currents computed in (6.68) (i.e., $I = I_1 + I_2 + \cdots + I_N$). Hence, we sum the rows *and* columns (the sum of all entries) of Z^{-1} to give the relationship between the *total* current through the conductor, I, and the voltage across the conductor, V, to give

$$I = \left(\sum_{i=1}^{N}\sum_{j=1}^{N}\left[\mathbf{Z}^{-1}\right]_{ij}\right) V \qquad (6.69)$$

We obtain the effective impedance of the entire conductor by inverting (6.69) to give

$$Z_{\text{total}} = \frac{1}{\sum_{i=1}^{N}\sum_{j=1}^{N}[\mathbf{Z}]_{ij}}$$

$$= R_{\text{total}} + j\omega L_{p,\text{total}} \tag{6.70}$$

where the real part of Z_{total} is the equivalent resistance of the land, R_{total}, and the imaginary part of Z_{total} is the product of $\omega = 2\pi f$ and the equivalent self partial inductance of the land, $L_{p,\text{total}}$. Note that $L_{p,\text{total}}$ is the sum of the external self partial inductance and the internal self partial inductance. As the frequency increases and the current crowds to the surface of the land, the magnetic flux internal to the land goes to zero, so the internal inductance should also go to zero, leaving the external self partial inductance of the land as the high-frequency inductance.

We will show computed results for typical lands. The thickness of a PCB land is 1.4 mils, where a mil is one-thousandth of an inch. The lands are typically etched from a copper-cladded board that is made from glass–epoxy or FR-4 material. The glass–epoxy substrate has a relative permeability of $\mu_r = 1$ (is not magnetic) and hence does not affect the inductance. It has a relative permittivity of about $\varepsilon_r = 4.7$, which does affect the capacitance. The copper cladding is said to be "1 ounce" since 1 ft^2 of this thickness weighs 1 oz. Typical land widths range from 5 to 30 mils. The following shows the result for a PCB land whose thickness is 1.4 mils, whose width is 15 mils, and whose length is 10 in. The current distribution over the land cross section will be shown for four frequencies: 100 kHz, 10 MHz, 100 MHz, and 1 GHz. The results for this case were obtained with NT = 16 and NW = 172. The width dimension of $w = 15$ mils is one skin depth ($\delta = 1/\sqrt{\pi f \mu_0 \sigma}$) at $f_{1\delta} = 30$ kHz and two skin depths at $f_{2\delta} = 120.34$ kHz. The thickness dimension of $t = 1.4$ mils is one skin depth at $f_{1\delta} = 3.45$ MHz and two skin depths at $f_{2\delta} = 13.8$ MHz. For the discretization of each conductor, the widths, Δw, and the thicknesses, Δt, of each subbar should be less than two skin depths *in order that the current over each subbar will be approximately uniformly distributed over it*, which was the basic assumption in the derivation of the subbar resistances and partial inductances. Hence, we should have $\Delta w = w/\text{NW} < 2\delta$ and $\Delta t = t/\text{NT} < 2\delta$. For NT = 16 and NW = 172, $\Delta t = \Delta w = 2.22$ μm. (This was the reason for choosing the NT and NW as we did.) Each dimension of the subbars, Δw and Δt, is two skin depths at 3.5 GHz. Hence, at the largest frequency of 1 GHz the current will be approximately uniformly distributed over each subbar. The distribution of the current over the conductor cross section is shown at the four frequencies in Fig. 6.14(a) through (d).

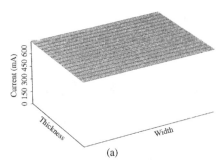

(a)

FIGURE 6.14(a). Current distribution over the cross section of a 1.4 mil × 15 mil land at 100 kHz.

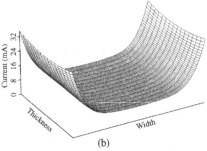

(b)

FIGURE 6.14(b). Current distribution over the cross section of a 1.4 mil × 15 mil land at 10 MHz.

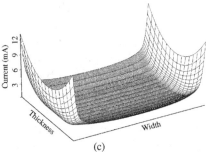

(c)

FIGURE 6.14(c). Current distribution over the cross section of a 1.4 mil × 15 mil land at 100 MHz.

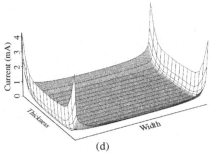

(d)

FIGURE 6.14(d). Current distribution over the cross section of a 1.4 mil × 15 mil land at 1 GHz.

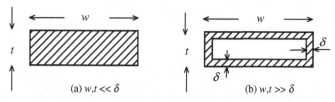

FIGURE 6.15. Current concentrating at the land surface in a thickness of dimension on the order of one skin depth.

Figure 6.14(a) shows that the current is uniformly distributed over the cross section at 100 kHz. We would expect that as the frequency of excitation increases, the current will move toward the surface, eventually lying predominantly in a thickness at the surface of dimension on the order of a skin depth, as illustrated in Fig. 6.15. Consequently, we would expect that the current would start to exhibit this concentration at the surface when the width or the thickness becomes greater than two skin depths: $2\delta < w$ or $2\delta < t$, whichever occurs first. For the width dimension of $w = 15$ mils, $f_{2\delta} = 120.34$ kHz, and for the thickness dimension of $t = 1.4$ mils, $f_{2\delta} = 13.8$ MHz. Hence, 100 kHz is such that the current distribution should remain uniformly distributed over the cross section. Since the width is much larger than the thickness, the first departure from a uniform distribution should occur when the width becomes on the order of two skin depths. Figure 6.14(b) for 10 MHz clearly shows that since the frequency is well above that for which the width is two skin depths but has not reached the point where the thickness is two skin depths or $f_{2\delta} = 13.8$ MHz, the current is crowding to the ends of the width dimension. Similarly, Fig. 6.14(c) for 100 MHz is above the point where the thickness is two skin depths, so crowding of the current is beginning to occur along the thickness dimension. Finally, Fig. 6.14(d) for 1 GHz shows the classic "bedpost" pattern, where the frequency is high enough that the current peaks sharply at all four corners of the land cross section.

We next show the frequency behavior of the net resistance and self partial inductance of the 1.4-mil × 15-mil land computed from (6.70). Figure 6.16(a) shows the behavior of the net resistance versus frequency. Figure 6.16(b) shows the *internal* inductance of the land computed from the current data via the method of Antonini et al. [19].

As the frequency increases such that the current lies in a thickness of approximately one skin depth at all four surfaces as in Fig. 6.15(b), we would expect the per-unit-length high-frequency resistance to asymptotically approach

$$r_{hf} = \frac{1}{\sigma\,(2\delta\,t + 2\delta\,w)}$$

$$= \frac{1}{2\sigma\,\delta\,(w + t)} \qquad \Omega/m \qquad (6.71a)$$

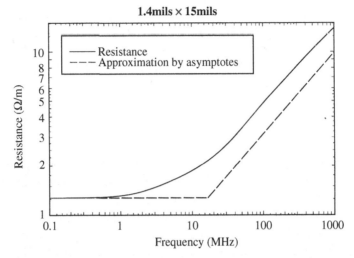

FIGURE 6.16(a). Net resistance of the 1.4 mil × 15 mil land vs. frequency.

The dc resistance, $r_{dc} = 1/\sigma w t$, and this high-frequency asymptote join at

$$2\delta = \frac{wt}{w + t} \tag{6.71b}$$

Hence, the high-frequency resistance should asymptotically approach a $\sqrt{f}$ increase since the skin depth $\delta = 1/\sqrt{\pi f \mu_0 \sigma}$ decreases as $\sqrt{f}$. Figure 6.16(a) exhibits this behavior. Similarly, as the frequency increases and the current crowds to the surface of the land, the magnetic flux internal to

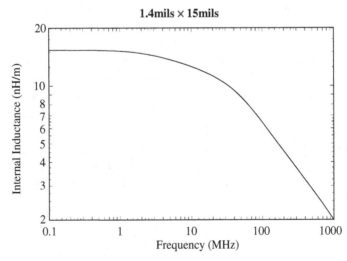

FIGURE 6.16(b). Internal inductance of the 1.4 mil × 15 mil land vs. frequency.

the land goes to zero as $\sqrt{f}$, so the internal inductance should also go to zero as $\sqrt{f}$. Figure 6.16(b) also exhibits this behavior. Because of the "peaking" of the current at the corners of the land, the high-frequency resistance in (6.71) is somewhat less than the actual high-frequency resistance [8].

Next, we investigate two identical lands having identical cross-sectional dimensions of $w = t = 50$ μm $\cong 2$ mils with various separations between the two lands. The total length of the two lands is 2.5 cm (about 1 in.). We compute the total resistance and partial inductance of each land from 1 to 100 MHz. We also plot the current distributions across the conductor cross sections for (1) an isolated conductor and (2) two identical conductors having various spacings between them when the two conductors carry (a) differential-mode currents and (b) common-mode currents. This demonstrates the proximity effect. The total resistance and partial inductance of each land will also be plotted for various land separations. We subdivide each land cross section into subbars of thickness $\Delta t = \Delta w = t/\text{NT} = w/\text{NW}$ for $\text{NT} = \text{NW} = 20$. For the discretization of each conductor, the widths, Δw, and the thicknesses, Δt, of each subbar should again be less than two skin depths *in order that the current over each subbar will be approximately uniformly distributed over it*, which was the basic assumption in the derivation of the subbar resistances and partial inductances (see Fig. 6.15). Hence, we should have $\Delta w = w/\text{NW} < 2\delta$ and $\Delta t = t/\text{NT} < 2\delta$. For $\text{NT} = \text{NW} = 20$, $\Delta t = \Delta w = 2.5$ μm and each dimension of the subbars, Δw and Δt, is two skin depths at 2.8 GHz. Each subbar will again be represented as shown in Fig. 6.13(b). The self partial inductances and the mutual partial inductances between all subbars will both be calculated using results from Hoer and Love [16] and given in (6.23) and (6.28), respectively.

The computations for an isolated conductor are as described before. For the case of two conductors, we write the relations between the impedances of the subbars of the two conductors as

$$
\begin{bmatrix} V_1 \\ V_1 \\ \vdots \\ V_1 \\ \cdots \\ V_2 \\ V_2 \\ \vdots \\ V_2 \end{bmatrix} = \underbrace{\begin{bmatrix} \mathbf{Z}_1 & \vdots & j\omega\,\mathbf{M}_{12} \\ \cdots & \cdots & \cdots \\ j\omega\,\mathbf{M}_{12} & \vdots & \mathbf{Z}_2 \end{bmatrix}}_{\mathbf{Z}} \begin{bmatrix} I_{11} \\ I_{12} \\ \vdots \\ I_{1N} \\ \cdots \\ I_{21} \\ I_{22} \\ \vdots \\ I_{2N} \end{bmatrix} \qquad (6.72)
$$

where $N = NW \times NT$. The $\mathbf{Z}_1$ and $\mathbf{Z}_2$ are each square with N rows and N columns each and contain the self impedances $Z = R + j\omega L_p$ of the subbars of that conductor *as well as* the mutual partial inductances $j\omega M_p$ between the subbars *of that conductor*. The $\mathbf{M}_{12}$ matrix is square with N rows and N columns and contains the mutual partial inductances *between subbars of the two conductors*. We first invert $\mathbf{Z}$ in (6.72), giving

$$\begin{bmatrix} I_{11} \\ I_{12} \\ \vdots \\ I_{1N} \\ \cdots \\ I_{21} \\ I_{22} \\ \vdots \\ I_{2N} \end{bmatrix} = \underbrace{\begin{bmatrix} \mathbf{Y}_{11} & \mathbf{Y}_{12} \\ \mathbf{Y}_{12} & \mathbf{Y}_{22} \end{bmatrix}}_{\mathbf{Y}=\mathbf{Z}^{-1}} \begin{bmatrix} V_1 \\ V_1 \\ \vdots \\ V_1 \\ \cdots \\ V_2 \\ V_2 \\ \vdots \\ V_2 \end{bmatrix} \tag{6.73}$$

Since the total current of each conductor is $I_1 = I_{11} + \cdots I_{1N}$ and $I_2 = I_{21} + \cdots I_{2N}$ and all subbars of each conductor are connected in parallel, we then sum the rows and columns of each of the four blocks of $\mathbf{Y} = \mathbf{Z}^{-1}$ to give a relation between the *total currents and voltages of the two conductors* as

$$\begin{bmatrix} I_1 \\ I_2 \end{bmatrix} = \begin{bmatrix} Y_{11} & Y_{12} \\ Y_{12} & Y_{22} \end{bmatrix} \begin{bmatrix} V_1 \\ V_2 \end{bmatrix} \tag{6.74a}$$

where each Y_{ij} is a scalar:

$$Y_{ij} = \sum_{n=1}^{N} \sum_{m=1}^{N} [\mathbf{Y}_{ij}]_{mn} \tag{6.74b}$$

Equation (6.74a) is inverted to give

$$\begin{bmatrix} V_1 \\ V_2 \end{bmatrix} = \begin{bmatrix} Z_{11} & Z_{12} \\ Z_{12} & Z_{22} \end{bmatrix} \begin{bmatrix} I_1 \\ I_2 \end{bmatrix} \tag{6.75}$$

We have two cases to consider: (1) differential-mode currents where the total currents through the two conductors are related as $I_2 = -I_1$, and (2) common-mode currents where the total currents through the two conductors are related as $I_2 = I_1$. From this we can determine the voltages across the two conductors for $I_1 = 1$ A as

$$V_{1,DM} = (Z_{11} - Z_{12})\, I_1$$
$$V_{2,DM} = (Z_{22} - Z_{12})\, I_2$$
$$V_{1,CM} = (Z_{11} + Z_{12})\, I_1$$
$$V_{2,CM} = (Z_{22} + Z_{12})\, I_2$$

(6.76)

where DM and CM denote differential- and common-mode voltages across the conductors, respectively. This gives the voltages across each of the two conductors for DM and CM excitation for a total current of 1 A through each of the two conductors. Hence the resistance and partial inductance of each conductor can be determined for DM and CM excitation as the real and imaginary parts of (6.76). (The self partial inductance of the conductor is the imaginary part divided by ω.) The voltages determined in (6.76) for 1-A DM and CM excitation can then be substituted into (6.73) to determine and plot the currents of the subbars for each of the conductors for DM and CM excitation.

Figure 6.17(a) shows the current distribution over the cross sections of the two conductors for a separation (edge to edge) of $s = 50\ \mu\text{m} \cong 2$ mils, differential-mode excitation, and an excitation frequency of 1 MHz, while Fig. 6.17(b) shows this for an excitation frequency of 100 MHz. Figure 6.18 repeats this for common-mode excitation. These plots show the expected crowding of the current toward the surfaces of the conductors when the cross-sectional dimensions become on the order of two skin depths. They also show

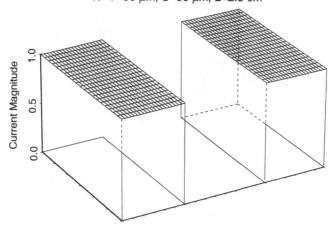

Differential-Mode Current Distribution, 1 MHz
W=T=50 µm, S=50 µm, L=2.5 cm

FIGURE 6.17(a). Current distribution over the conductor cross sections for differential-mode current and $s = 50\ \mu\text{m}$ and 1 MHz.

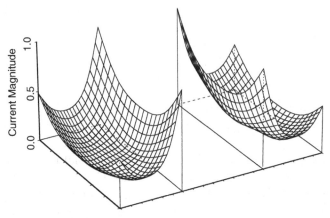

FIGURE 6.17(b). Current distribution over the conductor cross sections for differential-mode current and $s = 50$ μm and 100 MHz.

the proximity effect. For differential-mode excitation shown in Fig. 6.17(b) the currents tend to concentrate on the facing sides as is the case for wires. For common-mode excitation shown in Fig. 6.18(b), the currents tend to concentrate on opposide sides.

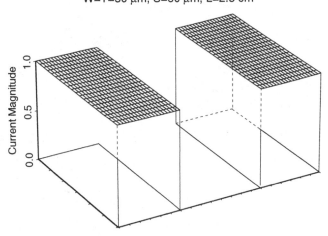

FIGURE 6.18(a). Current distribution over the conductor cross sections for common-mode current and $s = 50$ μm and 1 MHz.

Common-Mode Current Distribution, 100MHz
W=T=50 μm, S=50 μm, L=2.5 cm

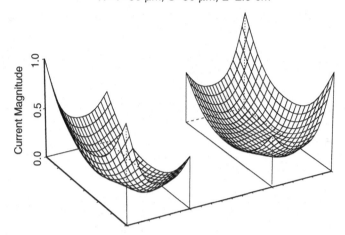

FIGURE 6.18(b). Current distribution over the conductor cross sections for common-mode current and $s = 50$ μm and 100 MHz.

Figure 6.19(a) shows the resistance for (1) an isolated conductor and (2) two conductors. Also shown is the high-frequency approximation in (6.71a). The dc and high-frequency asymptotes join at a frequency where $2\delta = wt/(w + t)$. Observe that for this close separation of the two conductors of $s = 50$ μm,

RESISTANCE
(T=W=50 μm, S=50 μm, Length=2.5 cm)

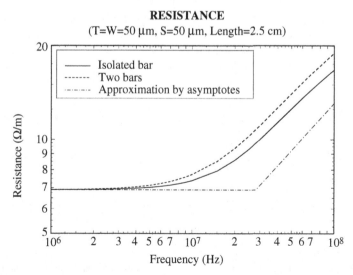

FIGURE 6.19(a). Total resistance for an isolated conductor and for two conductors.

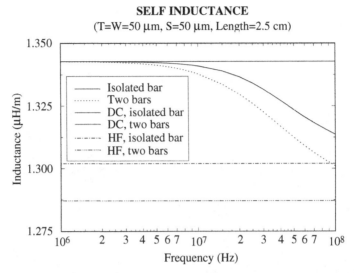

FIGURE 6.19(b). Total partial inductance for an isolated conductor and for two conductors.

exactly one conductor will fit between the two. Figure 6.19(b) shows the net self partial inductance of the isolated conductor as well as the two conductors. In addition, the dc and high-frequency limits of the self partial inductance are shown.

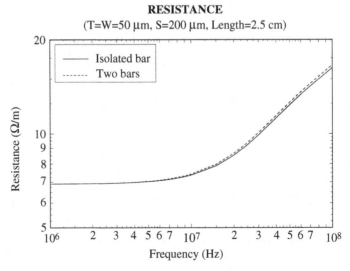

FIGURE 6.20(a). Total resistance for an isolated conductor and for two conductors for a separation of $s = 200$ μm.

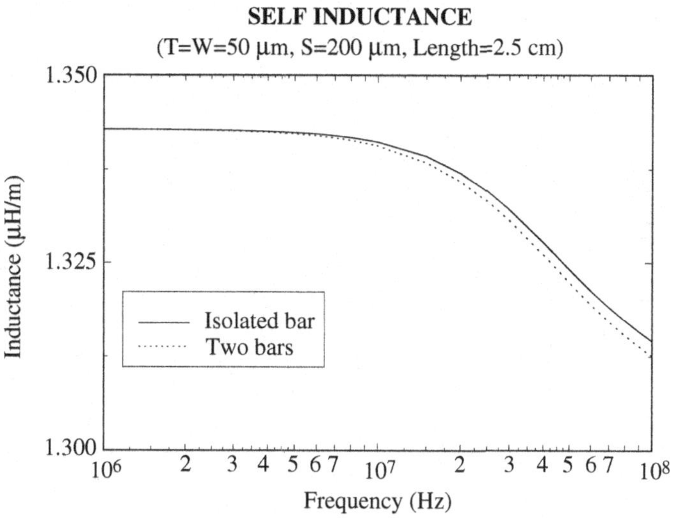

SELF INDUCTANCE

(T=W=50 μm, S=200 μm, Length=2.5 cm)

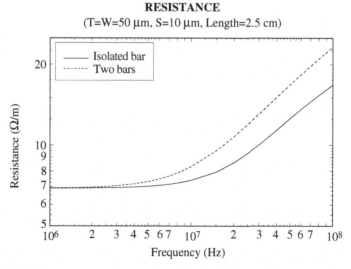

FIGURE 6.20(b). Total partial inductance for an isolated conductor and for two conductors for a separation of $s = 200$ μm.

Figure 6.20 shows the total resistance and partial inductance for an isolated conductor and for two conductors for a separation (edge to edge) of $s = 200$ μm. For these wide separations (four conductors will fit between the two) the resistance and inductance are not affected appreciably by the presence of the other conductor, as we would expect.

RESISTANCE

(T=W=50 μm, S=10 μm, Length=2.5 cm)

FIGURE 6.21(a). Total resistance for an isolated conductor and for two conductors for a separation of $s = 10$ μm.

SELF INDUCTANCE

(T=W=50 μm, S=10 μm, Length=2.5 cm)

FIGURE 6.21(b). Total partial inductance for an isolated conductor and for two conductors for a separation of $s = 10$ μm.

Figure 6.21 shows the total resistance and partial inductance for an isolated conductor and for two conductors for a separation (edge to edge) of $s = 10$ μm. For this very close separation, the resistance and inductance are affected significantly by the presence of the other conductor more than for $s = 50$ μm.

7

"LOOP" INDUCTANCE VS. "PARTIAL" INDUCTANCE

In the preceding chapters we have detailed the concept and calculation of "loop" inductance and "partial" inductance for various current-carrying structures consisting of conductors of circular, cylindrical cross section (wires), as well as conductors of rectangular cross section (PCB lands). In this final chapter we summarize the advantages and disadvantages of characterizing these structures with "loop" inductance or with "partial" inductance and give examples of the applications of partial inductance.

7.1 LOOP INDUCTANCE VS. PARTIAL INDUCTANCE: INTENTIONAL INDUCTORS VS. NONINTENTIONAL INDUCTORS

An important question that this book intends to resolve is: When should loop inductance be used to characterize a current-carrying, conductive structure, and when should partial inductance be used? A related question to be answered is: What are the advantages and disadvantages of loop inductance vs. partial inductance? There exists considerable misunderstanding throughout the electrical engineering community regarding "partial" inductance and where it is appropriate to use the concept to characterize the inductance effects of current-carrying structures. "Loop" inductance is a standard topic in undergraduate electrical engineering textbooks, but these textbooks do not contain any reference to "partial" inductance. Hence, electrical engineers are well trained in the understanding and calculation of loop inductance, but they have little or no understanding of the concept and uses of partial inductance. This unfortunate deficiency in the training of electrical engineers causes them erroneously to use formulas for loop inductance that do not apply to their situation of interest when they should instead use partial inductance formulas. *If one models a section of wire or PCB land with an inductance, they are inherently using partial inductance, whether they know it or not.* Loop inductance cannot be used to characterize a section of a conductor because as we have discussed, the induced emf in Faraday's law of induction that the loop inductance represents cannot be placed *uniquely* in any specific place in a closed current loop. The literature, both trade magazines and scholarly journals, is replete with examples of this misunderstanding and misuse of inductance, wherein the symbol for an inductor is used to model a section of wire or PCB land, yet a formula for "loop" inductance is used erroneously to compute the value of that inductance.

The inductive effects of a time-varying current inherent in Faraday's law of induction represent one of the most important parameters that determine the performance of today's electrical circuits and systems. Digital circuits and systems today have clock and data rates in the GHz range. The spectral content of these digital waveforms of trapezoidal pulse shape generally extends at least to the fifth harmonic of the repetition rate and depends on the pulse rise/fall times [5]. Frequencies of analog systems have also moved steadily into the GHz range. Interconnects such as wires and PCB lands were electrically short some 10 years ago and could be ignored in an analysis of the system performance. Today, the physical lengths of those same interconnects have not changed substantially but their electrical lengths have become a significant portion of a wavelength, and hence can no longer be ignored. These interconnects represent one of the most important parameters affecting digital as well as analog system performance. Most power distribution and signal interconnects in today's digital and analog systems must be modeled to predict the true performance of the system.

To understand the distinction in use between "loop" inductance and "partial" inductance, it is important to focus on "intentional" and "nonintentional" inductors. The solenoid and the toroid covered in Sections 4.2 are examples of "intentional" inductors. They are constructed intentionally to take advantage of the inductive effects inherent in Faraday's law of induction. One of the primary uses of intentional inductors such as the solenoid and the toroid are to block high-frequency signals while passing lower-frequency signals such as dc power. They are also used, along with capacitors, to construct bandpass filters that are so essential to radio communication. Lowpass and highpass filters are constructed as well using inductors in combination with capacitors. Similarly, bandreject filters remove unwanted signals. Shorted stubs consisting of two parallel lands shorted together at the far end are used to construct inductors that are suitable for use at microwave frequencies, where the parasitic effects of interwinding capacitance in conventional inductors would short out the inductor at these very high frequencies. These are examples of "intentional" inductors. For these structures we are interested only in the voltage at the terminals of the structure and are not interested in the voltages generated at points internal to the structure. Hence, "loop" inductance is useful in characterizing these structures for their typical uses.

Certain "transmission lines," such as the coaxial cable (Sections 4.1.3, and 4.7.3); the two-wire transmission line (Sections 4.6.1 and 4.7.2); one wire above a ground (Section 4.6.2); transmission lines composed of conductors of rectangular cross section such as the stripline, the microstrip line, and the PCB (Section 4.9); and multiconductor transmission lines (Section 4.8.2) are suitably characterized by per-unit-length "loop" inductances. Again, for these transmission-line structures we are only interested in the voltages at the terminals of the structure and are not interested in the voltages generated at points internal to the structure. Hence, "loop" inductance is useful in characterizing these structures for analyzing their typical use.

However, "nonintentional" inductances are generally undesired inductances and represent detrimental effects that must be incorporated into an analysis of the overall system to determine its performance degradation. It is generally not feasible or useful to attempt to characterize nonintentional inductances with "loop" inductance for a number of reasons, outlined in the following sections. Hence, partial inductance is the most appropriate characterization for nonintentional inductances.

7.2   TO COMPUTE "LOOP" INDUCTANCE, THE "RETURN PATH" FOR THE CURRENT MUST BE DETERMINED

Dc currents must form closed loops along conductors. The "loop" inductance characterizes that complete current loop at its terminals. For intentional inductors, the complete current loop is evident virtually "by design." Hence, the complete current loop for calculating the loop inductance is evident.

However, consider nonintentional inductors such as lands on a PCB that interconnect a source and a load. The "going down" path for a current from the source to the load is fairly easy to determine. But the return path for the current back to the source can take a number of alternative routes that are far from obvious. In today's highly dense and complicated PCBs it is virtually an impossible task to identify the return path for most currents. *If one cannot identify the complete path of the current loop, the loop inductance cannot be computed.* We therefore have no other recourse than to use partial inductance for characterizing the Faraday law inductive effect of a segment of the current loop such as a wire or a PCB land.

We ascribe the property of "inductance" to an intentional inductor such as a toroid or a solenoid, whether or not that inductor has any current flowing through it. But its utility exists only if a current passes through it. Similarly, when we ascribe a partial inductance to a segment of a conductor, that partial inductance is effective only when it has a current passing through it. That current must form a closed loop by some path that may not yet be obvious, nor do we need to determine "that path" when computing the partial inductance of a segment of the path. We must be assured that a return path for the current has been provided by the designer in some fashion that may not be readily obvious. If, by some omission in the circuit's physical construction, no path is provided for the current to return to its source, the partial inductance of a segment has no effect, for the same reason that an intentional inductor would have no effect: No current passes through it.

There is an important case where the return path of the current on a PCB is somewhat more obvious. If the current is allowed to return through one of the innerplanes buried in the board (either a "ground" distribution innerplane or a power distribution innerplane), it is well known that the current in that plane will tend to concentrate directly beneath the "going down" path. The return current will peak beneath the "going down" current and spread out in (on the surface of) the adjacent plane, giving a current density on the plane of [5]

$$J_s(x) = \frac{I}{\pi h \left(1 + (x/h)^2\right)} \qquad \text{A/m}$$

where h is the height of the "going down" current above the plane and x is the horizontal position along the plane, with $x = 0$ being directly beneath the "going down" current. This result was derived for a very ideal situation of a filamentary current above an infinite, perfectly conducting ground plane. Innerplanes in PCBs have various discontinuities in them and are of finite dimensions, so they are represented only approximately by this ideal case. For example, lands on a PCB that are above but near the edges of the innerplanes have substantial fringing fields and probably do not represent the case of an infinite ground plane. Similarly, the innerplanes may have gaps and other discontinuities in them. A power plane must have isolated sections to accommodate the various dc voltages of the system. Even the ground innerplanes may have gaps cut into them for various reasons. It is generally not recommended and is unnecessary to cut gaps in a ground innerplane, as this disrupts the return current paths [5]. But even for the ideal situation, which resembles a stripline or a microstrip line, computation of the loop inductance is a formidable task. Holloway and Kuester have made this calculation for the microstrip line [31]. It is worth noting that they do this using partial inductance concepts.

We can model all the conductor segments of a structure with their partial inductances (self and mutual). We can then solve the resulting lumped circuit using, for example, SPICE and hence determine the return paths for the currents without having to guess about the return path for a current. If the model does not have provision for at least one return path back to its source for a conductor segment, assigning a partial inductance to that segment will have no effect in the same fashion that not passing a current around the closed loop of an intentional inductor will eliminate any effect of that loop inductance. But to compute, a priori, a loop inductance, we *must* first determine the complete path for the loop current. To compute, a priori, the partial inductance of a conductor segment, we do *not* need to determine the complete current loop path.

7.3 GENERALLY, THERE IS NO UNIQUE RETURN PATH FOR ALL FREQUENCIES, THEREBY COMPLICATING THE CALCULATION OF A "LOOP" INDUCTANCE

Circuit designers provide *only one path* for the "going down" current. However, they tend to "leave it to the current" to select a return path back to the source. At dc and low frequencies, a current will return to its source along the path of lowest *resistance*. At higher frequencies, the current will return to its source along the path of lowest *impedance*, which is generally the path of lowest *inductance*: Resistance is an insignificant portion of the total impedance of a conductor at the higher frequencies [5]. A good example of this is the shielded wire, where the shield is above and "grounded to" a "ground" plane that was discussed in Chapter 1 and shown in Fig. 1.2. At dc and low frequencies, the "going down" current I takes its return path, I_G, through the massive ground plane, which obviously has a much lower dc resistance than other possible return paths. However, at higher frequencies, the current I takes its return path up through the shield, I_S, thereby minimizing the area and inductive impedance between the "going down" path and the return path [5]. Therefore, the return paths and hence the complete current loops are different for different frequencies for this structure.

Another example of this is the popular "gridded ground" system on a PCB shown in Fig. 7.1, where a "grid" of conductors is interconnected so as to provide a number of possible paths for the current to return to its source along [5].

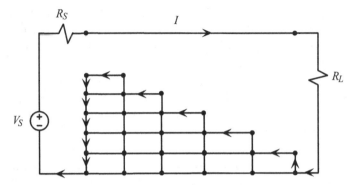

FIGURE 7.1. Ground grid for reducing the loop area of the loop current path.

This is done to avoid the radiated emissions of large current loops and is used in low-cost products to avoid more costly boards having innerplanes. It is also used even on PCBs with innerplanes, to minimize the (partial) inductance and associated "ground bounce" as well as common-mode currents generated by return currents that do not take the innerplane route [5]. At dc and low frequencies the "going down" current I returns along the path of lowest resistance using the simple "current-division" principle [1,2]. Simply model each wire segment of the grid with its dc resistance and compute (using SPICE or, if the circuit is simple, by hand) the path of lowest resistance. As the frequency of the current increases, the impedance of the complete path is governed by the inductance of the entire loop. From our previous calculations we know that the inductance of a current loop is related directly to its loop area. Hence, the return current "chooses" the path to minimize the total loop area giving the path of lowest loop impedance (inductance) nearest the "going down" current, as shown in Fig. 7.1. Modeling all the conductor segments of this structure with their resistances as well as their partial inductances (self and mutual), we can solve the resulting lumped-circuit model using SPICE and hence determine the lowest-impedance return current path. This was accomplished by Paul and Smith [32].

7.4 COMPUTING THE "GROUND BOUNCE" AND "POWER RAIL COLLAPSE" OF A DIGITAL POWER DISTRIBUTION SYSTEM USING "LOOP" INDUCTANCES

Signal integrity has become a paramount design consideration in today's high-speed digital systems. Signal integrity has to do with ensuring that the wave-shape and levels of those signals are maintained within tightly controlled limits to avoid logic errors and false switching when these levels rise or fall

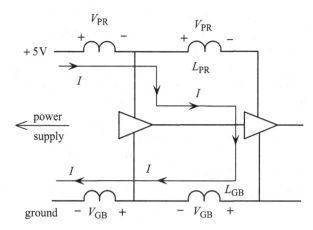

FIGURE 7.2. "Ground bounce" and "power rail collapse" in digital circuits.

into gray regions. Dc voltages are supplied to the modules in a digital system by lands routed on and within a PCB. For example, consider a two-conductor power distribution circuit for supplying dc voltages from a power supply to CMOS inverters as shown in Fig. 7.2. When the left inverter is in the high state, the right inverter is in the low state and current passes along the power distribution lands from the dc power supply to the power pin of the left in-verter, into the input of the right inverter, and returns to the power supply along the "ground" lands. When the left inverter is switched to the low state, the current of the power supply passes down through the right inverter and returns to the power supply. We have modeled the PCB lands connecting the power and the ground pins of the inverter modules to each other and to the dc power supply with inductances labeled as L_{PR} and L_{GB}. As the inverters switch, the currents through these lands go to or increase from zero or change direction, thereby inducing voltages across these inductors that are related to the rate of change of the currents through them. Voltages V_{PR} are developed across the inductors L_{PR} that may cause the voltages of the power pins of the modules to drop significantly, which is called *power rail collapse*. Similarly, voltages V_{GB} are developed across the inductors L_{GB}, causing the voltages of the two "ground" pins of the inverters to differ significantly, which is referred to as *ground bounce*. Both of these phenomena may cause logic errors, thereby degrading the signal integrity of the system.

There is considerable evidence that these voltages exist [33], but the essen-tial question is: What do we mean by these inductances? They certainly are not "loop" inductances since we know that an inductance of a closed current loop cannot be assigned uniquely to any place in that loop. These inductances are, in fact, partial inductances. Although not shown in this diagram, there are also mutual partial inductances between these self partial inductances that should be included if the conductors are close enough to each other. A diagram such as Fig. 7.2 is seen throughout the literature not only in "trade" magazines but also in scholarly journals. A closer inspection of those articles shows that many of the authors do not know how to correctly calculate the values for these inductances and erroneously use formulas for "loop" inductance. If we accept the fact that ground bounce and power rail collapse actually exist in digital circuits and are a severe problem, we have no recourse but to admit that "loop" inductance does not explain the phenomenon, and we must accept the utility of partial inductance concepts. A means for measuring these ground bounce and power rail collapse voltages using the concept of partial inductance was given in [33,34].

7.5   WHERE SHOULD THE "LOOP" INDUCTANCE OF THE CLOSED CURRENT PATH BE PLACED WHEN DEVELOPING A LUMPED-CIRCUIT MODEL OF A SIGNAL OR POWER DELIVERY PATH?

As we know from previous discussions, one cannot place the loop inductance uniquely in any specific position in the loop. You can only attribute partial inductances to specific segments of the closed current loop. For example, consider the parallel-wire transmission line shown in Fig. 7.3. In transmission-line analyses, we are only interested in the terminal voltages at the endpoints of the line. Hence, we can place the "loop" inductance in either conductor, as shown in Fig. 7.3, and obtain the same result for these terminal voltages. It is clear from Fig. 7.3 that the ground bounce and power rail collapse voltages cannot be determined uniquely using "loop" inductance.

Throughout the literature one sees equivalent-circuit models with lumped "inductances" representing segments of conductors. Upon closer scrutiny it

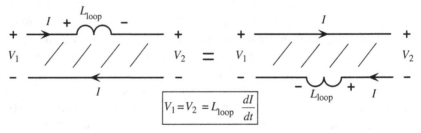

FIGURE 7.3. Loop inductance and the transmission line.

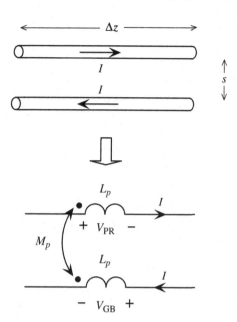

FIGURE 7.4. Modeling a two-wire transmission line with partial inductances.

is found that formulas for "loop" inductance are used to compute the values of those inductances. We can compute the "loop" inductance using partial inductances of the loop segments, but the reverse is not true. For example, we can model the two-wire transmission line in Fig. 7.3 using partial inductances as shown in Fig. 7.4. Using the dot convention [1,2] we obtain

$$V_{GB} = V_{PR} = L_p \frac{dI}{dt} - M_p \frac{dI}{dt}$$

$$= (L_p - M_p) \frac{dI}{dt} \tag{7.1}$$

The total voltage drop around the transmission line loop is the product of the "loop" inductance of the transmission line loop and the time derivative of the current. Hence, the total voltage around the loop is twice (7.1). Therefore, the transmission line "loop" inductance can be obtained from the partial inductances as

$$L_{\text{loop}} = 2 (L_p - M_p) \tag{7.2}$$

These self and mutual partial inductances for wires were obtained in Sections 5.3 and 5.4:

$$L_p = \frac{\psi_\infty}{I}$$

$$\cong \frac{\mu_0}{2\pi} \Delta z \left(\ln \frac{2\Delta z}{r_w} - 1 \right) \qquad \Delta z \gg r_w \qquad (7.3a)$$

and

$$M_p = \frac{\psi_\infty}{I}$$

$$= \frac{\mu_0}{2\pi} \Delta z \left[\ln \left(\frac{\Delta z}{s} + \sqrt{\left(\frac{\Delta z}{s}\right)^2 + 1} \right) - \sqrt{1 + \left(\frac{s}{\Delta z}\right)^2} + \frac{s}{\Delta z} \right]$$

$$\cong \frac{\mu_0}{2\pi} \Delta z \left(\ln \frac{2\Delta z}{s} - 1 \right) \qquad \Delta z \gg s \qquad (7.3b)$$

Combining these gives the loop inductance of the transmission line:

$$L_{loop} = 2 (L_p - M_p)$$

$$= 2\frac{\mu_0}{2\pi} \Delta z \left(\ln \frac{2\Delta z}{r_w} - 1 \right)$$

$$-2\frac{\mu_0}{2\pi} \Delta z \left(\ln \frac{2\Delta z}{s} - 1 \right)$$

$$= \frac{\mu_0}{\pi} \Delta z \ln \frac{s}{r_w} \qquad (7.4)$$

which was obtained directly by computing the magnetic flux threading the loop between the two wires in (4.73) in Section 4.6.1.

This result for the loop inductance of the transmission line in terms of partial inductances in (7.2) is rather obvious if we recall the physical meaning of the partial inductances as being the ratios of the magnetic flux between a wire and infinity and the current producing that flux. The quantity $(L_p - M_p)$ is the net magnetic flux through the loop formed by the two transmission-line conductors per unit of current as shown in Fig. 7.5. Adding the fluxes through the loop due to the two currents (equal but oppositely directed) gives the result in (7.2).

This partial inductance model of the transmission line in Fig. 7.4 also clearly shows an interesting observation that is not obtained from the transmission-line "loop" inductance circuit of Fig. 7.3. As we move the two wires closer, the value of the mutual partial inductance approaches the value of the self partial inductance (i.e., $M_p \to L_p$ as $s \to 0$), and hence the ground bounce and power rail collapse voltages approach zero (i.e., $V_{GB}, V_{PR} \to 0$ as $s \to 0$). This

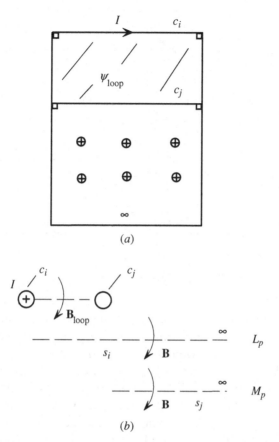

FIGURE 7.5. Loop inductance of a transmission line from partial inductances.

shows a routinely observed design rule that in order to reduce ground bounce and power rail collapse, the "going down" and return conductors should be placed as close as possible to each other. This useful design rule could not be determined using "loop" inductance, but it is frequently used without understanding the distinction between "loop" and "partial" inductances.

7.6 HOW CAN A LUMPED-CIRCUIT MODEL OF A COMPLICATED SYSTEM OF A LARGE NUMBER OF TIGHTLY COUPLED CURRENT LOOPS BE CONSTRUCTED USING "LOOP" INDUCTANCE?

The electromagnetic fields of all neighboring currents interact with each other to some degree, and to include all their effects, this coupling should be included in each circuit loop representation. Consider Fig. 7.6, where we have shown

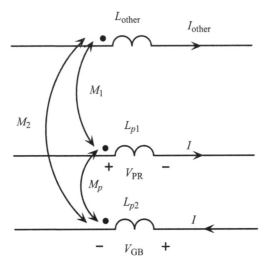

FIGURE 7.6. Including the effects of other currents.

a power distribution current loop in a digital system carrying current I, and also a conductor of a neighboring current loop on the PCB carrying current I_{other}. The neighboring current I_{other} as well as other neighboring currents *will affect the ground bounce and power rail collapse voltages.* This effect of neighboring currents could not be determined using "loop" inductances. However, their effect can be determined easily by modeling this situation with the self and mutual partial inductances of the conductors. To include the effect of the other current, we simply write the circuit equations using the dot convention as

$$V_{\text{GB}} = L_{p2}\frac{dI}{dt} - M_p\frac{dI}{dt} - M_2\frac{dI_{\text{other}}}{dt}$$

$$V_{\text{PR}} = L_{p1}\frac{dI}{dt} - M_p\frac{dI}{dt} + M_1\frac{dI_{\text{other}}}{dt}$$

With this circuit model we simply "turn the crank" and compute (perhaps with SPICE) the "ground bounce" voltage V_{GB} and the "power rail collapse" voltage V_{PR}, which are of considerable interest in signal integrity analyses for today's high-speed digital systems [5]. This would be a formidable if not impossible task using only "loop" inductances.

7.7   MODELING VIAS ON PCBS

A *via* (pronounced "veeya") is a hole drilled through a PCB to interconnect lands on the top and bottom surfaces as well as on innerplane layers within the

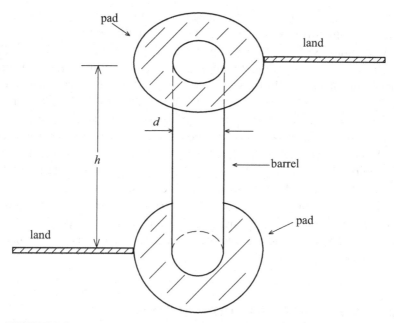

FIGURE 7.7. Via for interconnecting lands on a PCB that are on different layers.

PCB, as illustrated in Fig. 7.7. Circular "pads" attach the barrel to the lands. The via is an important feature in keeping the physical size of the PCBs from becoming prohibitive. However, it is also a significant factor affecting signal integrity since it represents a discontinuity in the transmission lines that are connected by it, thereby causing reflections that degrade the waveshape of the voltages and currents on the lands. A particularly simple inductance model of the via is as a simple wire of diameter d representing the barrel. Hence, the inductance of the via is simply the self partial inductance of a wire of diameter d and length h:

$$\frac{L_{\text{via}}}{h} = \frac{\mu_0}{2\pi} \left(\ln \frac{2h}{r_{\text{via}}} - 1 \right)$$

$$= \frac{\mu_0}{2\pi} \left(\ln \frac{4h}{d} - 1 \right) \qquad \text{H/m} \qquad (7.5a)$$

It is common to write this in nH and to use the dimensions for d and h in inches. Hence, the constant becomes $\mu_0/2\pi \to 5.08$ and we obtain

$$\frac{L_{\text{via}}}{h} = 5.08 \left(\ln \frac{4h}{d} - 1 \right) \qquad \text{nH/in.} \qquad (7.5b)$$

For example, for a board of standard thickness 62 mils and a via of radius 16 mils, which is equivalent to a No. 20 gauge wire, the inductance of the via is

5.32 nH/in., for a total via inductance of 0.33 nH. Mutual partial inductance between neighboring vias should be included in this model.

A popular signal integrity book gives the result as $L_{\text{via}} = 5.08\,h$ $\left[\ln(4h/d) + 1\right]$. Observe that there is a $+1$ in this result, whereas the correct partial inductance result in (7.5) has a -1 in it. For the previous dimensions, this gives a via inductance of 15.5 nH/in., a factor of 3 larger than the correct result in (7.5b). The authors of that book argued that their result was obtained using the per-unit-length inductance of a coaxial cable derived in (4.29), $L_{\text{via}} = (\mu_0/2\pi)\,h\,\ln(r_s/r_w)$, where the barrel of the via represents the inner wire of the cable of radius r_w and the "shield" (the "return path" for the current) is at a distance of $r_s = 2eh$ cylindrically about the barrel and $e = 2.71828\ldots$. Substituting $r_s = 2eh$ into the equation for the coaxial cable gives their result. A more defensible result is obtained with partial inductance concepts, and we would not need to determine a fictitious "return path" for the via current.

7.8   MODELING PINS IN CONNECTORS

Another aspect of system design that has the potential for degrading signal integrity are the numerous connectors in the system that make the inevitable connection between an off-board cable and the lands on the PCB. These connectors have numerous pins in them of radius r_{pin} and length l_{pin} that are inserted into a receptacle on the PCB. These essentially insert inductances into the signal propagation path that have the potential for degrading the quality of the signals being transferred through the connector. How shall we model these connector pins? The obvious choice is with partial inductances. Using the result for the self partial inductance of a wire in (7.3a) gives

$$L_{\text{pin}} = \frac{\mu_0}{2\pi} l_{\text{pin}} \left(\ln \frac{2l_{\text{pin}}}{r_{\text{pin}}} - \frac{3}{4}\right) \quad \text{H} \tag{7.6a}$$

This was cited in a textbook without recognition being given to it being a "partial" inductance. Where does the factor of $3/4$ arise? If we add the *internal inductance* of the wire, $\mu_0/8\pi$, to the self partial inductance in (7.3a), we obtain a total self partial inductance of

$$L_{\text{pin}} = \frac{\mu_0}{2\pi} l_{\text{pin}} \left(\ln \frac{2l_{\text{pin}}}{r_{\text{pin}}} - 1\right) + \frac{\mu_0}{8\pi} l_{\text{pin}}$$

$$= \frac{\mu_0}{2\pi} l_{\text{pin}} \left(\ln \frac{2l_{\text{pin}}}{r_{\text{pin}}} - \frac{3}{4}\right) \tag{7.6b}$$

Again mutual partial inductaries between neighbours pins should be included. As we computed earlier, the internal inductance is usually a negligible term and goes to zero as the frequency increases. This was not explained in the textbook. If all this were explained and the use of "partial" inductance of a wire were mentioned, the result would not seem to be "magic" and of unknown origin. For pins of rectangular cross section the textbook gives the formula

$$L_{\text{pin}} = \frac{\mu_0}{2\pi} l_{\text{pin}} \left(\ln \frac{4l_{\text{pin}}}{P} + \frac{1}{2} \right) \tag{7.7}$$

where P is the perimeter of the rectangular cross section of the pin: $P = 2(w + t)$. This result was given in equation (6.56c). Neither of these "mysterious formulas" in (7.6) and (7.7) were described as being "partial" inductances that we derived previously and were originally published by Grover [14] in 1946. In addition, the mutual inductance between two pins separated by a distance s is given as

$$M_p = \frac{\mu_0}{2\pi} l_{\text{pin}} \left[\ln \left(\frac{l_{\text{pin}}}{s} + \sqrt{\left(\frac{l_{\text{pin}}}{s} \right)^2 + 1} \right) - \sqrt{1 + \left(\frac{s}{l_{\text{pin}}} \right)^2} + \frac{s}{l_{\text{pin}}} \right]$$

$$\cong \frac{\mu_0}{2\pi} l_{\text{pin}} \left[\ln \left(\frac{2l_{\text{pin}}}{s} \right) - 1 \right] \qquad l_{\text{pin}} \gg s \tag{7.3b}$$

But, of course, this is simply the mutual partial inductance between the two current filaments that is derived in Chapter 5 and given in (5.21) and was also originally given by Grover [14]. So once again there has been widespread use of the concept of partial inductance without apparently knowing it or understanding the distinction between "loop" inductance and "partial" inductance.

7.9   NET SELF INDUCTANCE OF WIRES IN PARALLEL AND IN SERIES

Consider two wires of equal radii r_w that are connected in series as shown in Fig. 7.8. The lengths of the wires are l_1 and l_2 and their adjacent ends are separated by a distance of s. The equivalent circuit is also shown in Fig. 7.8. Summing the voltages developed across the two inductors of the equivalent circuit and using the dot convention gives

$$V = L_{p1} \frac{dI}{dt} + M_p \frac{dI}{dt} + L_{p2} \frac{dI}{dt} + M_p \frac{dI}{dt}$$

$$= (L_{p1} + M_p + L_{p2} + M_p) \frac{dI}{dt} \tag{7.8}$$

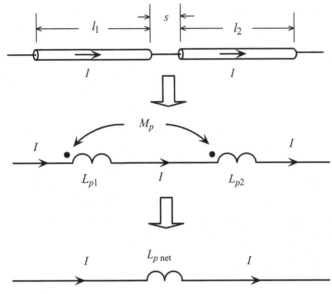

FIGURE 7.8. Two wires in series.

Hence, the net self partial inductance of the combination is

$$L_{p\,\text{net}} = L_{p1} + L_{p2} + 2M_p \qquad (7.9)$$

where the self partial inductances are obtained in Section 5.3 as

$$L_{pi} \cong \frac{\mu_0}{2\pi} l_i \left(\ln \frac{2l_i}{r_w} - 1 \right) \qquad l_i \gg r_w \qquad (5.18c)$$

The mututal partial inductance between two wires that are aligned but are offset by a distance s was obtained in Section 5.5 and Fig. 5.12 as

$$2M_p = L_{p(l_2+s+l_1)} - L_{p(l_1+s)} - L_{p(l_2+s)} + L_{ps} \qquad (5.28d)$$

In the special case where the two conductors are joined together, $s = 0$, the mutual inductance becomes

$$2M_p = L_{p(l_2+l_1)} - L_{pl_2} - L_{pl_1} \qquad s = 0 \qquad (5.28d)$$

Combining this result for $s = 0$ with (7.9) gives $L_{p\,\text{net}} = L_{p(l_2+l_1)}$, which makes sense.

Figure 7.9 shows two wires (possibly of different radii) connected in parallel. From the equivalent circuit and using the dot convention, we obtain the

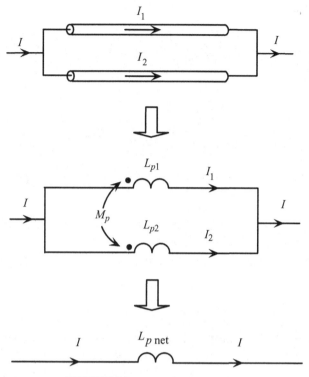

FIGURE 7.9. Two wires in parallel.

voltage across each conductor as

$$V = L_{p1} \frac{dI_1}{dt} + M_p \frac{dI_2}{dt}$$

$$= L_{p2} \frac{dI_2}{dt} + M_p \frac{dI_1}{dt} \tag{7.10}$$

Since the endpoints of the wires are connected, the two voltages across each wire must be equal. Writing this in matrix form gives

$$V \begin{bmatrix} 1 \\ 1 \end{bmatrix} = s \begin{bmatrix} L_{p1} & M_p \\ M_p & L_{p2} \end{bmatrix} \begin{bmatrix} I_1 \\ I_2 \end{bmatrix} \tag{7.11}$$

and s denotes the Laplace transform variable (essentially, the derivative operator here). Inverting this gives

$$\begin{bmatrix} I_1 \\ I_2 \end{bmatrix} = \frac{1}{s} \frac{1}{L_{p1}L_{p2} - M_p^2} \begin{bmatrix} L_{p2} & -M_p \\ -M_p & L_{p1} \end{bmatrix} \begin{bmatrix} 1 \\ 1 \end{bmatrix} V \tag{7.12}$$

Solving gives

$$\begin{bmatrix} I_1 \\ I_2 \end{bmatrix} = \frac{1}{s} \frac{1}{L_{p1}L_{p2} - M_p^2} \begin{bmatrix} L_{p2} - M_p \\ L_{p1} - M_p \end{bmatrix} V \qquad (7.13)$$

Adding the rows gives

$$\begin{aligned} I &= I_1 + I_2 \\ &= \frac{1}{s} \frac{L_{p1} + L_{p2} - 2M_p}{L_{p1}L_{p2} - M_p^2} V \end{aligned} \qquad (7.14)$$

Inverting this result gives the net partial inductance of the parallel combination:

$$L_{\text{net}} = \frac{L_{p1}L_{p2} - M_p^2}{L_{p1} + L_{p2} - 2M_p} \qquad (7.15)$$

If the two wires have identical lengths and radii, $L_{p1} = L_{p2} = L_p$, (7.15) reduces to

$$L_{\text{net}} = \frac{L_p + M_p}{2} \qquad (7.16)$$

It is generally thought that placing two wires in parallel gives a net inductance of the combination that is half that of one wire alone, since in electric circuit analysis, inductors in parallel combine like resistors in parallel [1,2]. This is not necessarily true because that usual assumption neglects to consider the mutual partial inductance between the two wires. The result in (7.16) shows that unless the two wires are placed relatively far apart, the net partial inductance of the combination will not equal half that of one wire. Placing the wires relatively far apart means that the mutual partial inductance approaches zero, $M_p \rightarrow 0$, as we have seen, and the result in (7.16) approaches $L_{p\,\text{net}} \rightarrow L_p/2$. On the other hand, moving the two wires closer together causes the mutual partial inductance to approach the value of the self partial inductance, $M_p \rightarrow L_p$, and the result in (7.16) approaches that of one wire, $L_{p\,\text{net}} \rightarrow L_p$. Hence, placing two wires in parallel and close together provides a net partial inductance that is not significantly less than using only one wire!

7.10 COMPUTATION OF LOOP INDUCTANCES FOR VARIOUS LOOP SHAPES

With the concept of partial inductances and the results derived previously for the self and mutual partial inductances of wires, it is a simple matter to derive the loop inductances of loops of various shapes as long as their perimeters consist of piecewise-linear segments. For example, consider the rectangular

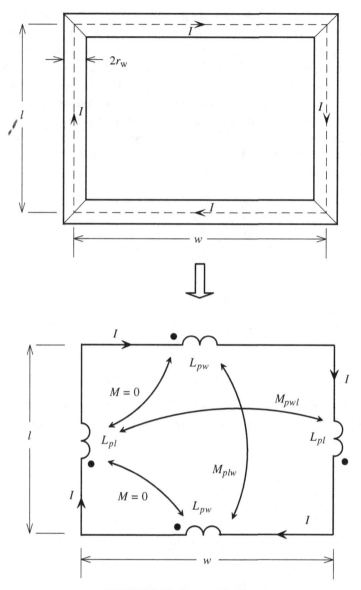

FIGURE 7.10. Rectangular loop.

loop composed of wires of equal radii r_w and side lengths of l and w shown in Fig. 7.10. Summing the voltages across the inductances in the equivalent circuit and using the dot convention for the mutuals gives the inductance of the loop as

$$L_{\text{loop}} = 2 \left(L_{pw} - M_{plw} \right) + 2 \left(L_{pl} - M_{pwl} \right) \qquad (7.17)$$

Notice the placement of the dots on the inductors. Parallel conductors should have the dots on the same ends since the self and mutual partial inductances give the magnetic fluxes between each conductor and infinity. To obtain the magnetic flux threading the loop between the two conductors, the dots should be on the same ends of parallel wires, thereby giving the magnetic flux through the surface between the two conductors carrying oppositely directed currents as the difference between the self and mutual partial inductances (see Figure 7.5 as well as Section 5.8 and Figures 5.5, 5.6, and 5.7). Substituting the self and mutual partial inductances from (5.18c) and (5.21b) into (7.17) yields, for $l, w \gg r_w$:

$$L_{loop} = \frac{\mu_0}{\pi} \left(w \ln \frac{2w}{r_w} - w - w \sinh^{-1} \frac{w}{l} + \sqrt{w^2 + l^2} - l \right.$$
$$\left. + l \ln \frac{2l}{r_w} - l - l \sinh^{-1} \frac{l}{w} + \sqrt{l^2 + w^2} - w \right) \qquad (7.18)$$

Simplfying this gives the same loop inductance obtained in Chapter 4 after a lengthy integration of the **B** field over the loop surface:

$$L_{loop} = \frac{\mu_0}{\pi} \left(w \ln \frac{2w}{r_w} + l \ln \frac{2l}{r_w} - w \sinh^{-1} \frac{w}{l} - l \sinh^{-1} \frac{l}{w} \right.$$
$$\left. + 2\sqrt{l^2 + w^2} - 2(w + l) \right) \qquad (4.18)$$

In the case of a square loop, $l = w$, (4.18) simplifies to the result obtained in Chapter 4 that was obtained after some tedious integration of the **B** field over the loop surface:

$$L_{square\ loop} = \frac{\mu_0}{\pi} \left[2l \ln \frac{2l}{r_w} - 2l \underbrace{\sinh^{-1}(1)}_{\ln(1+\sqrt{2})} + 2l\sqrt{2} - 4l \right]$$

$$= 2\frac{\mu_0}{\pi} l \left[\ln \frac{2l}{r_w} - \ln\left(1 + \sqrt{2}\right) + \sqrt{2} - 2 \right]$$

$$= 2\frac{\mu_0}{\pi} l \left[\ln \frac{l}{r_w} - 0.774 \right] \qquad l = w \gg r_w \qquad (4.20)$$

which matches Grover's result [14]. To these results we may add the internal inductances of the wire if necessary:

$$L_{loop,\ internal} = 2\frac{\mu_0}{8\pi} l + 2\frac{\mu_0}{8\pi} w \qquad (7.19)$$

Next, consider the equilateral triangle shown in Fig. 7.11 (see the discussion on placement of the dots in Section 5.8). Writing the voltage V across one of

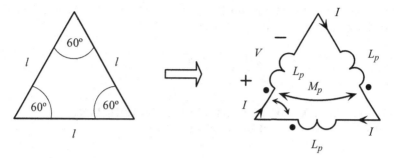

FIGURE 7.11. Equilateral triangle.

the inductors gives

$$V = L_p \frac{dI}{dt} - 2M_p \frac{dI}{dt}$$

$$= (L_p - 2M_p) \frac{dI}{dt} \tag{7.20}$$

The total voltage around the loop is three times (7.20), accounting for the voltages of all three sides. Hence, the net loop inductance as

$$L_{\text{loop}} = 3 \left(L_p - 2M_p \right) \tag{7.21}$$

Substituting the self partial inductances of the three wires from Chapter 5,

$$L_p \cong \frac{\mu_0}{2\pi} l \left(\ln \frac{2l}{r_w} - 1 \right) \qquad l \gg r_w \tag{5.18c}$$

and the mutual partial inductances between two inclined wires of equal length from Section 5.6, equation (5.46b),

$$M_p = \frac{\mu_0}{2\pi} \cos (60^\circ) \left(l \ln \frac{l + 2l}{l} \right)$$

$$= \frac{\mu_0}{2\pi} l \, (0.549) \tag{5.46b}$$

gives

$$L_{\text{loop}} = 3 \left(L_p - 2M_p \right)$$

$$= 3\frac{\mu_0}{2\pi} l \left(\ln \frac{2l}{r_w} - 1 - 2 \times 0.549 \right)$$

$$= 3\frac{\mu_0}{2\pi} l \left(\ln \frac{l}{r_w} + \ln 2 - 1 - 2 \times 0.549 \right)$$

$$= 3\frac{\mu_0}{2\pi} l \left(\ln \frac{l}{r_w} - 1.405 \right) \tag{7.22}$$

which matches Grover's result [14]. To this result we may add the internal inductances of the wire if necessary:

$$L_{\text{loop, internal}} = 3\frac{\mu_0}{8\pi}l \qquad (7.23)$$

7.11 FINAL EXAMPLE: USE OF LOOP AND PARTIAL INDUCTANCE TO SOLVE A PROBLEM

In this final section of the book we examine a typical example of computing the inductive coupling between two loops using (a) loop inductances and (b) partial inductances. Figure 7.12 shows the example dimensions. To simplify the numbers, each loop is chosen to be square with side dimensions 1 m × 1 m, and the two loops are offset by 1 m in the vertical dimension and by 1 m in the horizontal dimension. The loops are constructed of No. 20 gauge wires having radii of $r_w = 16$ mils. The first loop is driven by a 1-V sinusoidal source of frequency 10 MHz having a source resistance of 10 Ω. At a frequency of 10 MHz, a wavelength (in free space) is 30 m. Hence the dimensions of the loops and their separation can be considered to be electrically small, thereby allowing us to treat this problem as a lumped-circuit problem. The second loop also has a 10-Ω resistor inserted in it, and it is desired to compute the voltage induced across the terminals of that resistor, $V_{\text{out}}(t)$.

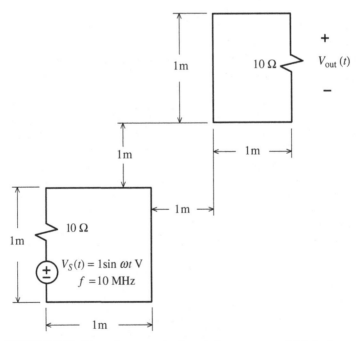

FIGURE 7.12. Example comparing loop inductance to partial inductance.

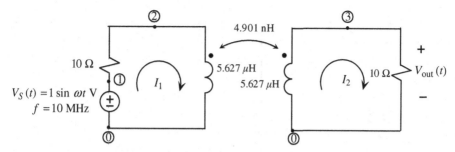

FIGURE 7.13. Example of Fig. 7.12 modeled with loop inductances.

We first compute the output voltage by modeling each loop as its loop self inductance and a mutual inductance between the two loops as shown in Fig. 7.13. The self inductance of each loop is computed from equation (4.20) as $L_{\text{loop}} = 5.627\ \mu\text{H}$. The mutual inductance between the two loops is computed from equation (4.111) as $M_{12} = 4.901$ nH. The complete model for the example using loop inductances is shown in Fig. 7.13. This can be solved by writing the (phasor) mesh current equations around the two loops giving [1,2]

$$\hat{V}_S = 1\angle 0° = \left(10 + j\omega\, 5.627 \times 10^{-6}\right)\hat{I}_1 - j\omega\, 4.901 \times 10^{-9}\hat{I}_2$$

$$0 = -j\omega\, 4.901 \times 10^{-9}\hat{I}_1 + \left(10 + j\omega\, 5.627 \times 10^{-6}\right)\hat{I}_2 \qquad (7.24)$$

Substituting $\omega = 2\pi f = 2\pi \times 10^7$ gives the phasor equations as

$$1\angle 0° = (10 + j353.58)\hat{I}_1 - j0.30795\hat{I}_2$$

$$0 = -j0.30795\hat{I}_1 + (10 + j353.58)\hat{I}_2 \qquad (7.25)$$

Solving this gives $\hat{I}_2 = 2.461 \times 10^{-6}\angle - 86.76°$ and $\hat{V}_{\text{out}} = 10\hat{I}_2 = 2.461 \times 10^{-5}\angle - 86.76°$ V. A simpler way to compute this result is by using PSPICE [2]. The nodes are numbered as shown in Fig. 7.13. The PSPICE program is

```
EXAMPLE
VS 1 0 AC 1 0
RS 1 2 10
L1 2 0 5.6273U
L2 3 0 5.6273U
K12 L1 L2 8.7097E-4
RL 3 0 10
.AC DEC 1 10MEG 10MEG
.PRINT AC VM(3) VP(3)
.END
```

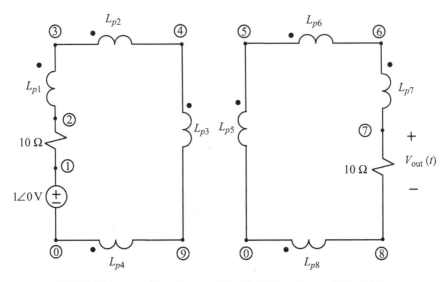

FIGURE 7.14. Modeling the example in Fig. 7.12 using partial inductances.

Note that PSPICE requires the description of mutual inductances in terms of their "coupling coefficients" as $k_{12} = M_{12}/\sqrt{L_1 L_2} = 8.7097 \times 10^{-4}$. The result is, as by hand calculation, $\hat{V}_{\text{out}} = V(3) = 2.461 \times 10^{-5} \angle - 86.76°$ V.

Next, we compute this result using partial inductances to model the segments of the loops and their interaction. The equivalent circuit is shown in Fig. 7.14. All of the self partial inductances are equal since the lengths of the sides of the loops are identical and equal to 1 m. These self partial inductances of each of the four sides of the two loops are computed from equation (5.18a) or approximately from (5.18c) and yield $L_p = 1.5 \ \mu\text{H}$. The inductances are labeled and have even or odd numbers. Mutual partial inductances between orthogonal segments are zero and hence there are mutual inductances only between even-numbered segments and only between odd-numbered segments. First, we compute the mutual partial inducances between parallel segments in the same loop using (5.21a): $M_{p13} = M_{p24} = M_{p57} = M_{p68} = 93.432$ nH. Next, we compute the mutual partial inductances between the vertical segments and between the horizontal segments of the two separate loops that are parallel but offset. We use (5.28) to perform that computation: $M_{p15} = M_{p37} = M_{p48} = M_{p26} = 35.524$ nH. Similarly, we obtain $M_{p17} = M_{p46} = 27.7175$ nH and $M_{p28} = M_{p35} = 45.7816$ nH.

The PSPICE program is

```
EXAMPLE
VS 1 0 AC 1 0
RS 1 2 10
```

```
L1 3 2 1.5U
L2 3 4 1.5U
L3 4 9 1.5U
L4 0 9 1.5U
L5 5 0 1.5U
L6 5 6 1.5U
L7 6 7 1.5U
L8 0 8 1.5U
RL 7 8 10
K13 L1 L3 0.062275
K15 L1 L5 0.023678
K17 L1 L7 0.018474
K35 L3 L5 0.030515
K37 L3 L7 0.023678
K57 L5 L7 0.062275
K24 L2 L4 0.062275
K68 L6 L8 0.062275
K26 L2 L6 0.023678
K48 L4 L8 0.023678
K28 L2 L8 0.030515
K46 L4 L6 0.018474
.AC DEC 1 10MEG 10MEG
.PRINT AC VM(7,8) VP(7,8)
.END
```

The result is $\hat{V}(7,8) = \hat{V}_{\text{out}} = 2.461 \times 10^{-5} \angle -86.76°$ V, which is precisely the same result as was obtained by using loop inductances!

If we examine the values of the mutual inductances for this problem, we find a seemingly curious result. All coupling between the two loops is transferred only through the mutual inductances, loop or partial. In the case of loop inductances in Fig. 7.13, this is solely through the mutual inductance between the two loops of $M_{12} = 4.901$ nH. In the case of partial inductances in Fig. 7.14, this coupling between the two loops occurs only through the mutual partial inductances between elements of the two different loops: M_{p15}, M_{p17}, M_{p35}, M_{p37}, M_{p28}, M_{p26}, M_{p48}, and M_{p46}. These have magnitudes that are on the order of 10^{-8}, which is an order of magnitude greater than the loop mutual inductance of Fig. 7.13. How can mutual partial inductances that differ by an order of magnitude from the loop mutual inductance M_{12} produce the same current in the second loop? The answer to this is that in the partial inductance circuit of Fig. 7.14, the effects of pairs of mutual partial inductances representing the coupling between the two loops *subtract* in the production of induced voltages across the segments of the perimeter of loop 2 to produce V_{out}. For example, the portion of the magnetic flux threading loop 2 due to the current in loop1 on segment 1 of that loop via the mutual partial inductance between that segment and segments 5 and 7 of loop 2 is $\psi_2 = (M_{p15} - M_{p17}) I_1$. This is sensible since the mutual partial inductance between two segments i and j, M_{pij}, gives the magnetic flux between segment j and infinity due to the current on segment i (see Fig. 5.7). Hence, the total flux between two parallel segments of loop 2 due to the current on another segment of loop 1 is the difference between the two mutual partial inductances (see Fig. 7.5). Hence, we may write the total magnetic flux through loop 2 (ψ_2 out of the page) due to the current around loop 1, I_1, as (use the right-hand rule)

$$\psi_2 = (M_{p35} - M_{p37}) I_1 - (M_{p15} - M_{p17}) I_1 + (M_{p28} - M_{p26}) I_1$$
$$- (M_{p48} - M_{p46}) I_1$$
$$= \underbrace{(M_{p35} - M_{p37} + M_{p17} - M_{p15} + M_{p28} - M_{p26} + M_{p46} - M_{p48})}_{M_{12}} I_1$$

Therefore, the loop mutual inductance between the two loops could be computed using partial mutual inductances as

$$M_{12} = M_{p35} - M_{p37} + M_{p17} - M_{p15} + M_{p28} - M_{p26} + M_{p46} - M_{p48}$$
$$= 4.901 \text{ nH}$$

giving precisely the same value for M_{12}. So the difficult and tedious derivation of the equation for the mutual loop inductance between the two loops, M_{12}, given in (4.110) in Section 4.10.1, could have been derived more easily in terms of the *prederived* mutual partial inductance formulas of Chapter 5.

It may appear that since the circuit for the partial inductance method in Fig. 7.14 is more involved than the circuit for the loop inductance method in Fig. 7.13, using loop inductances is preferable to using partial inductances. But when examined carefully, this is not the case. Solution of either circuit is trivial using SPICE or the personal computer version, PSPICE. The *heart of the solution is the values of the circuit elements of the circuit model*! For the loop inductance method one must compute the self inductances for each of the two loops as well as the mutual inductance between the two loops. The derivation of the equation for the loop self inductance, even a square loop, from the electromagnetic field equations is very involved: (see Section 4.1.1). Next, the derivation of the equation for the mutual inductance between two rectangular loops, even ones that lie in the same plane, from the electromagnetic field equations is also extremely involved and tedious: (see Section 4.10.1). You will not find these equations in handbooks or textbooks and must derive them yourself. Had we not already derived these self and mutual loop inductances *for this specific configuration*, you would be required to carry out these detailed derivations from the electromagnetic field equations. For every new problem *you* must *rederive* the formulas for that specific configuration! On the other hand, calculating the values of the self and mutual partial inductances in the partial inductance model of Fig. 7.14 is simple! We already derived the formulas for the self and mutual partial inductances of and between segments of straight wires: No more derivations need be done for a new configuration. Simply "build" a model of the problem by constructing it with piecewise-linear segments, compute the self and mutual partial inductances of and between the segments with the *prederived* formulas in Chapters 5 and 6, and then simply program PSPICE or any other lumped-circuit analysis tool to perform the circuit analysis calculations! So by using partial inductances, a circuit designer can build a circuit model *without ever having to deal with the complicated electromagnetic field equations to derive the values of those elements*! Aside from the very serious requirement in using loop inductances to identify the *complete current loop*, this is the essential beauty in using partial inductances over using loop inductances.

APPENDIX

FUNDAMENTAL CONCEPTS OF VECTORS

Fundamentally, the laws governing the calculation of capacitance and inductance are written in terms of *vectors* of the four electromagnetic field *vector quantities*, which are the electric field intensity vector $\mathbf{E}$, the electric flux density vector $\mathbf{D}$, the magnetic field intensity vector $\mathbf{H}$, and the magnetic flux density vector $\mathbf{B}$. Therefore, if we are to calculate and understand the notions of capacitance and inductance of a physical structure correctly, as well as use them correctly to construct a lumped-circuit model of that structure, we must understand some elementary properties of vectors and some elementary vector calculus ideas. Trying to avoid the use of vector calculus ideas by relying on one's daily "life experiences" to compute and properly interpret the meanings of capacitance and inductance of a structure has caused many of the incorrect results and misunderstanding, as well as the numerous erroneous applications that are seen throughout the literature and in conversations with engineering professionals.

We assume that the reader has a rudimentary familiarity with vectors, so this appendix is a review of those important concepts. The reader is referred to other textbooks on electromagnetics listed in the references for more details [3–6]. For the computation of inductance, this brief review will be sufficient.

Inductance: Loop and Partial, By Clayton R. Paul
Copyright © 2010 John Wiley & Sons, Inc.

A.1 VECTORS AND COORDINATE SYSTEMS

A vector, as distinguished from a scalar, contains two items of information about a physical quantity: its value and its direction of effect. A vector is shown in the figures as a line with an arrowhead to show that direction of effect and is denoted in the text as boldface (e. g. , **F**). The magnitude or length of a vector is denoted as F or as $F = |\mathbf{F}|$. To compute with vectors properly requires a coordinate system. We use primarily the rectangular (Cartesian) coordinate system that consists of three axes x, y, and z, as shown in Fig.A.1.

These axes are mutually orthogonal. In a rectangular coordinate system, a vector is described as

$$\mathbf{F} = F_x \mathbf{a}_x + F_y \mathbf{a}_y + F_z \mathbf{a}_z \qquad (A.1)$$

where the components of **F** along (projections of **F** *onto*) the x, y, and z axes are denoted as F_x, F_y, and F_z, respectively, and the unit vectors along the axes are denoted as $\mathbf{a}_x$, $\mathbf{a}_y$, and $\mathbf{a}_z$. These unit vectors are of unit length and are directed in the direction of increasing value of the coordinate axis.

There are other coordinate systems, such as the cylindrical and spherical coordinate systems described at the end of this appendix. Although a problem can be solved in any coordinate system, the choice of coordinate system used to solve the problem will simplify the solution considerably. The rectangular

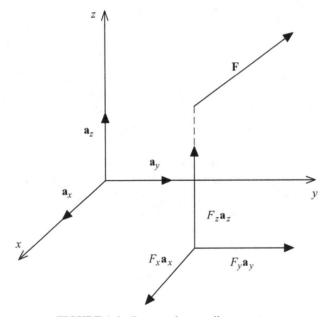

FIGURE A.1. Rectangular coordinate system.

coordinate system is more suitable for problems whose boundaries fit a rectangular shape. The cylindrical coordinate system is more suitable for problems whose boundaries fit a cylindrical shape, whereas the spherical coordinate system is more suitable for problems whose boundaries fit a spherical shape. The unit vectors of a rectangular coordinate system are mutually perpendicular at a point. Hence, the rectangular coordinate system is said to be an *orthogonal coordinate system*. The cylindrical and spherical coordinate systems discussed at the end of this appendix are similarly orthogonal coordinate systems. Vectors in any orthogonal coordinate system are added or subtracted by adding or subtracting their corresponding components:

$$\mathbf{A} \pm \mathbf{B} = (A_x \pm B_x)\, \mathbf{a}_x + (A_y \pm B_y)\, \mathbf{a}_y + (A_z \pm B_z)\, \mathbf{a}_z \qquad \text{(A.2)}$$

There are two ways of performing the multiplication of two vectors: the dot product and the cross product. The *dot product* of two vectors gives the result as a *scalar* and is defined by [3]

$$\begin{aligned} \mathbf{A}\cdot\mathbf{B} &= AB\cos\theta_{AB} \\ &= A_x B_x + A_y B_y + A_z B_z \end{aligned} \qquad \text{(A.3)}$$

where θ_{AB} is the angle between the two vectors as illustrated in Fig. A.2(a). In plain terms this gives (1) the product of the length of $\mathbf{A}$ and the *projection of* $\mathbf{B}$ *onto* $\mathbf{A}$, or (2) the product of the length of $\mathbf{B}$ and the *projection of* $\mathbf{A}$ *onto* $\mathbf{B}$. The result for the dot product in terms of the vector components in a rectangular coordinate system given in (A.3) is easy to remember: It is the sum of the products of the corresponding components of the two vectors. This will also be the case for the cylindrical and spherical coordinate systems. Two vectors are *perpendicular* if $\mathbf{A}\cdot\mathbf{B} = 0$. Also, the dot product of a vector with itself is its magnitude squared: $\mathbf{A}\cdot\mathbf{A} = |\mathbf{A}|^2$.

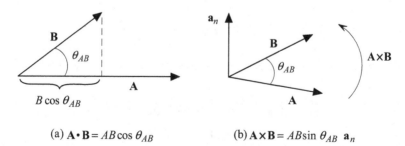

(a) $\mathbf{A}\cdot\mathbf{B} = AB\cos\theta_{AB}$ (b) $\mathbf{A}\times\mathbf{B} = AB\sin\theta_{AB}\,\mathbf{a}_n$

FIGURE A.2. Dot and cross product of two vectors.

The *cross product* of two vectors gives a *vector* and is defined by [3]

$$\mathbf{A} \times \mathbf{B} = AB \sin \theta_{AB} \mathbf{a}_n$$
$$= (A_y B_z - A_z B_y) \mathbf{a}_x + (A_z B_x - A_x B_z) \mathbf{a}_y + (A_x B_y - A_y B_x) \mathbf{a}_z$$

(A.4)

where θ_{AB} is the angle between the two vectors, as illustrated in Fig. A.2(b). The result gives a *vector* that is *perpendicular to the plane containing* **A** *and* **B**. The unit vector perpendicular to (normal to) this plane containing **A** and **B** is denoted as $\mathbf{a}_n$. Since there are two sides to this plane, which contains **A** and **B**, the direction of the unit normal is determined by the *right-hand rule*; that is, if the fingers of our right hand curl *from* **A** *to* **B**, the direction of the normal to this plane for **A**×**B** will be given by the thumb of our right hand. The reader should practice this since it is used throughout this book. The axes of the rectangular coordinate system are assumed to be ordered cyclically according to the convention of $x \to y \to z \to x \to y \to z \to \cdots$. In other words, if we cross the x axis into the y axis, we get the z axis: $\mathbf{a}_x \times \mathbf{a}_y = \mathbf{a}_z$. Note that, for example, $\mathbf{a}_y \times \mathbf{a}_x = -\mathbf{a}_z$. The vector result for the cross product in a rectangular coordinate system in terms of the vector components in (A.4) is easily remembered. Each component is of the form $(A_\beta B_\gamma - A_\gamma B_\beta) \mathbf{a}_\alpha$ in the order $\alpha \to \beta \to \gamma \to \alpha \to \beta \to \cdots$ according to the cyclic ordering of the axes. This rule for determining the cross product is the same in the cylindrical and spherical coordinate systems. Two vectors are *parallel* if $\mathbf{A} \times \mathbf{B} = 0$. Note that $\mathbf{A} \cdot \mathbf{B} = \mathbf{B} \cdot \mathbf{A}$ and the order in the dot product does not matter. However, the order in the cross product does matter: $\mathbf{A} \times \mathbf{B} = -\mathbf{B} \times \mathbf{A}$.

EXAMPLE

Two vectors lying in the yz plane are defined, as shown in Fig.A.3, as

$$\mathbf{A} = 3\mathbf{a}_y$$
$$\mathbf{B} = 2\mathbf{a}_y + \mathbf{a}_z$$

The lengths of the two vectors are $A = 3$ and $B = \sqrt{(2)^2 + (1)^2} = \sqrt{5}$. The dot product is

$$\mathbf{A} \cdot \mathbf{B} = (0\,\mathbf{a}_x + 3\,\mathbf{a}_y + 0\,\mathbf{a}_z) \cdot (0\,\mathbf{a}_x + 2\,\mathbf{a}_y + \mathbf{a}_z)$$
$$= 3 \times 2 + 0 \times 1$$
$$= 6$$

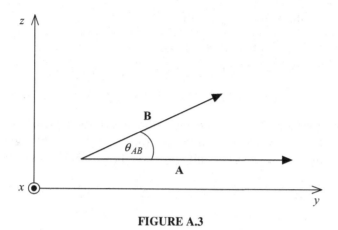

FIGURE A.3

From the dot product in (A.3),

$$\cos(\theta_{AB}) = \frac{\mathbf{A} \cdot \mathbf{B}}{A \, B}$$

$$= \frac{6}{3 \cdot \sqrt{5}}$$

$$= 0.894$$

Hence, the angle between the two vectors is $\theta_{AB} = \cos^{-1}(0.894) = 26.57°$. For these simple vectors, we can obtain this angle directly by trigonometry:

$$\theta_{AB} = \tan^{-1}\frac{1}{2}$$

$$= 26.57°$$

so we again obtain

$$\mathbf{A} \cdot \mathbf{B} = A \, B \cos\theta_{AB}$$

$$= 3 \times \sqrt{2^2 + 1^2} \times \cos(26.57°)$$

$$= 6$$

The cross product is

$$\mathbf{A} \times \mathbf{B} = (A_y B_z - A_z B_y)\,\mathbf{a}_x + (A_z B_x - A_x B_z)\,\mathbf{a}_y + (A_x B_y - A_y B_x)\,\mathbf{a}_z$$

$$= (3 - 0)\,\mathbf{a}_x + (0 - 0)\,\mathbf{a}_y + (0 - 0)\,\mathbf{a}_z$$

$$= 3\,\mathbf{a}_x$$

Directly, we obtain the same result:

$$\mathbf{A} \times \mathbf{B} = A \, B \, \sin \theta_{AB} \, \mathbf{a}_n$$
$$= 3 \times \sqrt{2^2 + 1^2} \times \sin(26.57°) \, \mathbf{a}_n$$
$$= 3 \, \mathbf{a}_n$$

Since both vectors lie in the yz plane, the unit normal *perpendicular to the plane containing* $\mathbf{A}$ and $\mathbf{B}$ is in the $\pm x$ direction. Using the right-hand rule and crossing $\mathbf{A}$ *to* $\mathbf{B}$ gives the unit normal in the positive x direction: $\mathbf{a}_n = \mathbf{a}_x$.

A.2   LINE INTEGRAL

The fundamental equations governing the electromagnetic field vectors (referred to collectively as Maxwell's equations) involve two basic integrals: the *line integral* and the *surface integral*. Hence, it is important that we understand what these mean and how to evaluate them. The vectors in the electromagnetic field equations are functions of the coordinate system variables x, y, and z, which is denoted by $\mathbf{F}(x, y, z)$ and hence are said to constitute a *field*. There are two possible types of fields: a *scalar* field and a *vector* field. An example of a *scalar field* is a plot of the temperature distribution in a room. Lines of constant temperature (a scalar) show the distribution of that field in the room. An example of a *vector field* would be the plot of flow rates and directions of the water flow in a river. The directions of these vectors show the direction of the water flow at that point, and the lengths of these vectors are proportional to the rates of flow at that point.

The *line integral* of a vector field is denoted as

$$\int_a^b \mathbf{F}(x, y, z) \cdot d\mathbf{l} = \int_a^b F(x, y, z) \, \cos \theta \, dl \tag{A.5}$$

The line integral means that we take the products of the projection of the vector $\mathbf{F}$ onto the path, $F \cos \theta$ (alternatively, the component of $\mathbf{F}$ tangent to the path), and the differential lengths, dl, along the path and sum them with an integral from the starting point a to the endpoint b, as illustrated in Fig. A.4.

An example of a line integral is the computation of the work required to push an object from one point to another when the force $\mathbf{F}$ is exerted on the object at an angle to the path as shown in Fig.A.5. The work done is $W = \int F \cos \theta \, dx = \int \mathbf{F} \cdot d\mathbf{l}$. The line integral is a very sensible result.

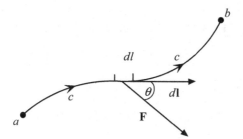

FIGURE A.4. Line integral.

There are two components of **F**: One component is *parallel* to the path and the other component is *perpendicular* to the path. Only the component *parallel to the path* should contribute to the result.

The actual computation of the line integral in a rectangular coordinate system is very simple. In a rectangular coordinate system a vector differential path length is

$$dl = dx\,\mathbf{a}_x + dy\,\mathbf{a}_y + dz\,\mathbf{a}_z \tag{A.6}$$

Hence

$$\mathbf{F} \cdot d\mathbf{l} = F_x\,dx + F_y\,dy + F_z\,dz \tag{A.7}$$

and the line integral becomes

$$\boxed{\begin{aligned} \int_a^b \mathbf{F} \cdot d\mathbf{l} &= \int_a^b F\cos\theta\,dl \\ &= \int_{x_a}^{x_b} F_x\,dx + \int_{y_a}^{y_b} F_y\,dy + \int_{z_a}^{z_b} F_z\,dz \end{aligned}} \tag{A.8}$$

where the path extends from (x_a, y_a, z_a) to (x_b, y_b, z_b) and each component of **F** is a function of x, y, and z: $F_x\,(x, y, z)$, $F_y\,(x, y, z)$, and $F_z\,(x, y, z)$. If

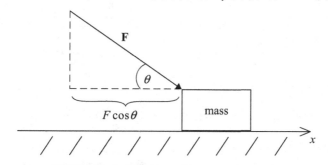

FIGURE A.5. Line integral in computing work.

the integral is taken around a closed path, it is denoted with a circle on the integral sign as $\oint_c \mathbf{F} \cdot d\mathbf{l}$ and c represents the contour of that closed path.

EXAMPLE

A vector field in the yz plane is given as

$$\mathbf{F}(x, y, z) = z\, \mathbf{a}_y$$

as shown in Fig. A.6. Determine the line integral of $\mathbf{F}$ along a straight-line path between the two points in the yz plane *from* point a at $(0,1,3)$ *to* point b at $(0,2,4)$. Observe that at all points in the yz plane the vector is directed in the y direction. However, its magnitude depends on z: for positive, increasing values of z, its magnitude (length) increases. For z negative, it is pointing in the $-y$ direction. Performing the line integral gives

$$\int_a^b \mathbf{F} \cdot d\mathbf{l} = \int_{x=0}^{0} \underbrace{F_x}_{0}\, dx + \int_{y=1}^{2} \underbrace{F_y}_{z}\, dy + \int_{z=3}^{4} \underbrace{F_z}_{0}\, dz$$

$$= \int_{y=1}^{2} z\, dy$$

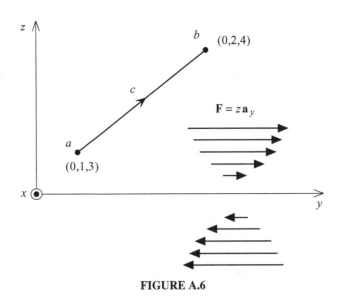

FIGURE A.6

$$= \int_{y=1}^{2} (y+2) \, dy$$

$$= \frac{7}{2}$$

and we have substituted the equation of the path, $z = y + 2$.

A.3   SURFACE INTEGRAL

The *surface integral* is

$$
\int_{s} \mathbf{F}(x, y, z) \cdot d\mathbf{s} = \int_{s} \mathbf{F}(x, y, z) \cdot \mathbf{a}_{n} \, ds
$$
$$
= \int_{s} F(x, y, z) \, \cos \theta \, ds \qquad \text{(A.9)}
$$

The surface integral gives the integral of the products of the components of **F** that are *perpendicular to the surface s* and the differential surface elements *ds* as shown in Fig. A.7. The unit normal perpendicular to the surface is denoted as $\mathbf{a}_{n}$, and the differential surface area is $d\mathbf{s} = ds \, \mathbf{a}_{n}$. The surface integral gives the *flux of the vector field* **F** *through the surface s*. This is like shining a light through an opening. There are two components of **F**: One component is *parallel* to the surface and the other component is *perpendicular*

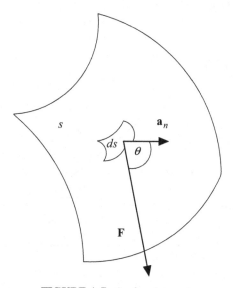

FIGURE A.7. Surface integral.

to the surface. Only the component of the light flux that is *perpendicular* to the opening contributes to the net light flux passing through that opening. If the surface s is a closed surface, the surface integral is denoted with a circle on the integral sign: $\oint_s \mathbf{F} \cdot d\mathbf{s}$. Hence, the surface integral in (A.9) is said to give the net *flux* of the vector field through the surface s.

Observe that there is a major difference between the line integral and the surface integral. The line integral involves the components of $\mathbf{F}$ that are *parallel* to (tangent to) the path, whereas the surface integral involves the components of $\mathbf{F}$ that are *perpendicular* to the surface.

The evaluation of the surface integral in a rectangular coordinate system is very simple. The vector differential surface is

$$d\mathbf{s} = dy\,dz\,\mathbf{a}_x + dx\,dz\,\mathbf{a}_y + dx\,dy\,\mathbf{a}_z \qquad (A.10)$$

Note that the components of this are the differential surface areas whose unit normals are perpendicular to them (e. g., $dy\,dz\,\mathbf{a}_x$). Hence, the surface integral simplifies, in a rectangular coordinate system, to

$$\int_s \mathbf{F}(x, y, z) \cdot d\mathbf{s} = \int_{S_x} F_x\,dy\,dz + \int_{S_y} F_y\,dx\,dz + \int_{S_z} F_z\,dx\,dy \qquad (A.11)$$

EXAMPLE

A wedge-shaped surface lies in the yz plane as shown in Fig. A.8. Determine the flux of the vector field

$$\mathbf{F} = (x + 2)\,\mathbf{a}_x$$

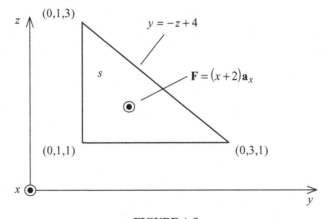

FIGURE A.8

through the surface. The surface integral becomes

$$\int_s \mathbf{F}(x, y, z) \cdot d\mathbf{s} = \int_{S_x} F_x \, dy \, dz$$

$$= \int_{z=1}^{3} \int_{y=1}^{y=-z+4} \left(\underbrace{x}_{0} + 2 \right) dy \, dz$$

$$= \int_{z=1}^{3} \int_{y=1}^{y=-z+4} 2 \, dy \, dz$$

$$= \int_{z=1}^{3} (-2z + 6) \, dz$$

$$= 4$$

We have substituted $x = 0$ over the surface into $F_x = x + 2$ and the equation of the top part of the wedge, $y = -z + 4$, in the limit of one of the integrals.

A.4   DIVERGENCE

The line and surface integrals apply over regions of space. The following vector calculus results, the *divergence* and the *curl*, are the *point forms* of these integrals which apply to points in space and are differential relations that give the relationships between the field vectors at points in space.

The *divergence* of a vector field gives the net *outflow* or *flux* of a vector field from a point, hence the name *divergence*, and is defined by

$$\nabla \cdot \mathbf{F}(x, y, z) = \underbrace{\lim}_{\Delta v \to 0} \frac{\oint_s \mathbf{F} \cdot d\mathbf{s}}{\Delta v} \qquad (A.12)$$

This is illustrated in Fig. A.9. If we surround a point by a *closed* surface s that contains a differential volume Δv, compute the net flux of $\mathbf{F}$ *out of* the closed surface per unit of volume enclosed by s, and then let the surface and enclosed volume shrink to zero, the limit of that is the *divergence* of $\mathbf{F}$ at that point. Essentially, this gives an indication of any sources of $\mathbf{F}$ that are located at the point. If the divergence of $\mathbf{F}$ is negative at the point, we say that a *sink* exists at that point. So the divergence indicates whether there is a net *outflow* of $\mathbf{F}$ at that point. If we puncture an inflated ballon, we get a divergence of the air contained in that ballon.

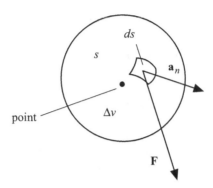

FIGURE A.9. Divergence of a vector field.

The "del operator," ∇, is somewhat equivalent to a derivative in scalar calculus and is an "operator" defined by [3]

$$\nabla = \mathbf{a}_x \frac{\partial}{\partial x} + \mathbf{a}_y \frac{\partial}{\partial y} + \mathbf{a}_z \frac{\partial}{\partial z} \qquad (A.13)$$

Using the del operator, we obtain the divergence of a vector field in a rectangular coordinate system as

$$\nabla \cdot \mathbf{F}(x, y, z) = \frac{\partial F_x}{\partial x} + \frac{\partial F_y}{\partial y} + \frac{\partial F_z}{\partial z} \qquad (A.14)$$

It is very important to observe that the divergence of a vector field gives a *scalar* quantity as the result.

EXAMPLE

A vector field is described by

$$\mathbf{F} = x\,\mathbf{a}_x + y\,\mathbf{a}_y + z\,\mathbf{a}_z$$

as plotted in Fig. A.10. Determine the divergence of the field. The divergence of this field is

$$\nabla \cdot \mathbf{F}(x, y, z) = \frac{\partial F_x}{\partial x} + \frac{\partial F_y}{\partial y} + \frac{\partial F_z}{\partial z}$$

$$= 1 + 1 + 1 = 3$$

Since this result is independent of x, y, and z, there is a net outflow of the vector at every point in the space. This is a sensible result since the field is constant over any sphere of radius $r = \sqrt{x^2 + y^2 + z^2}$ centered at the origin of the coordinate system and is directed normal to the surface of that sphere. Hence, from the basic definition of the divergence given in (A.12) we can

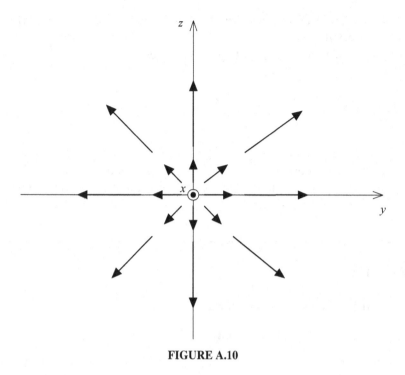

FIGURE A.10

calculate directly

$$\nabla \cdot \mathbf{F}\,(x,\, y,\, z) = \underbrace{\lim}_{\Delta v \to 0}\ \frac{\oint_s \mathbf{F} \cdot d\mathbf{s}}{\Delta v}$$

$$= \frac{r \times 4\pi r^2}{4/3\pi\, r^3}$$

$$= 3$$

A.4.1   Divergence Theorem

We can interchange certain surface and volume integrals with the *divergence theorem* [3]:

$$\boxed{\oint_s \mathbf{F} \cdot d\mathbf{s} = \int_v (\nabla \cdot \mathbf{F})\ dv} \tag{A.15}$$

This result provides that if we integrate the divergence of **F** throughout some volume v, we can obtain the same result by performing the surface integral of **F** over the *closed* surface s that contains the volume v. This is a very sensible result if we think about what these quantities mean. According to (A.12), the divergence $\nabla \cdot \mathbf{F}$ gives the net outflow or flux of $\mathbf{F}$ throughout the volume Δv per unit of that volume. Rewriting (A.12) gives

$$\oint_s \mathbf{F} \cdot d\mathbf{s} = \underbrace{\lim_{\Delta v \to 0}} \left[\nabla \cdot \mathbf{F}(x, y, z) \ \Delta v \right]$$

$$= \int_v (\nabla \cdot \mathbf{F}) \, dv \qquad \text{(A.12)}$$

Hence, it makes sense that we can obtain the net flux out of the closed surface s that encloses that volume, $\oint_s \mathbf{F} \cdot d\mathbf{s}$, by performing the volume integral of $\nabla \cdot \mathbf{F}$ throughout that volume.

EXAMPLE

Verify the divergence theorem for the vector field

$$\mathbf{F} = x \, \mathbf{a}_x + y \, \mathbf{a}_y + z \, \mathbf{a}_z$$

for the square volume whose corners are at (0,0,0), (0,0,1), (0,1,0),(0,1,1), (1,0,0), (1,0,1), (1,1,0), and (1,1,1) as illustrated in Fig. A.11. The surface integral over the closed surface s is

$$\oint_s \mathbf{F} \cdot d\mathbf{s} = \underbrace{\int_{z=0}^1 \int_{y=0}^1 F_x \, dy \, dz}_{\text{front}} - \underbrace{\int_{z=0}^1 \int_{y=0}^1 F_x \, dy \, dz}_{\text{back}} - \underbrace{\int_{z=0}^1 \int_{x=0}^1 F_y \, dx \, dz}_{\text{left}}$$

$$+ \underbrace{\int_{z=0}^1 \int_{x=0}^1 F_y \, dx \, dz}_{\text{right}} - \underbrace{\int_{y=0}^1 \int_{x=0}^1 F_z \, dx \, dy}_{\text{bottom}} + \underbrace{\int_{y=0}^1 \int_{x=0}^1 F_z \, dx \, dy}_{\text{top}}$$

$$= \underbrace{\int_{z=0}^1 \int_{y=0}^1 \underbrace{x}_{1} \, dy \, dz}_{\text{front}} - \underbrace{\int_{z=0}^1 \int_{y=0}^1 \underbrace{x}_{0} \, dy \, dz}_{\text{back}} - \underbrace{\int_{z=0}^1 \int_{x=0}^1 \underbrace{y}_{0} \, dx \, dz}_{\text{left}}$$

$$+ \underbrace{\int_{z=0}^1 \int_{x=0}^1 \underbrace{y}_{1} \, dx \, dz}_{\text{right}} - \underbrace{\int_{y=0}^1 \int_{x=0}^1 \underbrace{z}_{0} \, dx \, dy}_{\text{bottom}} + \underbrace{\int_{y=0}^1 \int_{x=0}^1 \underbrace{z}_{1} \, dx \, dy}_{\text{top}}$$

$$= 1 - 0 - 0 + 1 - 0 + 1$$

$$= 3$$

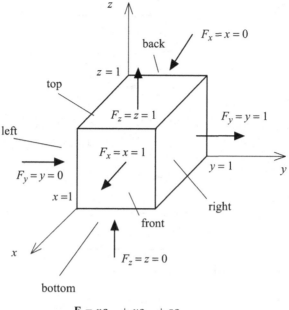

$$\mathbf{F} = x\mathbf{a}_x + y\mathbf{a}_y + z\mathbf{a}_z$$

FIGURE A.11

Notice that the surface integral determines the *net flux leaving the closed surface*. A vector component points into one side and out of the other side. Hence, half the integrals are positive and half the integrals are negative. Observe also that each integrand is 0 or 1 over a surface and the dimensions of each side are 1. Therefore, the integral over a side is either 0 or1. Since

$$\nabla\cdot\mathbf{F}\,(x, y, z) = \frac{\partial F_x}{\partial x} + \frac{\partial F_y}{\partial y} + \frac{\partial F_z}{\partial z}$$
$$= 1 + 1 + 1 = 3$$

the right-hand side of the divergence theorem in (A.15) also gives the same result:

$$\int_v (\nabla\cdot\mathbf{F})\,dv = \int_{x=0}^{1}\int_{y=0}^{1}\int_{z=0}^{1} 3\,dx\,dy\,dz$$
$$= 3$$

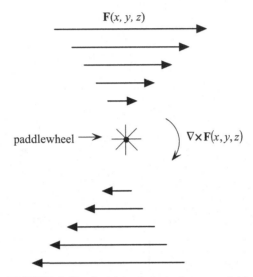

FIGURE A.12. Curl (circulation) of a vector field.

A.5   CURL

While the divergence gives the net outflow or flux of a vector field from a point, the *curl* of a vector field gives the *net circulation or rotation of the field about a point*. For example, consider the vector field shown in Fig. A.12. This field might represent the flow of the water in a river. If we insert a small paddlewheel as shown, the flow pattern will cause the paddlewheel to rotate in the clockwise direction. If we turned the paddlewheel such that its axis was parallel to the field lines, it would not rotate.

Figure A.13 shows how we might define the *circulation* of a vector field in one plane. Define a flat surface s in that plane and the associated contour c enclosing it. Define the unit normal to that plane as $\mathbf{a}_n$, with its direction

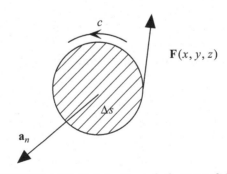

FIGURE A.13. Defining the curl of a vector field.

according to the right-hand rule with respect to the direction of c around that surface perimeter. The net circulation at the point per unit of the enclosed surface area *in this plane* would be

$$\textit{circulation per unit area} = \mathbf{a}_n \left(\lim_{\Delta s \to 0} \frac{\oint_c \mathbf{F} \cdot d\mathbf{l}}{\Delta s} \right) \tag{A.16}$$

By performing the line integral of $\mathbf{F}$ around the contour c enclosing the surface Δs and dividing by that surface, we get a measure of the circulation (in this case in the counterclockwise direction). A direction is given to that circulation by the unit vector $\mathbf{a}_n$ normal to the surface. The direction of the unit normal is obtained in accordance with the right-hand rule. Since the result is circulation or rotation of the field, we should obtain the total circulation or rotation in three orthogonal planes. The result gives the curl of the vector field as

$$\nabla \times \mathbf{F}(x, y, z) = \mathbf{a}_x \left(\lim_{\Delta s_{yz} \to 0} \frac{\oint_{c_{yz}} \mathbf{F} \cdot d\mathbf{l}}{\Delta s_{yz}} \right) + \mathbf{a}_y \left(\lim_{\Delta s_{xz} \to 0} \frac{\oint_{c_{xz}} \mathbf{F} \cdot d\mathbf{l}}{\Delta s_{xz}} \right)$$
$$+ \mathbf{a}_z \left(\lim_{\Delta s_{xy} \to 0} \frac{\oint_{c_{xy}} \mathbf{F} \cdot d\mathbf{l}}{\Delta s_{xy}} \right)$$

$$\tag{A.17}$$

where Δs_{xy}, for example, is a flat surface in the xy plane which is perpendicular to $\mathbf{a}_z$, and c_{xy} denotes the contour around the enclosed surface Δs_{xy}.

Applying the del operator that is defined in (A.13) gives a mechanical way of determining the curl in a rectangular coordinate system [3]:

$$\nabla \times \mathbf{F}(x, y, z) = \left(\frac{\partial F_z}{\partial y} - \frac{\partial F_y}{\partial z} \right) \mathbf{a}_x + \left(\frac{\partial F_x}{\partial z} - \frac{\partial F_z}{\partial x} \right) \mathbf{a}_y$$
$$+ \left(\frac{\partial F_y}{\partial x} - \frac{\partial F_x}{\partial y} \right) \mathbf{a}_z$$

$$\tag{A.18}$$

Observe that each of these components can be remembered easily using the cyclic rule for the cross product, the cyclic ordering of the three axes, and the definition of the del operator given in (A.13). For example, each component of the curl is of the form $(\partial F_\gamma / \partial \beta - \partial F_\beta / \partial \gamma) \, \mathbf{a}_\alpha$, where the ordering is $\alpha \to \beta \to \gamma \to \alpha \cdots$.

EXAMPLE

Determine the curl of the vector field

$$\mathbf{F} = z\,\mathbf{a}_y$$

that is illustrated in Fig. A.14. First we see clearly that there will be circulation and the rotation will be clockwise with the unit normal being in the negative x direction. Substituting into (A.18) yields

$$\nabla \times \mathbf{F}\,(x,\,y,\,z) = \left(\underbrace{\frac{\partial F_z}{\partial y}}_{0} - \underbrace{\frac{\partial F_y}{\partial z}}_{1} \right)\mathbf{a}_x + \left(\underbrace{\frac{\partial F_x}{\partial z}}_{0} - \underbrace{\frac{\partial F_z}{\partial x}}_{0} \right)\mathbf{a}_y$$

$$+ \left(\underbrace{\frac{\partial F_y}{\partial x}}_{0} - \underbrace{\frac{\partial F_x}{\partial y}}_{0} \right)\mathbf{a}_z$$

$$= -\mathbf{a}_x$$

as expected.

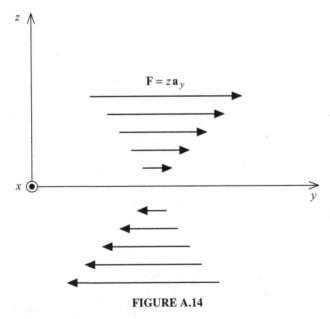

FIGURE A.14

A.5.1   Stokes's Theorem

Similar to the divergence theorem, *Stokes's theorem* allows us to interchange a surface integral and a line integral [3]:

$$\boxed{\oint_c \mathbf{F} \cdot d\mathbf{l} = \int_s (\nabla \times \mathbf{F}) \cdot d\mathbf{s}} \qquad \text{(A.19)}$$

Stokes's theorem provides that the surface integral of the curl of $\mathbf{F}$ over an open surface s will give the same result as performing the line integral of $\mathbf{F}$ around the contour c that encloses that open surface. As was the case for the divergence theorem, Stokes's theorem is a very sensible result. According to (A.17), the curl of a vector field, $\nabla \times \mathbf{F}$, gives the net circulation or rotation of a field around a contour that encloses a differential surface per unit of that enclosed surface. Rewriting the x component of (A.17) gives

$$\oint_{c_{yz}} \mathbf{F} \cdot d\mathbf{l} = \lim_{\Delta s_{yz} \to 0} \left\{ \left[\nabla \times \mathbf{F}(x, y, z) \right]_x \Delta s_{yx} \right\}$$

$$= \int_{s_{yz}} (\nabla \times \mathbf{F}) \cdot d\mathbf{s}$$

Hence, it makes sense that by integrating the curl over the surface with a surface integral we will obtain the same result as the line integral around the contour enclosing that surface would give.

EXAMPLE

Verify Stokes's theorem for the vector field

$$\mathbf{F} = z \, \mathbf{a}_y$$

and the closed contour c and its enclosed surface s shown in Fig. A.15. The curl of $\mathbf{F}$ is

$$\nabla \times \mathbf{F} = \left(\underbrace{\frac{\partial F_z}{\partial y}}_{0} - \underbrace{\frac{\partial F_y}{\partial z}}_{1} \right) \mathbf{a}_x + \left(\underbrace{\frac{\partial F_x}{\partial z}}_{0} - \underbrace{\frac{\partial F_z}{\partial x}}_{0} \right) \mathbf{a}_y + \left(\underbrace{\frac{\partial F_y}{\partial x}}_{0} - \underbrace{\frac{\partial F_x}{\partial y}}_{0} \right) \mathbf{a}_z$$

$$= -\mathbf{a}_x$$

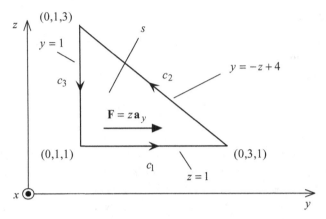

FIGURE A.15

Hence, the right-hand side of Stokes's theorem is

$$\int_s (\nabla \times \mathbf{F}) \cdot ds = \int_{z=1}^{3} \int_{y=1}^{y=-z+4} \underbrace{(-1)\,\mathbf{a}_x}_{\nabla \times \mathbf{F}} \cdot \underbrace{(\mathbf{a}_x\,dy\,dz)}_{ds}$$

$$= \int_{z=1}^{3} \int_{y=1}^{y=-z+4} (-1)\,dy\,dz$$

$$= -2$$

Since $\mathbf{F} \cdot d\mathbf{l} = F_y\,dy = z\,dy$, the left-hand side of Stokes's theorem is

$$\oint_c \mathbf{F} \cdot d\mathbf{l} = \underbrace{\int_{y=1}^{3} F_y\,dy}_{c_1} + \underbrace{\int_{y=3}^{1} F_y dy}_{c_2} + \underbrace{\int_{y=1}^{1} F_y dy}_{c_3}$$

$$= \int_{y=1}^{3} \underbrace{z}_{1}\,dy + \int_{y=3}^{1} \underbrace{z}_{-y+4}\,dy + \int_{y=1}^{1} z\,dy$$

$$= 2 - 4 + 0$$

$$= -2$$

which is the same.

A.6   GRADIENT OF A SCALAR FIELD

Perhaps one of the best illustrations of the use of the *gradient* is a topographical map. Contours of constant elevation (above sea level) are shown as closed contours. We might denote this as the scalar field EL (x, y, z). Think of this scalar function as depicting a three-dimensional map with the x and y coordinates giving the horizontal position over the Earth's surface, and the z axis giving the elevation of each point above sea level. The closer the contours of constant elevation are to each other, the steeper the slope (i. e., the greater the change in elevation with a change in horizontal distance). If we wanted to chart a course for hiking that would avoid the steep slopes, we would choose a path between points on adjacent contours of constant elevation with those contours being as widely separated as possible. In doing so, we would make the vertical distance we move as long a horizontal distance as possible. Also, to make the trip as expeditious as possible we would choose a route that is perpendicular to those contours.

Denote some general scalar field as $f(x, y, z)$. A differential change in the function (the scalar field) as we move between contours of constant value of f is

$$df = \frac{\partial f(x, y, z)}{\partial x}dx + \frac{\partial f(x, y, z)}{\partial y}dy + \frac{\partial f(x, y, z)}{\partial z}dz \qquad \text{(A.20)}$$

Using the del operator in (A.13):

$$\nabla = \mathbf{a}_x \frac{\partial}{\partial x} + \mathbf{a}_y \frac{\partial}{\partial y} + \mathbf{a}_z \frac{\partial}{\partial z} \qquad \text{(A.13)}$$

we define the *gradient* of f as

$$\nabla f = \frac{\partial f(x, y, z)}{\partial x}\mathbf{a}_x + \frac{\partial f(x, y, z)}{\partial y}\mathbf{a}_y + \frac{\partial f(x, y, z)}{\partial z}\mathbf{a}_z \qquad \text{(A.21)}$$

Note that the gradient of a scalar field $f(x, y, z)$, $\nabla f(x, y, z)$, gives a *vector* as the result. Recalling the vector differential path length in (A.6),

$$d\mathbf{l} = dx\,\mathbf{a}_x + dy\,\mathbf{a}_y + dz\,\mathbf{a}_z \qquad \text{(A.6)}$$

we can write (A.20) in terms of the gradient as

$$df = \nabla f \cdot d\mathbf{l} \tag{A.22}$$

which you should verify.

Now we interpret the meaning of the gradient. The differential change in (A.22) is

$$df = \nabla f \cdot d\mathbf{l}$$
$$= |\nabla f| \, dl \, \cos \theta \tag{A.22}$$

where θ is the angle between the gradient vector, ∇f, and the differential path length vector, $d\mathbf{l}$. The rate of change of the scalar field along this path is

$$\frac{df}{dl} = |\nabla f| \cos \theta \tag{A.23}$$

If we want to move in the direction of the maximum rate of change of the scalar field (i. e., perpendicular to the contours of constant f), the path taken must be perpendicular to the gradient vector (i. e., $\theta = 90°$):

$$\left. \frac{df}{dl} \right|_{max} = |\nabla f| \tag{A.24}$$

Therefore, *the gradient vector gives both the direction and the magnitude of the maximum space rate of change of the scalar field.*

EXAMPLE

Show that the gradient of the scalar field $f(x, y, z) = x + y$ is normal to the lines of constant f. The scalar field is plotted in Fig. A.16. The gradient is

$$\nabla f = \frac{\partial f}{\partial x} \mathbf{a}_x + \frac{\partial f}{\partial y} \mathbf{a}_y$$
$$= \mathbf{a}_x + \mathbf{a}_y$$

which is plotted in Fig. A.16. Obviously, the gradient is perpendicular to the lines of constant f, and it also points in the direction of the maximum rate of change of f.

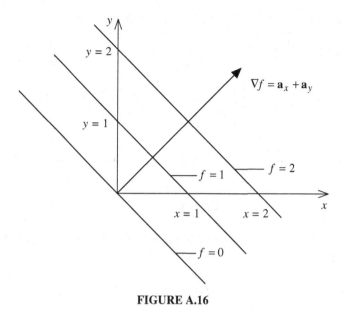

FIGURE A.16

A.7   IMPORTANT VECTOR IDENTITIES

An important vector identity that will prove very useful in defining the concept of partial inductance is

$$\nabla \cdot (\nabla \times \mathbf{F}) = 0 \qquad\qquad (A.25)$$

Note that it would make no sense to write $\nabla \times (\nabla \cdot \mathbf{F})$ because the divergence $\nabla \cdot \mathbf{F}$ gives a *scalar* and we cannot take the curl of a scalar. With our understanding of the meaning of curl and divergence, this identity is sensible. The curl of a vector field, $\nabla \times \mathbf{F}$, gives the net *circulation* or *rotation* of the field, whereas the divergence of a field, $\nabla \cdot \mathbf{F}$, gives the net *outflow* or *flux* of the field from a point. We have two situations to consider: (1) If the vector field has circulation at a point, $\nabla \times \mathbf{F} \neq 0$, it can have no divergence (net outflow of the field) at that point and (A.25) is satisfied; (2) on the other hand, if the vector field has no circulation at a point, $\nabla \times \mathbf{F} = 0$, the divergence of this is zero.

A simple way to prove this important identity is to carry out the operation in a rectangular coordinate system using symbols. For example,

$$\nabla \times \mathbf{F} = \left(\frac{\partial F_z}{\partial y} - \frac{\partial F_y}{\partial z} \right) \mathbf{a}_x + \left(\frac{\partial F_x}{\partial z} - \frac{\partial F_z}{\partial x} \right) \mathbf{a}_y + \left(\frac{\partial F_y}{\partial x} - \frac{\partial F_x}{\partial y} \right) \mathbf{a}_z$$

If we next take the divergence of this result, we obtain

$$\nabla \cdot (\nabla \times \mathbf{F}) = \frac{\partial}{\partial x}\left(\frac{\partial F_z}{\partial y} - \frac{\partial F_y}{\partial z}\right) + \frac{\partial}{\partial y}\left(\frac{\partial F_x}{\partial z} - \frac{\partial F_z}{\partial x}\right) + \frac{\partial}{\partial z}\left(\frac{\partial F_y}{\partial x} - \frac{\partial F_x}{\partial y}\right)$$

$$= \frac{\partial^2 F_z}{\partial x \partial y} - \frac{\partial^2 F_y}{\partial x \partial z} + \frac{\partial^2 F_x}{\partial y \partial z} - \frac{\partial^2 F_z}{\partial y \partial x} + \frac{\partial^2 F_y}{\partial z \partial x} - \frac{\partial^2 F_x}{\partial z \partial y}$$

$$= 0$$

Another useful vector identity is that *the curl of the gradient of a scalar field is zero*:

$$\boxed{\nabla \times \nabla f\,(x, y, z) = 0} \tag{A.26}$$

Integrating this over some open surface s and using Stokes's theorem on the result gives

$$\int_s [\nabla \times \nabla f] \cdot ds = \oint_c (\nabla f) \cdot d\mathbf{l}$$

$$= \oint_c \frac{\partial f}{\partial x} dx + \frac{\partial f}{\partial y} dy + \frac{\partial f}{\partial z} dz$$

$$= \oint_c df$$

$$= 0$$

The result is due to integrating df around a *closed path*. This identity can be directly proven by carrying out the operations in (A.26) symbolically in a rectangular coordinate system:

$$\nabla \times \nabla f = \nabla \times \left(\frac{\partial f}{\partial x}\mathbf{a}_x + \frac{\partial f}{\partial y}\mathbf{a}_y + \frac{\partial f}{\partial z}\mathbf{a}_z\right)$$

$$= \left(\frac{\partial}{\partial y}\frac{\partial f}{\partial z} - \frac{\partial}{\partial z}\frac{\partial f}{\partial y}\right)\mathbf{a}_x + \left(\frac{\partial}{\partial z}\frac{\partial f}{\partial x} - \frac{\partial}{\partial x}\frac{\partial f}{\partial z}\right)\mathbf{a}_y$$

$$+ \left(\frac{\partial}{\partial x}\frac{\partial f}{\partial y} - \frac{\partial}{\partial y}\frac{\partial f}{\partial x}\right)\mathbf{a}_z$$

$$= 0$$

A.8   CYLINDRICAL COORDINATE SYSTEM

A point in a *cylindrical coordinate system* is defined by the three variables r, ϕ, and z, as illustrated in Fig. A.17. The coordinate r is the radial distance of the

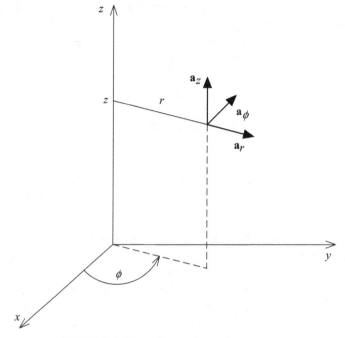

FIGURE A.17. Cylindrical coordinate system.

point from the z axis (parallel to the xy plane), the coordinate ϕ is the angular displacement (in *radians* with $0 \leq \phi \leq 360°$) of the projection of the point on the xy plane measured counterclockwise from the positive x axis, and the coordinate z is the distance of the projection of the point along the z axis. The corresponding three unit vectors $\mathbf{a}_r$, $\mathbf{a}_\phi$, and $\mathbf{a}_z$ are directed in the direction of increasingvalue of the variable and are mutually perpendicular. Hence, the cylindrical coordinate system, like the rectangular coordinate system, is an *orthogonal coordinate system*.

A vector in cylindrical coordinates is again described in terms of its unit vectors as

$$\mathbf{A} = A_r\mathbf{a}_r + A_\phi\mathbf{a}_\phi + A_z\mathbf{a}_z \qquad (A.27)$$

Two vectors are again added or subtracted by adding or subtracting their corresponding components:

$$\mathbf{A} \pm \mathbf{B} = \left(A_r \pm B_r\right)\mathbf{a}_r + \left(A_\phi \pm B_\phi\right)\mathbf{a}_\phi + \left(A_z \pm B_z\right)\mathbf{a}_z \qquad (A.28)$$

The dot product of two vectors is, again, the sum of the products of the corresponding components:

$$\boxed{\begin{aligned}\mathbf{A}\cdot\mathbf{B} &= AB\cos\theta_{AB}\\ &= A_r B_r + A_\phi B_\phi + A_z B_z\end{aligned}} \tag{A.29}$$

The cross product of two vectors is, again,

$$\boxed{\begin{aligned}\mathbf{A}\times\mathbf{B} &= AB\sin\theta_{AB}\,\mathbf{a}_n\\ &= \left(A_\phi B_z - A_z B_\phi\right)\mathbf{a}_r + \left(A_z B_r - A_r B_z\right)\mathbf{a}_\phi + \left(A_r B_\phi - A_\phi B_r\right)\mathbf{a}_z\end{aligned}}$$

$$\tag{A.30}$$

Note that the coordinates are ordered $r \to \phi \to z \to r \to \phi \to z \to r \to$ $\cdots$ such that $\mathbf{a}_r \times \mathbf{a}_\phi = \mathbf{a}_z$. Note that $\mathbf{a}_\phi \times \mathbf{a}_r = -\mathbf{a}_z$ and $\mathbf{a}_r \times \mathbf{a}_z = -\mathbf{a}_\phi$. The vector result for the cross product in a cylindrical coordinate system in terms of the vector components is, again, easily remembered. Each component is of the form $\left(A_\beta B_\gamma - A_\gamma B_\beta\right)\mathbf{a}_\alpha$ in the order $\alpha \to \beta \to \gamma \to \alpha \to \beta \to \cdots$ according to the cyclic ordering of the coordinates $r \to \phi \to z \to r \to \phi \to z \to r \to \cdots$.

The algebra results above are the same as for the rectangular coordinate system. However, the vector calculus results will be different from those for a rectangular coordinate system since one of the variables of the cylindrical coordinate system, ϕ, does not have the dimensions of distance. Differential changes in the coordinates give differential arc lengths dr, $r\,d\phi$, and dz, as illustrated in Fig. A.18. Note that the ϕ variable is the only one of the three whose units are not a length. (The units of ϕ are *radians*.)For a differential change in ϕ, $d\phi$, the corresponding change in arc length for a radius of r is $r\sin d\phi \cong r\,d\phi$ using the small-angle approximation for the sine. Hence, a vector differential arc length is

$$d\mathbf{l} = dr\,\mathbf{a}_r + r\,d\phi\,\mathbf{a}_\phi + dz\,\mathbf{a}_z \tag{A.31}$$

and the line integral is

$$\boxed{\int_a^b \mathbf{F}(r,\phi,z)\cdot d\mathbf{l} = \int_{r_a}^{r_b} F_r\,dr + \int_{\phi_a}^{\phi_b} F_\phi\,r\,d\phi + \int_{z_a}^{z_b} F_z\,dz} \tag{A.32}$$

A vector differential surface is

$$d\mathbf{s} = (r\,d\phi\,dz)\,\mathbf{a}_r + (dr\,dz)\,\mathbf{a}_\phi + (dr\,r\,d\phi)\,\mathbf{a}_z \tag{A.33}$$

Each of these components is formed by the products of the two sides of each differential surface in Fig. A.18 that is perpendicular to the unit vector for

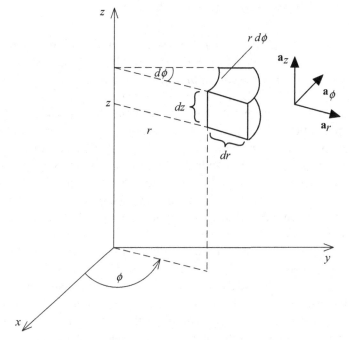

FIGURE A.18. Differential elements in a cylindrical coordinate system.

that side. For example, the side perpendicular to $\mathbf{a}_r$ has sides of length dz and $r\,d\phi$, while the side perpendicular to $\mathbf{a}_\phi$ has sides of length dz and dr. Hence, the surface integral is

$$\int_s \mathbf{F}(r, \phi, z) \cdot d\mathbf{s} = \int_{s_r} \mathbf{F}_r\, r\, d\phi\, dz + \int_{s_\phi} \mathbf{F}_\phi\, dr dz + \int_{s_z} \mathbf{F}_z\, dr\, r\, d\phi \qquad \text{(A.34)}$$

The divergence and the curl are a bit more complicated than for the rectangular coordinate system. The derivations of these are given in reference [3,6] and become

$$\nabla \cdot F(r, \phi, z) = \frac{1}{r}\frac{\partial (rF_r)}{\partial r} + \frac{1}{r}\frac{\partial F_\phi}{\partial \phi} + \frac{\partial F_z}{\partial z} \qquad \text{(A.35)}$$

$$\nabla \times \mathbf{F}(r, \phi, z) = \left(\frac{1}{r}\frac{\partial F_z}{\partial \phi} - \frac{\partial F_\phi}{\partial z}\right)\mathbf{a}_r + \left(\frac{\partial F_r}{\partial z} - \frac{\partial F_z}{\partial r}\right)\mathbf{a}_\phi$$
$$+ \left[\frac{1}{r}\frac{\partial (rF_\phi)}{\partial r} - \frac{1}{r}\frac{\partial F_r}{\partial \phi}\right]\mathbf{a}_z \qquad \text{(A.36)}$$

A.9   SPHERICAL COORDINATE SYSTEM

A point in a *spherical coordinate system* is defined by the three variables r, θ, and ϕ, as illustrated in Fig. A.19. The coordinate r is the radial distance of the point from the *origin of the coordinate system*, the coordinate θ (in *radians* with $0 \leq \theta \leq 180°$) is the angular displacement from the positive z axis, and the coordinate ϕ is the angular displacement (in *radians* with $0 \leq \phi \leq 360°$) of the projection of the point on the xy plane measured counterclockwise from the positive x axis. Note that the r in a spherical coordinate system is different from the r in a cylindrical coordinate system. The corresponding three unit vectors $\mathbf{a}_r$, $\mathbf{a}_\theta$, and $\mathbf{a}_\phi$ are directed in the direction of increasing value of the variable and are mutually perpendicular. Hence, the spherical coordinate system, like the rectangular and cylindrical coordinate systems, is an *orthogonal coordinate system*.

Some textbooks denote the radius dimension in a cylindrical coordinate system as ρ instead of r to distinguish it from the radius r in a spherical coordinate system. It is usually rather simple to distinguish between the two. If r is the distance perpendicular to the z axis and parallel to the xy plane, this is the cylindrical coordinate system. If r is the distance from the origin of the coordinate system, this is the spherical coordinate system.

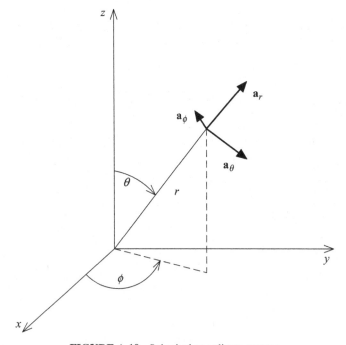

FIGURE A.19. Spherical coordinate system.

A vector in spherical coordinates is again described in terms of its unit vectors as

$$\mathbf{A} = A_r \mathbf{a}_r + A_\theta \mathbf{a}_\theta + A_\phi \mathbf{a}_\phi \qquad \text{(A.37)}$$

Two vectors are again added or subtracted by adding or subtracting their corresponding components:

$$\mathbf{A} \pm \mathbf{B} = \left(A_r \pm B_r\right) \mathbf{a}_r + \left(A_\theta \pm B_\theta\right) \mathbf{a}_\theta + \left(A_\phi \pm B_\phi\right) \mathbf{a}_\phi \qquad \text{(A.38)}$$

The dot product of two vectors is, again, the sum of the products of the corresponding components:

$$\begin{aligned} \mathbf{A}\cdot\mathbf{B} &= AB\cos\theta_{AB} \\ &= A_r B_r + A_\theta B_\theta + A_\phi B_\phi \end{aligned} \qquad \text{(A.39)}$$

The cross product of two vectors is, again,

$$\begin{aligned} \mathbf{A}\times\mathbf{B} &= AB\sin\theta_{AB}\,\mathbf{a}_n \\ &= \left(A_\theta B_\phi - A_\phi B_\theta\right)\mathbf{a}_r + \left(A_\phi B_r - A_r B_\phi\right)\mathbf{a}_\theta + (A_r B_\theta - A_\theta B_r)\,\mathbf{a}_\phi \end{aligned}$$

$$\text{(A.40)}$$

Note that the coordinates are ordered $r \to \theta \to \phi \to r \to \theta \to \phi \to r \to \cdots$ such that $\mathbf{a}_r \times \mathbf{a}_\theta = \mathbf{a}_\phi$. Note that $\mathbf{a}_\theta \times \mathbf{a}_r = -\mathbf{a}_\phi$ and $\mathbf{a}_r \times \mathbf{a}_\phi = -\mathbf{a}_\theta$. The vector result for the cross product in a spherical coordinate system in terms of the vector components is, again, easily remembered. Each component is of the form $\left(A_\beta B_\gamma - A_\gamma B_\beta\right)\mathbf{a}_\alpha$ in the order $\alpha \to \beta \to \gamma \to \alpha \to \beta \to \cdots$, according to the cyclic ordering of the coordinates $r \to \theta \to \phi \to r \to \theta \to \phi \to r \to \cdots$.

The algebra results above are the same as for the rectangular and the cylindrical coordinate systems. However, the vector calculus results will be different from those for a rectangular coordinate system since two of the variables of the spherical coordinate system, θ and ϕ, do not have the dimensions of distance. Differential changes in the coordinates give differential arc lengths dr, $r\,d\theta$, and $r\sin\theta\,d\phi$, as illustrated in Fig.A.20. Hence, a vector differential arc length is

$$d\mathbf{l} = dr\,\mathbf{a}_r + r\,d\theta\,\mathbf{a}_\theta + r\sin\theta\,d\phi\,\mathbf{a}_\phi \qquad \text{(A.41)}$$

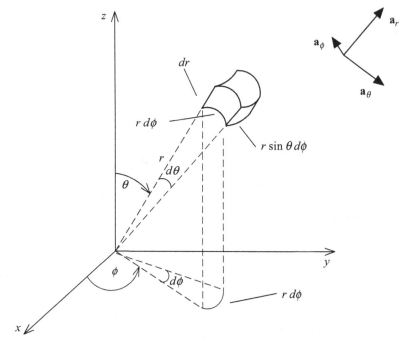

FIGURE A.20. Differential elements in a spherical coordinate system.

and the line integral is

$$\int_a^b \mathbf{F}(r, \theta, \phi) \cdot d\mathbf{l} = \int_{r_a}^{r_b} F_r \, dr + \int_{\theta_a}^{\theta_b} F_\theta \, r \, d\theta + \int_{\phi_a}^{\phi_b} F_\phi \, r \sin \theta \, d\phi \quad \text{(A.42)}$$

A vector differential surface is

$$d\mathbf{s} = (r \, d\theta \, r \sin \theta \, d\phi) \, \mathbf{a}_r + (dr \, r \sin \theta \, d\phi) \, \mathbf{a}_\theta + (dr \, r \, d\theta) \, \mathbf{a}_\phi \quad \text{(A.43)}$$

Each of these components is formed by the products of the two sides of each differential surface in Fig. A.20 that is perpendicular to the unit vector for that side. For example, the side perpendicular to $\mathbf{a}_r$ has sides of length $r \, d\theta$ and $r \sin \theta \, d\phi$, while the side perpendicular to $\mathbf{a}_\theta$ has sides of length dr and $r \sin \theta \, d\phi$. Hence the surface integral is

$$\int_s \mathbf{F}(r, \theta, \phi) \cdot d\mathbf{s} = \int_{S_r} F_r \, r \, d\theta \, r \sin \theta \, d\phi$$
$$+ \int_{S_\theta} F_\theta \, dr \, r \sin \theta \, d\phi + \int_{S_\phi} F_\phi \, dr \, r d\theta \quad \text{(A.44)}$$

The divergence and the curl are again a bit more complicated. The derivations of these are given in references [3,6] and become

$$\nabla \cdot \mathbf{F}(r, \theta, \phi) = \frac{1}{r^2} \frac{\partial \left(r^2 F_r\right)}{\partial r} + \frac{1}{r \sin \theta} \frac{\partial \left(\sin \theta \, F_\theta\right)}{\partial \theta} + \frac{1}{r \sin \theta} \frac{\partial F_\phi}{\partial \phi} \qquad (A.45)$$

$$\nabla \times \mathbf{F}(r, \theta, \phi) = \frac{1}{r \sin \theta} \left[\frac{\partial \left(F_\phi \sin \theta\right)}{\partial \theta} - \frac{\partial F_\theta}{\partial \phi}\right] \mathbf{a}_r$$
$$+ \frac{1}{r} \left[\frac{1}{\sin \theta} \frac{\partial F_r}{\partial \phi} - \frac{\partial \left(r F_\phi\right)}{\partial r}\right] \mathbf{a}_\theta + \frac{1}{r} \left[\frac{\partial \left(r F_\theta\right)}{\partial r} - \frac{\partial F_r}{\partial \theta}\right] \mathbf{a}_\phi$$

$$(A.46)$$

TABLE OF IDENTITIES, DERIVATIVES, AND INTEGRALS USED IN THIS BOOK

Identities

(1)
$$\ln \frac{a + \sqrt{x^2 + a^2}}{-b + \sqrt{x^2 + b^2}} = \sinh^{-1} \frac{a}{x} - \sinh^{-1} -\frac{b}{x}$$

$$= \sinh^{-1} \frac{a}{x} + \sinh^{-1} \frac{b}{x} \tag{5.24}$$

(2)
$$\ln \left(x + \sqrt{x^2 + 1} \right) = -\ln \left(-x + \sqrt{x^2 + 1} \right)$$

(3)
$$\sinh^{-1} x \equiv \ln \left(x + \sqrt{x^2 + 1} \right)$$

$$= -\sinh^{-1}(-x) \tag{D700.1}$$

(4)
$$K(k) = \int_{\zeta=0}^{\pi/2} \frac{d\zeta}{\sqrt{1 - k^2 \sin^2 \zeta}} \tag{D773.1}$$

(5)
$$E(k) = \int_{\zeta=0}^{\pi/2} \sqrt{1 - k^2 \sin^2 \zeta} \; d\zeta \tag{D774.1}$$

(6)
$$\tan^{-1} \theta_1 \pm \tan^{-1} \theta_2 = \tan^{-1} \frac{\theta_1 \pm \theta_2}{1 \mp \theta_1 \theta_2} \qquad \theta_1, \theta_2 \geq 0$$

(7)
$$\tan^{-1}(x + y) + \tan^{-1}(x - y) = \tan^{-1} \frac{2x}{1 - x^2 + y^2}$$

(8)
$$\tan^{-1}(x + y) - \tan^{-1}(x - y) = \tan^{-1} \frac{2y}{1 + x^2 - y^2}$$

Inductance: Loop and Partial, By Clayton R. Paul
Copyright © 2010 John Wiley & Sons, Inc.

(9)
$$\ln\frac{b}{a} = \ln\left(\frac{w}{a}+1\right)$$

$$\cong \frac{w}{a} \qquad w \ll a \tag{D601}$$

(10)
$$\int_{\theta=0}^{\pi} \sqrt{1-k^2\cos^2\theta}\,d\theta = 2\int_{\theta=0}^{\pi/2}\sqrt{1-k^2\cos^2\theta}\,d\theta$$

$$= 2\int_{\theta=0}^{\pi/2}\sqrt{1-k^2\sin^2\theta}\,d\theta$$

(11)
$$\int_{\theta=0}^{\pi}\frac{1}{\sqrt{1-k^2\cos\theta}}\,d\theta = 2\int_{\theta=0}^{\pi/2}\frac{1}{\sqrt{1-k^2\cos\theta}}\,d\theta$$

$$= 2\int_{\theta=0}^{\pi/2}\frac{1}{\sqrt{1-k^2\sin\theta}}\,d\theta$$

(12)
$$(1-x)^{-1/2} \cong 1 + \frac{1}{2}x + \cdots \tag{D1}$$

(13)
$$\sinh^{-1}x = \ln\left(x+\sqrt{x^2+1}\right)$$

$$= -\sinh^{-1}(-x)$$

$$= -\ln\left(-x+\sqrt{x^2+1}\right) \tag{D700.1}$$

(14)
$$\sinh^{-1}\frac{x}{a} = -\sinh^{-1}\left(-\frac{x}{a}\right)$$

$$= \ln\left[\frac{x}{a}+\sqrt{\left(\frac{x}{a}\right)^2+1}\right]$$

$$= \ln\left(x+\sqrt{x^2+a^2}\right) - \ln a \tag{D700.1}$$

(15)
$$\ln\left[\frac{l}{d}+\sqrt{\left(\frac{l}{d}\right)^2+1}\right] = \ln\frac{2l}{d} + \frac{1}{4}\left(\frac{d}{l}\right)^2 - \frac{3}{32}\left(\frac{d}{l}\right)^4$$

$$+\cdots \qquad \frac{l}{d} > 1$$

$$= \frac{l}{d} - \frac{1}{6}\left(\frac{l}{d}\right)^3 + \frac{3}{40}\left(\frac{l}{d}\right)^5$$

$$-\cdots \qquad \frac{l}{d} < 1 \tag{D602.1}$$

(16)
$$\sqrt{1 + \left(\frac{d}{l}\right)^2} = 1 + \frac{1}{2}\left(\frac{d}{l}\right)^2 - \frac{1}{8}\left(\frac{d}{l}\right)^4 + \frac{1}{16}\left(\frac{d}{l}\right)^6$$

$$-\cdots \qquad \frac{d}{l} \leq 1$$

$$= \frac{d}{l}\sqrt{\left(\frac{l}{d}\right)^2 + 1}$$

$$= \frac{d}{l} + \frac{1}{2}\left(\frac{l}{d}\right) - \frac{1}{8}\left(\frac{l}{d}\right)^3 + \frac{1}{16}\left(\frac{l}{d}\right)^5$$

$$-\cdots \qquad \frac{l}{d} \leq 1 \tag{D5.3}$$

(17)
$$\ln\left(x + \sqrt{x^2 + a^2}\right) - \ln\left(-y + \sqrt{y^2 + a^2}\right) = \ln\left(y + \sqrt{y^2 + a^2}\right)$$
$$- \ln\left(-x + \sqrt{x^2 + a^2}\right)$$

(18)
$$\ln\frac{a + \sqrt{x^2 + a^2}}{-b + \sqrt{x^2 + b^2}} = \sinh^{-1}\frac{a}{x} - \sinh^{-1}\left(-\frac{b}{x}\right)$$

$$= \sinh^{-1}\frac{a}{x} + \sinh^{-1}\frac{b}{x}$$

(19)
$$\tanh^{-1} x = \frac{1}{2}\ln\frac{1+x}{1-x} \qquad x^2 < 1 \tag{D702}$$

Derivatives

(1)
$$\frac{d}{dr}\ln\left[\frac{a}{r} + \sqrt{\left(\frac{a}{r}\right)^2 + 1}\right] = -\frac{a}{r\sqrt{a^2 + r^2}}$$

(2)
$$\frac{d}{dx}\sinh^{-1}\frac{a}{x} = \frac{d}{dx}\operatorname{csc}h^{-1}\frac{x}{a}$$

$$= \frac{-a}{|x|\sqrt{x^2 + a^2}} \tag{D728.8}$$

(3)
$$\frac{d}{dr}\ln\left(a + \sqrt{a^2 + r^2}\right) = -\frac{a}{r\sqrt{a^2 + r^2}} + \frac{1}{r}$$

(4)
$$\frac{\partial}{\partial u}\tan^{-1} u = \frac{1}{1 + u^2} \tag{D512.4}$$

(5)
$$\frac{d\left(\frac{u}{v}\right)}{dx} = \frac{v\frac{du}{dx} - u\frac{dv}{dx}}{v^2} \tag{D65}$$

Integrals

(1)
$$\frac{\partial}{\partial y} \int_a^b f(x, y)\, dx = \int_a^b \frac{\partial f(x, y)}{\partial y}\, dx$$
(D69.3)

(2)
$$\int \frac{1}{\left(a^2 + x^2\right)^{3/2}}\, dx = \frac{x}{a^2 \sqrt{a^2 + x^2}}$$
(D200.3)

(3)
$$\int \frac{1}{a^2 + x^2}\, dx = \frac{1}{a} \tan^{-1} \frac{x}{a}$$
(D120.1)

(4)
$$\int \frac{dx}{\left(ax^2 + b\right) \sqrt{fx^2 + g}} = \frac{1}{\sqrt{b}\sqrt{ag - bf}} \tan^{-1} \frac{x\sqrt{ag - bf}}{\sqrt{b}\sqrt{fx^2 + g}}$$
(D387)

(5)
$$\int_0^\pi \frac{(a - b\cos x)\, dx}{a^2 + b^2 - 2ab\cos x} = \begin{cases} \dfrac{\pi}{a} & a > b > 0 \\ 0 & b > a > 0 \end{cases}$$
(D859.124)

(6)
$$\int \frac{1}{\sqrt{x^2 + a^2}}\, dx = \ln\left(x + \sqrt{x^2 + a^2}\right)$$
(D200.01)

(7)
$$\int \ln\left(x^2 + a^2\right)\, dx = x\ln\left(x^2 + a^2\right) - 2x + 2a\tan^{-1} \frac{x}{a}$$
(D623)

(8)
$$\int \frac{x}{a^2 + x^2}\, dx = \frac{1}{2} \ln\left(a^2 + x^2\right)$$
(D121.1)

(9)
$$\int \frac{dx}{x\sqrt{x^2 + a^2}} = -\frac{1}{a} \ln\left| \frac{a + \sqrt{x^2 + a^2}}{x} \right|$$
(D221.01)

(10)
$$\int \sinh^{-1} \frac{x}{a}\, dx = x\sinh^{-1} \frac{x}{a} - \sqrt{x^2 + a^2} \qquad a > 0$$
(D730)

(11)
$$\int \frac{dx}{\left(ax^2 + bx + c\right)^{3/2}} = \frac{4ax + 2b}{\left(4ac - b^2\right)\left(ax^2 + bx + c\right)^{1/2}}$$
(D380.003)

(12)
$$\int \frac{x\, dx}{\left(ax^2 + bx + c\right)^{3/2}} = -\frac{2bx + 4c}{\left(4ac - b^2\right)\left(ax^2 + bx + c\right)^{1/2}}$$
(D380.013)

(13)
$$\int \frac{\sqrt{x^2 + a^2}}{x}\, dx = \sqrt{x^2 + a^2} - a\ln \frac{a + \sqrt{x^2 + a^2}}{x}$$
(D241.01)

(14)
$$\int \frac{x}{\sqrt{x^2 + a^2}}\, dx = \sqrt{x^2 + a^2}$$
(D201.01)

(15)
$$\int \ln ax\, dx = x\ln ax - x$$
(D610.01)

(16)
$$\ln\left[\frac{l}{r_w}+\sqrt{\left(\frac{l}{r_w}\right)^2+1}\right]=\ln\frac{2l}{r_w}+\frac{1}{4}\left(\frac{r_w}{l}\right)^2-\frac{3}{32}\left(\frac{r_w}{l}\right)^4$$

$$+\cdots\qquad\frac{l}{r_w}\gg1$$
(D602.1)

(17)
$$\int\frac{dx}{\sqrt{x^2+bx+c}}=\ln\left(2\sqrt{x^2+bx+c}+2x+b\right)$$
(D380.001)

(18)
$$\int\ln\left(a+\sqrt{a^2+x^2}\right)dx=x\ln\left(a+\sqrt{a^2+x^2}\right)$$
$$-x+a\ln\left(x+\sqrt{a^2+x^2}\right)$$
(D740)

(19)
$$\int\sqrt{a^2+x^2}dx=\frac{a^2}{2}\ln\left(x+\sqrt{a^2+x^2}\right)+\frac{x}{2}\sqrt{a^2+x^2}$$
(D230.01)

(20)
$$\int x\ln\left(a+\sqrt{a^2+x^2}\right)dx=-\frac{x^2}{4}+\frac{a}{2}\sqrt{a^2+x^2}$$
$$+\frac{x^2}{2}\ln\left(a+\sqrt{a^2+x^2}\right)$$
(D740.1)

(21)
$$\int\ln\left(x+\sqrt{a^2+x^2}\right)dx=x\ln\left(x+\sqrt{a^2+x^2}\right)-\sqrt{a^2+x^2}$$
(D625)

(22)
$$\int x\sqrt{a^2+x^2}dx=\tfrac{1}{3}\left(a^2+x^2\right)^{3/2}$$
(D231.01)

(23)
$$\int_{x=0}^{2\pi}\ln\left(1+b^2-2b\cos x\right)dx=\begin{cases}4\pi\ln b & b>1\\0 & b<1\end{cases}$$
(D865.73)

(24)
$$\int x\ln x\,dx=\frac{x^2}{2}\ln x-\frac{x^2}{4}$$
(D610.1)

(25)
$$\int x\ln\left(x^2+a^2\right)dx=\frac{1}{2}\left(x^2+a^2\right)\ln\left(x^2+a^2\right)-\frac{1}{2}x^2$$
(D623.1)

(26)
$$\int\tan^{-1}\frac{x}{a}dx=x\tan^{-1}\frac{x}{a}-\frac{a}{2}\ln\left(a^2+x^2\right)$$
(D525)

(27)
$$\int x^2\ln\left(a^2+x^2\right)dx=\frac{x^3}{3}\ln\left(a^2+x^2\right)-\frac{2}{9}x^3+\frac{2}{3}xa^2-\frac{2}{3}a^3\tan^{-1}\frac{x}{a}$$
(D623.2)

(28)
$$\int x\tan^{-1}\frac{a}{x}dx=\frac{ax}{2}+\frac{x^2+a^2}{2}\tan^{-1}\frac{a}{x}$$
(D528.1)

(29) $$\int x^3 \ln\left(x^2 + a^2\right) dx = \frac{x^4 - a^4}{4} \ln\left(x^2 + a^2\right) - \frac{x^4}{8} + \frac{x^2 a^2}{4}$$ (D623.3)

(30) $$\int \tan^{-1} \frac{a}{x} dx = x \tan^{-1} \frac{a}{x} + \frac{a}{2} \ln\left(x^2 + a^2\right)$$ (D528)

(31) $$\int x^2 \tan^{-1} \frac{a}{x} dx = \frac{x^3}{3} \tan^{-1} \frac{a}{x} + \frac{ax^2}{6} - \frac{a^3}{6} \ln\left(x^2 + a^2\right)$$ (D528.2)

(32) $$\int \ln x \, dx = x \ln x - x$$ (D610)

REFERENCES AND FURTHER READINGS

[1] C.R. Paul, *Analysis of Linear Circuits*, McGraw-Hill, New York, 1989.

[2] C.R. Paul, *Fundamentals of Electric Circuit Analysis*, Wiley, New York, 2001.

[3] C.R. Paul and S.A. Nasar, *Introduction to Electromagnetic Fields*, McGraw-Hill, New York, second edition, 1987, and third edition, 1998.

[4] C.R. Paul, *Electromagnetics for Engineers: With Applications to Digital Systems and Electromagnetic Compatibility*, Wiley, Hoboken, NJ, 2004.

[5] C.R. Paul, *Introduction to Electromagnetic Compatibility*, second edition, Wiley-Interscience, Hoboken, NJ, 2006.

[6] C.T.A. Johnk, *Engineering Electromagnetic Fields and Waves*, second edition, Wiley, New York, 1988.

[7] H.B. Dwight, *Tables of Integrals and Other Mathematical Data*, fourth edition, Macmillan, New York, 1961.

[8] C.R. Paul, *Analysis of Multiconductor Transmission Lines*, second edition, Wiley-Interscience, Hoboken, NJ, 2008.

[9] W.B. Boast, *Vector Fields*, Warren B. Boast, Ames, Iowa, 1964.

[10] W.R. Smythe, *Static and Dynamic Electricity*, third edition, revised printing, Hemisphere Publishing Company, New York, 1989.

[11] E. Weber, *Electromagnetic Fields, Theory and Applications:* Vol. I, *Mapping of Fields*, Wiley, New York, 1950.

[12] W. Kaplan, *Advanced Calculus*, Addison-Wesley, Reading, MA, 1952.

Inductance: Loop and Partial, By Clayton R. Paul
Copyright © 2010 John Wiley & Sons, Inc.

[13] R.M. Fano, L.J. Chu, and R.B. Adler, *Electromagnetic Fields, Energy, and Forces,* Wiley, New York, 1960, second printing 1963.

[14] F.W. Grover, *Inductance Calculations,* Dover Publications (Instrument Society of America), New York, 1973, (first published in 1946).

[15] A.E. Ruehli, "Inductance Calculations in a Complex Integrated Circuit Environment," *IBM Journal of Research and Development,* pp. 470–481, September 1972.

[16] C. Hoer and C. Love, "Exact Inductance Equations for Rectangular Conductors with Applications to More Complicated Geometries," *Journal of Research of the National Bureau of Standards: C. Engineering and Instrumentation,* vol. 69C, no. 2, pp. 127–137, April–June 1965.

[17] G.A. Campbell, "Mutual Inductance of Circuits Composed of Straight Wires," *Physical Review,* vol. 5, pp. 452–458, June 1915.

[18] C.L. Holloway and E.F. Kuester, "Dc Internal Inductance for a Conductor of Rectangular Cross Section," *IEEE Transactions on Electromagnetic Compatibility,* vol. 51, no. 2, pp. 338–344, May 2009.

[19] G. Antonini, A. Orlandi, and C.R. Paul, "Internal Impedance of Conductors of Rectangular Cross Section," *IEEE Transactions on Microwave Theory and Techniques,* vol. 47, no. 7, pp. 979–985, July 1999.

[20] E.B. Rosa and F.W. Grover, "Formulas and Tables for the Calculation of Mutual and Self Inductances (Revised)," *Bulletin of the National Bureau of Standards,* vol. 8, no. 1, **169**, pp. 1–231, 1912.

[21] P.L. Kalantarov and L.A. Tseitlin, *Raschet Induktivnostie,* Energoatomizdat, Leningrad, Russia, 1986.

[22] P. Silvester, *Modern Electromagnetic Fields,* Prentice-Hall, Englewood Cliffs, NJ, 1968.

[23] J.C. Maxwell, *A Treatise on Electricity and Magnetism,* Vol. II, second edition, Clarendon Press, Oxford, 1881.

[24] W.D. Stevenson, Jr., *Elements of Power System Analysis,* second edition, McGraw-Hill, New York, 1962.

[25] E.B. Rosa, "Calculation of the Self-Inductance of Single-Layer Coils," *Bulletin of the National Bureau of Standards,* vol. 2, no. 2, **31**, pp. 161–187, 1906.

[26] E.B. Rosa and L. Cohen, "On the Self-Inductance of Circles," *Bulletin of the National Bureau of Standards,* vol. 4, no. 1, **75**, pp. 149–159, 1907–1908.

[27] E.B. Rosa, "The Self and Mutual Inductance of Linear Conductors," *Bulletin of the National Bureau of Standards,* vol. 4, no. 2, **80**, pp. 301–344, 1907–1908.

[28] E.B. Rosa and L. Cohen, "Formulae and Tables for the Calculation of Mutual and Self-Inductance," *Bulletin of the National Bureau of Standards,* vol. 5, no. 1, **93**, pp. 1–132, 1908–1909.

[29] A. Gray, *The Theory and Practice of Absolute Measurements in Electricity and Magnetism,* Vol. II, Part I, Macmillan, London and New York, 1893.

[30] T.J. Higgins, "Formulas for the Geometric Mean Distances of Rectangular Areas and of Line Segments," *Journal of Applied Physics*, vol. 14, pp. 188–195, April 1943.

[31] C.L. Holloway and E.F. Kuester, "Net and Partial Inductance of a Microstrip Ground Plane," *IEEE Transactions on Electromagnetic Compatibility*, vol. 40, no. 1, pp. 33–45, February 1998.

[32] C.R. Paul and T.S. Smith, "Effect of Grid Spacing on the Inductance of Ground Grids," *1991 IEEE International Symposium on Electromagnetic Compatibility*, Cherry Hill, NJ, August 1991.

[33] C.R. Paul, "Modeling of Electromagnetic Interference Properties of Printed Circuit Boards," *IBM Journal of Research and Development*, vol. 33, no. 1, pp. 33–50, January 1989.

[34] C.R. Paul, "What Do We Mean by 'Inductance'? Part I: Loop Inductance," *IEEE EMC Society Magazine*, Fall 2007, pp. 95–101, and "What Do We Mean by 'Inductance'? Part II: Partial Inductance," *IEEE EMC Society Magazine*, Winter 2008, pp. 72–79.

INDEX

Inductance: Loop and Partial, By Clayton R. Paul
Copyright © 2010 John Wiley & Sons, Inc.

Printed in the United States
By Bookmasters